Flowering plants

IN WEST AFRICA

—

MARGARET STEENTOFT, F.L.S.

Formerly of the University of Ibadan and lately
Associate Professor of the Institute of Biological Education
Danmarks Lærerhøjskole, Copenhagen

T0275762

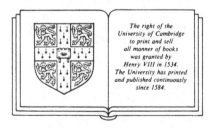

The right of the
University of Cambridge
to print and sell
all manner of books
was granted by
Henry VIII in 1534.
The University has printed
and published continuously
since 1584.

CAMBRIDGE UNIVERSITY PRESS

Cambridge

New York   New Rochelle   Melbourne   Sydney

CAMBRIDGE UNIVERSITY PRESS
Cambridge, New York, Melbourne, Madrid, Cape Town, Singapore, São Paulo

Cambridge University Press
The Edinburgh Building, Cambridge CB2 8RU, UK

Published in the United States of America by Cambridge University Press, New York

www.cambridge.org
Information on this title: www.cambridge.org/9780521261920

First published 1988
This digitally printed version 2008

A catalogue record for this publication is available from the British Library

Library of Congress Cataloguing in Publication data

Steentoft, Margaret.
 Flowering plants in West Africa.

 Bibliography: p.
 Includes index.
 1. Botany—Africa, West. 2. Angiosperms—Africa,
West. I. Title.
QK393.S743 1988   583.13′0966   87–24223

ISBN 978-0-521-26192-0 hardback
ISBN 978-0-521-06312-8 paperback

Contents

Introduction

When I wrote *Introduction to the flowering plants of West Africa* (1965), I was particularly aware of the needs of post-O level GCE students and their teachers, whether in schools or in the first year of some form of higher education or further training. In the intervening twenty years or so, not only has a concentrated research effort in West Africa vastly increased our knowledge of West African plants, but uses for, and attitudes to, this natural resource have changed. There have been two periods of drought and famine south of the Sahara, their severity directly and mainly attributable to economic policies, whether wrong in themselves, wrongly applied or simply non-existent.

A knowledge of species biology is not merely the basis for the academic study of a discipline such as taxonomy, but is the essential foundation for decision-making in rural economics in both its aspects, development and conservation. Famines will recur while population pressure, deforestation and traditional farming starved of investment continue. Yet India is self-supporting in food production and has no foreign debt. The difference between the two subcontinents is striking, and the expertise of a wide range of specialists is going to be needed if the lot of millions of people south of the Sahara is to be ameliorated.

With this in mind, the information presented in this book has been made adaptable to a variety of purposes for a variety of readers, in teaching–learning situations both formal and informal, for self-instruction and reference, and as a model for the investigation of families and species of flowering plants other than those selected here. Only the availability (and ability to use) the *Flora of West tropical Africa* (2nd edn) and a dictionary of botanical terminology is assumed. The families (and species) chosen have been selected for their ecological or economic importance. The families are arranged in the same sequence as in the *Flora of West tropical Africa* and the species names taken from that flora, unless newer ones exist. These have been added in parentheses, the reference being given only for those not occurring in the other major African floras (see General Bibliography).

Terminology is also largely that of the *Flora of West tropical Africa* but is extended where more recent work affords useful information. Ptyxis (the folding of individual leaves in the bud, as opposed to vernation, the placing of leaves in the bud) is an example (Cullen, 1978), while the description of leaf venation is being systematised, and that of fruits is very much needed. Descriptions of fruits based on temperate examples are sadly inadequate in the tropics. Roth (1977) has studied some tropical species.

Each family includes a section on field recognition, a necessary preliminary to many kinds of study, and the associated bibliography is confined to recent publications, in which reference to earlier work is made. Keeping up to date with the literature is best achieved by consulting the annual *AETFAT Index*, first published in 1954, and publications of the Royal Botanic Gardens, Kew (see References below), which will eventually go on-line.

Interspecific relationships are dealt with in Chapter 1, vegetation, a mixture of many individuals of various species, in Chapter 2, while, in each family, pollination and fruit and seed dispersal are reviewed. These topics are all aspects of species biology, perhaps those aspects most readily accessible to non-botanists. Taken together, they explain why, when the human species interferes with a population of another species, not only that species but many others are affected, often to their detriment.

Grateful acknowledgment is made of the facilities provided, over the last three years, by the Director of the Royal Botanic Gardens, Kew, and the Curator of the Forest Herbarium, Oxford, and of the help provided by numerous specialists, in particular F. N. Hepper at Kew.

References

AETFAT Index. L'Association pour l'étude taxonomique de la flore d'Afrique tropicale. Bruxelles: Laboratoire de botanique systematique, Université libre de Bruxelles. Since 1975 in the series *Travaux du Laboratoire de botaniqe systématique et de Phytogéographie*, Université libre de Bruxelles. Annual.

Cullen, J. (1978). A preliminary survey of ptyxis (vernation) in the Angiosperms. *Notes from the Royal Botanic Garden, Edinburgh*, **37**, pp. 161–214.

Roth, I. (1977). Fruits of angiosperms. In *Handbuch der Pflanzenanatomie*, vol. 10, part 1. Berlin: Borntraeger.

Steentoft Nielsen, M. (1965). *Introduction to the flowering plants of West Africa*. London: University of London Press.

The Kew Record of taxonomic literature. Kew: Royal Botanic Gardens. Annually from 1971 to 1981, now monthly.

Symbols and abbreviations

⸚ hypogynous
÷ perigynous
⸓ epigynous
⊕ actinomorphic
.l. zygomorphic
♂ male
♀ female
⚥ hermaphrodite
∅ diameter
± more or less, sometimes
A androecium
Ab abaxial
Ad adaxial
C corolla
CAM Crassulacean acid metabolism
EFN Extrafloral nectary
G gynaecium ($\overline{\text{G}}$ ovary inferior; $\underline{\text{G}}$ ovary superior)
(Gr.) Greek
(L.) Latin
K calyx
P perianth (applied where there are tepals, not distinct calyx and corolla)

The species described in Chapter 2 and in Field recognition sections may be assumed to have alternate leaves unless otherwise stated.

1

Species associations

Every individual interacts in a variety of ways with a number of other individuals. Individuals of the same species, if situated spatially and temporally near enough to each other, compete, though ramets may compete or act as reservoirs of supply for each other according to circumstance. This is a simplified view of intraspecific relationships.

Species associations are interspecific relationships, and the following account must needs take the form of a rather broad review, since information derived directly from observation in West Africa is quite limited.

Interspecific organismal associations were termed symbioses by De Bary (1879), and a number of kinds of such relationship is now recognised. The named kinds are obviously different from each other, but there may well exist more such relationships, differing from each other in subtle and, as yet, undiscovered ways.

It is proposed to define symbiosis as provision by permanent or transient physical contact of one species for another. At least one nutrient or service of significance for the survival of at least one of the participants is involved and some degree of obligacy is present in the relationship.

This definition is wider than that of Boucher, James & Keeler (1982), but narrower than that of Starr (1975). It includes mutualism, in which the associate species confer benefit (exchange nutrients/services) on each other.

Non-mutualistic symbiosis

Two early recognised symbioses, both much to the disadvantage of the host flowering plant, are its parasitic symbioses with fungi and bacteria, an obligate relationship for these organisms for at least part of their life cycle, and its totally obligate symbioses with predatory herbivores. There are also four West African genera of flowering plants, belonging to the families Balanophoraceae, Convolvulaceae, Lauraceae and Orobanchaceae, which are obligate root parasites of other flowering plants. Both parasites and predators benefit at the expense of the host, depriving the host of water and nutrients.

Loss of nutrients can include loss of light by shading because of, for example, a heavy liane canopy or covering of leaf surface with algae. Climbers are generally considered to be commensals of their supporting plants (phorophytes), advantage accruing to the climber but no harm being done to the phorophyte; but, in practice, only trunk climbers avoid shading phorophyte foliage (Newton, 1971).

Epiphytes have, in general, also been regarded as commensals, but deleterious effects on the host and penetration of host tissue, either by epiphyte root hairs or by the epiphyte's fungal partner have been observed. Some kind of parasitism is involved at least in cases where a degree of obligacy exists. Epiphytes of the crown also compete for light.

In so far as lianes and epiphytes appear casually on their phorophytes, they compete with them without forming symbioses. The phorophyte may possess physical means of reducing competition, such as a population of ants (see below), or have chemical defences.

The hemiparasitic flowering plants, though green, have been shown to be harmful to their hosts among crop species, and may be presumed to be so in other species. Room (1973) has written a series of papers on the relationship of *Tapinanthus bangwensis* (Loranthaceae) with its host, the cocoa tree. It grows obliquely into the xylem of its host, and the cocoa branch distal to the hemiparasite dies while the hemiparasite continues to absorb the products of the host's photosynthesis. Room adds four species of Loranthaceae (*Englerina gabonensis, Tapinanthus buntingii, T. belvisii* and *T. farmari*) to the list of cocoa hemiparasites in the *Flora of West tropical Africa* (1958), where numerous other hosts are mentioned. Balle & Hallé (1961) present an account of the family in Côte d'Ivoire and other flora accounts have since appeared (see General bibliography, p. 331).

The Scrophulariaceae contains at least two genera which are root hemiparasites of other flowering plants: *Striga* spp., on guinea corn etc. and *Sopubia ramosa* on *Imperata cylindrica* (Williams, 1960). Members of the Olacaceae, Opiliaceae and Santalaceae are also root hemiparasites of a variety of woody species (Kuijt, 1969).

Hemiparasites can be vegetative mimics of their host, thus sharing, or avoiding, the attention of herbivores with their host, but their flowers are, by

Fig. 1.1. Liane canopies. A. Trunk liane (*Hüllenlianen*). B. Inner crown liane (*Gedekte lianen*). C. Outer crown liane (*Wipfellianen*). D. Curtain liane (*Lianenvorhänge*). (From Vareschi, 1980.)

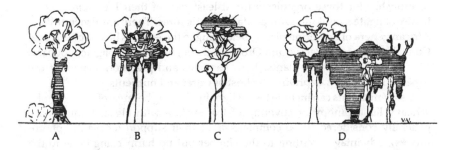

A   B   C   D

contrast, often quite brilliant, e.g. those of *Tapinanthus* spp., which are polli-nated by birds (Feehan, 1985).

An entirely different relationship, but one also involving mimetic deceit (Wiens, 1978), exists between flowers and animals. The plant deceives the animal and benefits by pollination or seed dispersal, while the animal not only receives little reward, but instead uses energy on non-productive activi-ties. The flower or fruit imitated (the model) offers a genuine reward to the animal, while the mimic merely offers false cues which attract the animal and stimulate it to behave in a manner beneficial to the plant.

This kind of relationship must be reckoned as parasitism or predation, a kind of consumption by the plant of the animal, and, as with the more obvious form recognised above, it must be balanced, so that the parasite/pred-ator species is adequately supplied without the parasitised/predated species being eliminated.

A much larger class of food deceivers, well represented in West Africa, is made up of species with unisexual flowers, in which the flowers of one sex produce rewards for the pollinator but the flowers of the other sex do not; they merely mimic the reward-producing flower in appearance. Bawa (1977) has shown that ♀ flowers of an American *Cupania* sp. (Sapindaceae) imitate ♂ flowers adequately by means of staminodes alone, and that the same species of insect visits flowers of both sexes. Stigmas may fulfil the same function. Other West African families showing this form of mimicry (unless ♀ flowers are eventually shown to produce a stigmatic exudate or other attractant) include the Annonaceae, Cucurbitaceae, Euphorbiaceae and Palmae.

Brood site deception, in which the insect is persuaded that the flower pro-vides a suitable site for egg-laying, with provision of food for the larvae, occurs in the Araceae, Aristolochiaceae and Asclepiadaceae (Yeo, 1973). The flowers smell of dung or rotting flesh and are often trap flowers, in which the insect is constrained to remain a sufficient length of time for pollination to have taken place.

Similar examples concerning flowers with parts imitating fungi in which fungus gnats (mycetophilids) lay their eggs, have been described in the fami-lies Araceae and Aristolochiaceae in Europe (Vogel, 1973).

The opposite of brood site deception, when a butterfly is persuaded that its potential brood site is already occupied, is described in America for *Passiflora* spp. (Gilbert, 1975), where the base of the leaf has two small butterfly-egg-shaped bodies of yellow tissue, which signal to the butterfly that the potential brood site is already occupied. Since no eggs are laid, the plant is saved from damage by caterpillars, and the butterfly must fly further in search of a brood site.

Three forms of deception found in orchids provoke insect behaviour during which pollination occurs.

1. Antagonist deception, in which the orchid induces an insect to 'attack'

it by resembling one of the insect's natural enemies.

2. Sexual deception, in which the orchid invites pseudocopulation by resembling the mate of the potential visitor.

3. Prey deception, in which orchid flowers mimic the prey of, for example, hunting wasps, which lay eggs where there appears to be food for the larvae.

The adequacy of these mechanisms depends on the inability of insect imagos to learn and have yet to be observed among West African orchids.

Mimicry in connection with seed dispersal also occurs. The disperser is persuaded that the diaspore is nutritionally desirable, either by its resembling a one-coloured red or blue fleshy fruit, or by its resembling a two-coloured arilloid seed, arilloid being a general term for seed flesh. Seeds disguised in this way are generally of poor nutritive value relative to the effort required of the disperser, though not necessarily without nutritional value. Examples in the tropics occur in the families Amaryllidaceae, Euphorbiaceae, Meliaceae, Musaceae and Papilionaceae.

Mutualisms

When both partners benefit from the association, a mutualistic symbiosis exists on a basis of exchange, for example of water and minerals in exchange for organic nitrogen, or of food (pollen or nectar) in return for the service of a pollinator. Mutualisms with fungi, bacteria and animals are exceedingly common, the rule rather than the exception, and it is highly probable that every taxonomic species of flowering plant takes part in one or more such associations.

Boucher *et al.* (1982) distinguish between indirect and direct mutualisms, the latter group comprising the relationships traditionally regarded as mutualistic which will be discussed below.

Indirect mutualisms are illustrated by a model of relationship between three species (A, B and C). In one case, A and C are assumed to compete with B, thus indirectly benefiting each other, while in the second case it is assumed that AB and BC enjoy mutualistic relationships and thus A and C benefit each other indirectly through B. It is probable that such relationships will be found to be quite common, and possible examples are pointed out in what follows. In the case of the ant–plant mutualism to which coccids are the third party, a quite different relationship exists (see below).

Direct mutualisms exist between flowering plants on the one hand and fungi, bacteria and animals on the other.

The flowering-plant–fungus relationship takes the form of fungus-inhabited roots, mycorrhizae. The host in three of these groups is photoautotrophic.

The sterile septate mycelia of members of the basidiomycetes or ascomycetes form an externally sheathing mycorrhiza on the roots of woody species

sheathing mycorrhiza), while the non-septate mycelium of phycomycetes of the family Endogonaceae forms an internal intercellular and intracellular mycorrhiza (arbuscular mycorrhiza) in a very wide range of species. Mycelia of both kinds extend a long way into the surrounding litter or soil and are effective in passing on nutrients to the host, which also achieves an improved water economy. In particular, nutrients required in relatively large quantities (such as phosphorus) and minerals of low mobility, both of which soon become depleted in the root zone, are acquired from further afield than the root net can reach. The fungus also produces growth substances, which stimulate the root to branch and assume its characteristic mycorrhizal form. In return, the fungus is supplied with organic carbon components, including vitamins and other growth factors.

The fungus, in sheathing mycorrhizae, penetrates between the cells of the root, forming the so-called Hartig net. Hyphae also penetrate the root cells, and intimate contact is established, though the hyphae are rarely digested. The sheath of hyphae acts as a phosphorus reservoir.

There are rather few examples of sheathing mycorrhizae in the tropics though there are numerous ones in temperate countries. Redhead (1980) gives a list of caesalpiniaceous species with ectomycorrhizae and refers to literature relating to the occurrence of mycorrhizae. The fungi concerned exploit leaf litter, which decomposes very rapidly in the tropics, by means of multihyphal mycelial strands. Both host and fungus are obligate symbionts, and must occur together, so that the rarity of sheathing mycorrhizas may not be difficult to explain.

Afforestation with exotic tree species such as pines has shown that the necessary fungi are not always present in West African soils, some being unable to withstand the temperatures of the dry season. Inoculation of seedling trees with suitable fungal strains has to take place in the nursery, before planting out.

Arbuscular mycorrhizae are of cosmopolitan occurrence, and both woody and herbaceous species may form them and so commonly do so that it is quicker to point out where they are not known. In general, hydrophytes and helophytes lack them, also some species of physiologically dry (salt) and physically dry habitats. All crop plants possess them, including the root-nodule-bearing members of the Papilionaceae, except for the grasses, which only form arbuscular mycorrhizae under conditions of acute phosphorus shortage (Mosse, Hayman & Arnold, 1973).

The sedge and rushes are mainly helophytes but they appear to be exceptionally well able to absorb phosphorus even at very low external concentrations.

The mycelia of Endogonaceae invade root cells, forming invaginations in the plasma membranes of the host cell so that fungus and host are in close contact. The fungus is supplied with organic carbon compounds in exchange for a supply of phosphorus, which is not, however, stored by the fungus. The invaginations are eventually digested by the host.

Table 1.1. *West African species with arbuscular mycorrhizae*

Anacardiaceae	Meliaceae (contd)
Anacardium occidentale	*K. ivorensis*
Annonaceae	*Lovoa trichilioides*
Annonidium mannii	Mimosaceae
Cleistopholis patens	*Albizia adianthifolia*
Apocynaceae	*A. zygia*
Alstonia boonei	*Cylicodiscus gabunensis*
Rauwolfia vomitoria	*Leucaena glauca**
Bignoniaceae	*Pentaclethra macrophylla*
*Jacaranda mimosifolia**	*Samanea saman**
Bombacaceae	*Xylia xylocarpa**
Ceiba pentandra	Moraceae
Caesalpiniaceae	*Antiaris africana*
*Cassia siamea**	*Bosquiea angolensis*
Gossweilerodendron	*Chlorophora excelsa*
balsamiferum	*Musanga cecropioides*
Hylodendron gabunense	*Myrianthus arboreus*
*Peltophorum ferrugineum**	*Treculia africana*
Combretaceae	Ochnaceae
Terminalia ivorense	*Lophira alata*
T. superba	Papilionaceae
Compositae	*Dalbergia latifolia**
Vernonia conferta	*D. sissoo**
Ebenaceae	*Gliricida sepium**
Diospyros insculpta	Rubiaceae
D. mespiliformis	*Nauclea diderrichii*
Euphorbiaceae	*Randia cladantha*
*Hevea brasiliensis**	Rutaceae
Macaranga barteri	*Fagara* spp.
Ricinodendrom heudelotii	Simaroubaceae
Hypericaceae	*Hannoa klaineana*
Harungana madagascariensis	Sterculiaceae
Lauraceae	*Cola gigantea* var.
Persea gratissima	*glabrescens*
Loganiaceae	*Hildegardia barteri*
Anthocleista nobilis	*Mansonia altissima**
Meliaceae	*Triplochiton scleroxylon*
*Azadirachta indica**	Ulmaceae
Carapa procera	*Trema guineensis*
Entandrophragma angolensis	Verbenaceae
Guarea cedrata	*Gmelina arborea**
G. thompsonii	*Tectona grandis**
Khaya grandifoliola	*Vitex* spp.

Data from Redhead (1960).
Asterisk denotes introduced species.

The hyphae penetrate both soil and litter, and are also tolerant of phosphate fertilisers. This is an advantage in the cultivation of legume crops, because the application of phosphate fertiliser improves nodulation.

The Endogonaceae are obligate symbionts, but there is some doubt about how obligate the relationship is for the flowering plant. Certainly in soils of low fertility, and this probably means in most of West Africa, such mycorrhizae may be expected. Some crop plants in particular, e.g. *Citrus* spp., are highly dependent on mycorrhizae.

Two advantages of possessing mycorrhizae, in nutrient and water supplies, have been mentioned, but other benefits may also accrue to the flowering plant, including increased tolerances to aluminium and high temperature, and of protection from, or compensation for, disease. Soil nutrients may also be protected from leaching by being trapped in the fungus. In the mixed communities which make up vegetation, much remains to be done in defining the role of mycorrhizae in competition. The saprotrophic abilities of the hyphae outside the flowering plant are poorly understood.

Although regarded as examples of direct mutualism, mycorrhizae may prove to participate in indirect mutualisms, connecting individuals of two or more flowering plant species. In an example from North America (Woods & Beck, 1964), seven stumps of red maple (*Acer rubrum*) were treated with ^{45}Ca and ^{32}P and surrounding trees monitored for radioactivity. This was found in numerous trees of 19 other species up to 24 feet (*c.* 7.5 m) away from the treated stumps.

Mycorrhizae are also formed between orchids and sterile septate mycelia of the genus *Rhizoctonia*, which is highly pathogenic to other flowering plants. These fungi can decompose a wide variety of even complex carbohydrates, and such substances as lignin as well. However, they do not attack the cell walls of their orchid host. The orchid is partially saprotrophic, but ultimately the intracellular portions of the fungus are digested, and there is a net gain of carbohydrate by the orchid, which at least in this respect, parasitises the fungus. The fungus receives organic nitrogen and vitamins in return.

Older and larger green orchids seem to be less dependent on symbiosis, but all orchids, green and non-green, are dependent on a fungal infection of the swollen seed, if this is to germinate, non-green orchids remaining in an obligate relationship with the fungus. The source of carbon for the permanently saprotrophic orchid is either the soil or another green plant, in both cases via the fungus.

The few species of non-orchidaceous achlorophyllous plants in West Africa belong to the families Burmanniaceae, Gentianaceae and Triuridaceae. According to Furman & Trappe (1971), these plants are probably also to be regarded as parasitic upon their fungal partner, as has been shown for the European saprophytic orchid *Monotropa hypopytis*, which was found to be connected to forest tree species via its fungus, a *Boletus* sp. The 'saprophyte' is epiparasitic upon the green plant, since nutrient transfer from this plant to the 'saprophyte' has been shown.

Permanent physical interspecific connections via mycorrhizae between two or more individuals of different flowering plant species are now well known and should be looked for in West Africa. Another kind of connection, by root grafts (Kozlowski & Cooling, 1961) is also a possibility.

Mutualisms between flowering plants and bacteria take the form of nodules formed on roots or in leaves, and the bacteria are nitrogen-fixing. The nodules developed on legume roots accommodating *Rhizobium* spp. are probably best known. Of the leguminous species examined, 90% of Mimosaceae and Papilionaceae are reported to be nodulated but only 20% of the Caesalpiniaceae. The Mimosaceae and Papilionaceae may also possess arbuscular mycorrhizas. *Rhizobium*-like nodulation of species other than legumes is reviewed by Becking (1975), and, in these, nitrogen fixation has also been demonstrated.

Of the kind of nodule accommodating an actinomycete of the genus *Frankia*, only one example is known in West Africa. This is *Casuarina equisetifolia*, a species introduced into West Africa, but elsewhere occurring naturally near the sea, on land which is occasionally flooded by seawater. Each nodule is lobed, and each lobe develops a negatively gravitropic root, the appearance of each bunch of roots being quite characteristic. The nodules are perennial (as opposed to the more or less seasonal ones of *Rhizobium*), and about 14 other genera have been shown to possess this form of nodule. *Casuarina equisetifolia* has one of the highest rates of nitrogen fixation among those species for which this value is known – 229 kg N_2/ha per year, a measurement which has yet to be made in West Africa.

Leaf nodules have long been known in Rubiaceae in West Africa, and in these several species of bacteria have been demonstrated, notably *Chromobacterium lividum* and *Klebsiella rubiacearum*. These are also nitrogen-fixing and are present in vegetative buds, the new leaf primordia being infected by the older leaves surrounding them. The nodules appear to form in substomatal chambers.

Lersten & Horne (1976) reviewed the occurrence of leaf nodules among the Rubiaceae, also in West Africa, and commented on the claims made for a number of non-rubiaceous genera (*Cassia occidentalis*, *Cienfuegosia* spp. and *Laguncularia racemosa*), all of which remain to be substantiated.

A slightly different case is that of the West African *Dioscorea sansibarensis*, which has leaf tip glands reported to harbour bacteria (Chevalier, 1936).

The role of nodule symbionts in fixing nitrogen, made available to the flowering plant host in the form of organic nitrogen and exchanged for organic carbon compounds, parallels the phosphorus–organic carbon exchange relationship of flowering plants with the fungi of mycorrhizae. These relationships are assumed to be advantageous to both parts, but the extent of the advantage remains to be determined.

These kinds of nutritional relationships certainly affect our idea of how the individual flowering plant functions in its community. In particular, the indirect mutualisms operating between individuals of the same and different species via mycorrhizas need further investigation.

Mutualistic relationships with animals

Pollination

The accomplishment of pollination is incidental to the satisfaction of some animal need. Some West African examples are described by Steentoft (1988). Faegri & Pijl (1980), Proctor & Yeo (1973), Free (1970) and Real (1983) provide the fullest general accounts to date.

Animals are attracted by visual and/or odoriferous signals which promise reward, often pollen or nectar though other forms are now known. Nectar-like stigmatic exudates occur in sweet potato and several other families (Martin & Telek, 1971), while floral oils (needed as larval food), oils, gums and resins (attractive to bees) and food bodies (attractive to beetles) also occur.

Attraction to brood sites is well known in figs, in the fruit of which agaonid wasps breed, but is also found in cocoa, where midges which breed in the decaying pods are always on hand to perform pollination. Similarly in jackfruit (*Artocarpus heterophylla*) (in Java), it has been found that pollinating Diptera breed in the fallen ♂ inflorescences, and so are always found near the tree.

A case of indirect mutualism in connection with pollination concerning a West African genus is reported by Schemske (1981). Two *Costus* spp. in tropical America share a pollinator, a species of bee. The flowers of the two *Costus* spp. resemble each other closely, so that the species appear to 'co-operate' in providing the bee with an adequate number of flowers and thus retain its services. Both *Costus* spp. benefit indirectly through the bee.

Vibratile or 'buzz' pollination was first described for *Cassia* spp. and *Solanum wendlandii*. It occurs in species with apically dehiscent anthers (opening by pores or slits but without valves) and dry pollen, and 6–8% of the world flora is estimated to possess such a mechanism. Bees able to behave appropriately extract pollen from these anthers by vibrating their indirect flight muscles, these very strong vibrations being transmitted through their bodies and via their contacts with the anthers to the pollen, causing it to be ejected through the pore. The vibrations are audible, hence the name.

When pollination has occurred, and the stigma ceases to be receptive and the pollen is moribund, visual and odoriferous changes occur in flowers, which no longer offer rewards. Pollinators then tend to avoid them and after a period of further growth and development, mature fruit and seeds appear.

Dispersal

Successful dispersal from the point of view of the plant requires removal of diaspores from the allelopathic or competitive influence of the parent, or from its seed shadow, to which predators might be attracted.

Some dispersal is non-symbiotic, being incidental to the animal's activities, as the diaspore is not attractive. Adhesive fruits and seeds, removed externally

upon an animal's fur or feathers, are dispersed in this way, while the seeds
consumed by grazing animals are transported internally.

Where, however, seeds are transported in connection with the consumption
of attractive flesh, a mutualistic relationship exists between the disperser and
the diaspore species.

The flesh may be contributed by all or some part of the pericarp, by the
placenta or by the seed, and it is advertised by colour or aroma, sometimes
both. Seeds and stones are usually treated in one of two ways: either they
are rejected relatively near the site of collection, while the flesh is consumed
(as birds treat berries of Loranthaceae) or the whole fruit is transported some
distance in the gut of the consumer, the seeds reappearing in dung. The
period of transport in the gut is generally regarded as improving germination,
both in terms of numbers of seeds germinating and in terms of speed of
germination. The effect of a multiple planting in rich compost is certainly
achieved. Improved germination may result partly from scarification of the
testa.

Seeds consumed during grazing by cows are deposited in dung in portions
of half a million or more, but frugivorous animals eat fewer and larger fruits
of particular kinds.

Dispersal by birds is reviewed by Snow (1981), Killick (1959), Alexandre
(1978) adding some West African examples. Only birds with so-called 'hard'
gizzards destroy the seeds or stones of the fruits they consume.

The role of bats has been studied by Ayensu (1974), and examples include
pawpaw and guava. Alexandre (1978) adds a further species, *Strombosia glaucescens*.

Small mammals are mentioned, in connection with dispersal, by Killick
(1959), and Alexandre (1978) identifies rodents as dispersing *Octoknema borea-
lis*; but in the main the role of mammals other than bats is only just beginning
to be appreciated. In the Taï Forest of Côte d'Ivoire, Alexandre (1978) found
monkeys responsible for the dispersal of *Dacryodes klaineana*, *Diospyros sanza-
minika* and *Pycnanthus angolensis*, and it seems probable that the fruits of *Coelo-
caryon oxycarpum*, so similar to those of *Pycnanthus*, are dispersed in the same
way. Examples of species dispersed by baboons and elephants are listed in
Tables 1.2 and 1.3.

Alexandre (1978) estimated that, in the Taï Forest, 30% of upper storey
and emergent species (40% of individuals) were elephant dispersed, and that
such fruits were at least 5 cm Ø, lay on the ground singly or in groups and
were dull in colour but had strong smelling, often resinous, flesh, or even
a resinous shell. Drupes and large berries are prominent, also the pods of
Caesalpiniaceae and Mimosaceae with pulpa.

Plant–ant mutualisms

If all the large trees of a species are removed from a stretch of forest, not
only are they destroyed, but other species in that area may be destroyed

with them and the chances of survival of many more impaired. Because of the involvement of the tree species in numerous mutualistic relationships, few of which are obvious to the casual observer, a chain of destruction ensues.

Several kinds of ant–plant mutualism exist and they illustrate some of the complexity and variety possible among animal–plant relationships (Beattie, 1985).

Pollination by ants has been suggested in few species, confirmed for only one (arctic) species. Cocoa, cashew and *Capsicum* are among the suggested crop plants, and Hagerup (1943) has suggested two genera of Euphorbiaceae. Hagerup worked in the southern sahel, and the general review offered by Buckley (1982) also tends to support the idea that ant pollination may be found to be more common in semi-desert and desert communities.

Table 1.2. *Families, discussed in Chapters 3 to 40, with species germinating in baboon and elephant droppings*

Anacardiaceae	Guttiferae[a]
Annonaceae	Labiatae[b]
Apocynaceae	Meliaceae[b]
Bignoniaceae	Mimosaceae
Bombacaceae	Moraceae*
Caesalpiniaceae	Palmae
Cucurbitaceae[b]	Papilionaceae[b]
Cyperaceae	Rubiaceae*
Euphorbiaceae*	Sapotaceae
Gramineae[b]	Verbenaceae

Data from Lieberman, Hall & Swaine (1979) and Alexandre (1978).
[a] Elephants only.
[b] Baboons only. Species marked with an asterisk are represented by 100 or more seedlings.

Table 1.3. *Species of families, not discussed in Chapters 3 to 40, germinating from baboon droppings*

Boraginaceae	Flacourtiaceae
Cordia guineensis	*Flacourtia flavescens**
C. rothii	Menispermaceae
Capparidaceae	*Triclisia subcordata*
Capparis brassii	Myrtaceae *Eugenia coronata*
C. erythrocarpos	Opiliaceae
Crataeva adansonii	*Opilia celtidifolia*
Ebenaceae	Plumbaginaceae
Diospyros abyssinica	*Plumbago zeylanica*
D. mespiliformis	Tiliaceae
Ficoidaceae	*Grewia carpinifolia**
Trianthema portulacastrum	*G. villosa**

Data from Lieberman et al. (1979).
Species marked with an asterisk are represented by 100 or more seedlings.

Ants produce a variety of substances affecting the viability of pollen and its growth. In particular myrmicacin (β-hydroxydecanoic acid), which is secreted on to the body surface of several ant species, inhibits pollen tube growth. It is a potent antibiotic, potentially of value in the crowded conditions in the ant nest.

Ants can also act as floral robbers, and Guerrant & Fiedler (1981) propose various means by which such robbery may be deterred. Some plant species possess an ant guard, which deters other floral robbers (see below). Little is known of these relationships in West Africa.

In seed dispersal, ants play one of two roles. Harvester ants seek edible diaspores, especially grain and other seeds low in defensive chemicals, and store them in underground granaries ready for consumption. Grain and seeds may, however, be dropped or abandoned and thus dispersed and may sprout (*Anogeissus leiocarpus*), or seedlings may be removed from the granary and deposited in the area around the nest. The nutrition of granivorous ants in Côte d'Ivoire is discussed in Levieux & Diomande (1978).

As seed flesh harvesters, ants collect inedible seeds for the sake of the attractive and nutritious food body associated with each seed. The flesh is often an elaiosome, pale in colour and containing a high proportion of lipids, and acting as a handle by which the seed can be grasped. Morphologically

Table 1.4. *Species germinating in elephant droppings, or otherwise observed to be dispersed by elephants*

Burseraceae
  Canarium schweinfurthii[a]
  Dacryodes klaineana[a]
Capparidaceae
  Capparis erythrocarpos[b]
  Buchholzia coriacea
Humiriaceae
  Sacoglottis gabonensis
Irvingiaceae
  Irvingia gabonensis
  Klainedoxa gabonensis
Loganiaceae
  Strychnos spp.
Octoknemataceae
  Okoubaka aubrevillei
Pandanaceae
  Panda oleosa
Passifloraceae
  Adenia lobata

Rhamnaceae
  Ziziphus spp.[b]
Rosaceae
  Parinari excelsa
  P. sp.[c]
  Acioa sp.
  Hirtella butayei[d]
Rutaceae
  Afraegle paniculata
Simaroubaceae
  Gymnostemon zaizou
Tiliaceae
  Desplatsia chrysochlamys
  Duboscia viridiflora
  Grewia spp.
Zygophyllaceae
  Balanites aegyptiaca[b]
  B. wilsoniana

Data from Alexandre (1978).
[a] Also bird dispersed.
[b] Cited by Alexandre (1978).
[c] *Magnistipula fleuryana*.
[d] *Magnistipula butayei*.

the elaiosome is often a caruncle (*Ricinus*) or strophiole (*Polygala*, also Euphorbiaceae) (Pijl, 1972). In Labiatae the elaiosome stems from the receptacle, and in the Compositae it is an outgrowth of the pseudonut.

In other cases, the food body may have quite a different morphological origin as, for example, in *Rottboellia*, where it is formed from the tissue of the spikelet axis, and in *Datura* where it is proteinaceous, and probably part of the placenta.

In Malaysia, food bodies of diverse morphological nature have been reported in a number of genera known in West Africa: *Cleome, Clerodendrum, Cyanastrum, Desmodium, Lochnera, Sterculia* and *Turnera*. Some of these need further investigation in West Africa.

Once the food body is eaten, the seed is removed from the nest and may ultimately germinate. It has at least escaped the attentions of predators and the consequences of fire while in the nest, and is then deposited on the ant midden, in ground of high fertility where seedling establishment is promoted.

The remaining class of plant–ant mutualisms in West Africa is concerned with the defence of the flowering plant against predators or parasites. The plant offers food or nest sites (domatia) attractive to the ants, which then, by their naturally aggressive behaviour, defend their territory, the plant, against would-be herbivores and, indeed, against lianes and epiphyllic algae. Expenditure of plant resources on ant food, whether plant juices, food bodies or extra-floral nectar, appears to be worth while to the plant, improved fruit and seed set and reduced leaf damage and nectar robbery resulting from ant occupation.

Plant juices are expended on ants through an intermediary relationship with coccids, which suck the juices and secrete honeydew (sweet fluid faeces), which is then consumed by the ants. Coccids are tended by the ants in nests on the host plant, in specially formed domatia.

Food bodies are oil- or protein-rich tissues on young plant growth, collected by ants as an item of diet. O'Dowd (1982) has demonstrated the presence of pearl bodies, globular uni- or multi-cellular bodies on the surface of young tissue, in balsa (*Ochroma pyramidale*). These look pearl-like because of the presence of lipids. About 50 genera show the development of pearl bodies in one or more species, and the families where pearl bodies should be sought include the Acanthaceae, Ampelidaceae, Bombaceae, Malvaceae, Melastomataceae, Sterculiaceae and Urticaceae.

Extrafloral nectaries (EFNs) produce the equivalent of floral nectar, but usually it is more proteinaceous and more often contains non-protein amino acids than does floral nectar.

The glands producing this nectar are commonly on leaves (Caesalpiniaceae, Ebenaceae and Mimosaceae), stipules (Ampelidaceae, Malpighiaceae), or on the base of the petiole (*Dioscorea, Impatiens*), the base of the leaf blade (*Turnera*), here and on pedicels (*Bixa*), bracts (*Costus*), or on the abaxial sides of sepals (*Cleome, Ipomoea*), or on petioles (also *Ipomoea*).

Schnell, Cusset & Quenum (1963) identify a number of tropical families, for which they describe examples where EFNs occur. Included are the Asclepiadaceae, Bignoniaceae, Cucurbitaceae, Euphorbiaceae, Labiatae, Malvaceae, Papilionaceae, Passifloraceae and Verbenaceae. The Combretaceae should also be included.

The occurrence of domatia in African plants is summarised by Huxley (1986) and an illustrated account can be found by Schnell & Grout de Beaufort (1966). Domatia occur on woody plants and are constructed either on a temporary plant part (leaf, leaf pouch or stipule), i.e. deciduous domatia, or in a hollowed-out stem, i.e. permanent domatia. In the former case, the ants must move the homopterans (coccids) they tend at intervals.

The presence of coccids has not always been ascertained, but certainly at least either EFNs (or other food bodies) or coccids are present, sometimes two or all three of these.

Janzen (1972) studied *Barteria fistulosa* (now *B. nigritiana* subsp. *fistulosa*), a small under-storey tree of lowland forest occurring from S. Nigeria to Zaïre. The ant *Pachysima aethiops* inhabits internodal domatia (with coccids and fungus) and defends the tree against herbivores, lianes and epiphyllic algae. *Barteria* has EFNs in the form of glandular teeth on the leaves. The fungus observed in the domatia probably grows saprophytically on nest debris, and Janzen infers that it is consumed by ants. *Pachysima* patrols mainly young leaves.

Barteria and *Pachysima* are mutualists, *Barteria* providing nest sites, nectar and prey in return for defence. The ants are saprovores feeding on coccid faeces but providing them in return with shelter, and thus also mutualists, but the coccids are consumers of *Barteria* juices. The fungus exploits the debris of all three organisms and is harvested by *Pachysima*.

Schotia africana (now *Leonardoxa africana*), studied by McKey (1984) is a tree of roughly the same size as *Barteria* (*c.* 15 m) and occurs mainly along watercourses from S. Nigeria to Gabon. It exists in both myrmecophyte and nonmyrmecophyte populations. The former in Cameroun is inhabited by two species of ants, both of which occupy (separate) swollen internodes lacking coccids and fungi. The ant *Petalomyrmex phylax* patrols only young leaves, which produce little or no nectar, and feeds only on the minute herbivores attempting to attack the leaves. Older leaves are tougher and possess chemical defences such as tannins, and also produce nectar. These leaves, which live two or more years, are patrolled by another species of ant, *Cataulacus mckeyi*. These ants mostly only patrol the basal pair of leaflets of each leaf and neither attack herbivores nor clean the shoots of epiphytic growths. *Cataulacus* also attacks *Petalomyrmex*, though it cannot enter a *Petalomyrmex* nest as the entrance is too small.

The relationship between *Schotia* and *Petalomyrmex* is mutualistic, consisting of the provision of nest site and prey for the ant in return for defence, but *Cataulacus* is a herbivore, consuming extrafloral nectar and using nest sites (without making return) and a predator of *Petalomyrmex*.

The three-dimensional mosaic of ant species discovered to occupy cocoa farms (Majer, 1976) could well encompass a variety of relationships, of which the two described above are examples. Since swollen shoot virus is transmitted

Fig. 1.2. Ant domatia. A. *Barteria fistulosa* node × $\frac{3}{5}$. B. *Schotia africana* stem × $\frac{3}{5}$. C. *Nauclea vanderguchtii* stem tip × $\frac{3}{5}$. D–G. *Cola marsupium*. D. Base of underside of leaf showing openings to domatia × 1. E. Upper side of leaf × $\frac{1}{3}$. F. Distal domatia, upper side of leaf × 1. G. Central domatia, upper side of leaf × 1. (From Schnell & Grout de Beaufort, 1966, Figs. 2C, 1B, 6G, 7F–I.)

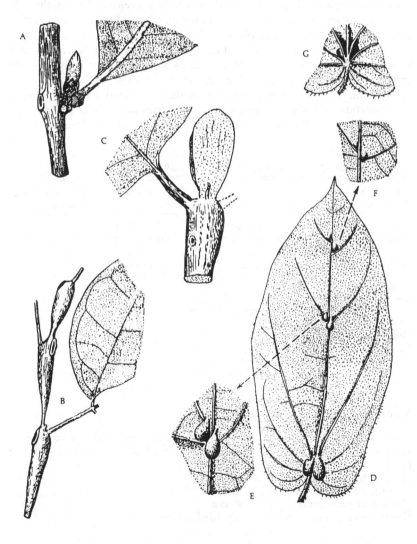

by the coccids, and spread by the ants concerned, an understanding of the relationship involved is very important.

Two other relationships between plants and ants are reported in America and Asia, respectively, but appear to be absent from West Africa. In America, leaf-cutting ants prey upon plants; they remove foliage, which they further macerate in their nests, defaecate on it and inoculate it with fungus spores, thus creating ant gardens. The ants then eat the fungus. In Asia, ant epiphytes are particularly common in nutrient-poor areas. The epiphytes produce large tubers, which are occupied by ants. These are either scavengers or tend coccids and feed on honeydew. Debris tends to collect in the nest and acts as a source of nutrients for the epiphyte, which may even send roots through the nest. The ants thus perform some of the functions of the extensive terrestrial root system which the epiphytes lack.

The existence of a second kind of domatia, for mites, and the relationship which may exist between plants and mites are but poorly known. There appears to be no literature specifically relevant to West Africa.

In summary, the flowering plant may be disadvantaged (harmed) through the predation of herbivores and by the parasitism not only of bacteria and fungi, but of other flowering plants. Lianes and epiphytes are almost certainly

Table 1.5. *West African flowering plant species with ant domatia*

Deciduous domatia		Domatia in	EFNs/coccids present
Ebenaceae	*Diospyros conocarpa*	Whole leaf	
Rubiaceae	*Gardenia imperialis*	Leaf pouch	+/+
	Psychotria 3 spp.[a]	Stipule	/+
Sterculiaceae	*Cola marsupium*	Leaf pouch	
Permanent domatia			
Caesalpiniaceae	*Schotia africana*[b]		+/
Passifloraceae	*Barteria fistulosa*[c]		+/+
Rubiaceae	*Canthium* 4 spp.		/+
	Cuvieria 7 spp.[d]		/+
	Nauclea vanderguchtii		/+
	Rothmannia 3 spp.		
	Uncaria africana		/+
Verbenaceae	*Clerodendrum* 2 spp.		
	Vitex thyrsiflora		

Data from Huxley (1986), Bequaert (1922) and Schnell & Grout de Beaufort (1966).
[a] Not with certainty West African spp.
[b] Now *Leonardoxa africana*.
[c] Now *Barteria nigritiana* subsp. *fistulosa*.
[d] Presence of coccids and fungus recorded for the genus.

harmful to their phorophytes to some degree, and flowering plant parasites and hemiparasites are certainly harmful to their (flowering plant) hosts.

Flowering plants may, however, take advantage of animals through mimicry, and, if the flowering plant concerned is a 'saprophyte' or orchid, advantage may also be taken of fungi. Mutualistic relations exist between flowering plants and animals (pollination and dispersal), fungi (phosphorus) and with bacteria (nitrogen) and ants (defence).

Thus, when a plant species population is eliminated, or even merely reduced or otherwise disadvantaged by a deleterious change in the habitat, the species population of at least one, usually many, other species will also be affected negatively. As a rule, several populations of other species will be affected directly, many more at second hand.

Bibliography

Alexandre, D.-Y. (1978). Le rôle disseminateur des eléphants en forêt de Taï, Côte d'Ivoire. *La terre et la vie*, **32**, pp. 47–72.

Ayensu, E. S. (1974). Plant and bat interactions in West Africa. *Annals of the Missouri Botanic Garden*, **61**, pp. 702–27.

Balle, S. & Hallé, N. (1961). Les Loranthacées de la Côte d'Ivoire. *Adansonia* sér. 2, **1**, pp. 208–65.

Bawa, K. D. (1977). Reproductive biology of *Cupania guatamalensis* Radlk. (Sapindaceae). *Evolution*, **31**, pp. 52–63.

Beattie, A. J. (1985). *The evolutionary ecology of ant–plant mutualisms*. Cambridge: Cambridge University Press.

Becking, J. H. (1975). Root nodules in non-legumes. In *The development and function of roots*, ed. J. G. Torrey & D. T. Clarkson, pp. 507–66. New York: Academic Press.

Bequaert, J. (1922). Ants of the American Congo Expedition IV. *Bulletin of the American Museum of Natural History*, **45**, pp. 333–583.

Boucher, D. H., James, S. & Keeler, K. H. (1982). The ecology of mutualism. *Annual Review of Ecology and Systematics*, **13**, pp. 315–47.

Buckley, R. C. (1982). Ant–plant interactions: a world review. In *Ant–plant interactions in Australia*, ed. R. C. Buckley. The Hague: Junk.

Chevalier, A. (1936). Contribution à l'étude de quelques espèces africaines du genre *Dioscorea*. *Bulletin du Musée d'histoire naturelle, Paris*, **8**, pp. 520–51.

Dafni, A. (1984). Mimicry and deception in pollination. *Annual Review of Ecology and Systematics*, **15**, pp. 259–78.

De Bary, A. (1879). *Die Erscheinung der Symbiose*. Strasburg: Karl J. Trubner.

Faegri, K. & Pijl, L. van der (1980). *The principles of pollination ecology*. Oxford: Pergamon Press.

Feehan, J. (1985). Explosive flower opening in ornithophily: a study of pollination mechanisms in some Central African Loranthaceae. *Botanical Journal of the Linnean Society*, **90**, pp. 129–44.

Free, J. B. (1970). *Insect pollination of crops*. New York: Academic Press.

Furman, T. E. & Trappe, J. M. (1971). Phylogeny and ecology of mycotrophic achlorophyllous Angiosperms. *Quarterly Review of Biology*, **46**, pp. 219–25.

Gilbert, L. E. (1975). Ecological consequences of coevolved mutualism between butterflies and plants. In *Coevolution of animals and plants*, ed. L. E. Gilbert &

P. H. Raven, pp. 210–40. Austin: University of Texas Press.

Guerrant, E. O. & Fiedler, P. L. (1981). Flower defences against nectar pilferage by ants. *Reproductive botany* (Supplement to *Biotropica*), **13**, pp. 25–33.

Hagerup, O. (1943). Myrebestøvning. *Botanisk tidsskrift*, **46**, pp. 116–23.

Huxley, C. R. (1986). Evolution of benevolent ant/plant relationships. In *Insects and the plant surface*, ed. B. E. Juniper & R. S. Southwood, pp. 257–82. London: Edward Arnold.

Janzen, D. H. (1972). Protection of *Barteria* (Passifloraceae) by *Pachysima* ants (Pseudomyrmecinae) in a Nigerian rain forest. *Ecology*, **53**, pp. 885–92.

Jones, C. E. & Little, R. J. (1983). *Handbook of experimental pollination ecology*. New York: Van Nostrand Reinhold.

Killick, H. J. (1959). The ecological relationships of certain plants in the forest and savanna of central Nigeria. *Journal of Ecology*, **47**, pp. 115–27.

Kozlowski, T. T. & Cooling, J. H. (1961). Natural root grafting in forest trees. *Wisconsin College of Agriculture and Forestry Research Note*, **56**.

Kuijt, J. (1969). *The biology of parasitic flowering plants*. Berkeley, California: University of California Press.

Lersten, N. & Horne, H. T. (1976). Bacterial leaf nodule symbiosis in Angiosperms with emphasis on Rubiaceae and Myrsinaceae. *Botanical Review*, **42**, pp. 145–214.

Levieux, J. & Diomande, T. (1978). The nutrition of granivorous ants. I. *Insectes sociaux*, **25**, pp. 127–40.

Liebermann, D., Hall, J. B. & Swaine, M. D. (1979). Seed dispersal by baboons in the Shai Hills, Ghana. *Ecology*, **60**, pp. 65–75.

Majer, J. D. (1976). The maintainance of the ant mosaic in Ghana cocoa farms. *Journal of Applied Ecology*, **13**, pp. 123–44.

Martin, F. W. & Telek, L. (1971). The stigmatic secretion of the sweet potato. *American Journal of Botany*, **58**, pp. 317–22.

McKey, D. (1984). Interaction of the ant–plant *Leonardoxa africana* (Caesalpiniaceae) with its obligate inhabitants in a rain forest in Cameroun. *Biotropica*, **16**, pp. 81–99.

Mikola, P. (ed.) (1980). *Tropical mycorrhiza research*, Oxford: Oxford University Press.

Mosse, B., Hayman, D. S. & Arnold, D. J. (1973). Plant growth responses to vesicular-arbuscular mycorrhiza. V. *New Phytologist*, **72**, pp. 809–15.

Newton, L. (1971). Epiphytes in West Africa. *Epiphytes*, **3**, pp. 69–70.

O'Dowd, D. J. (1982). Pearl bodies as ant food: an ecological role for some leaf emergences of tropical plants. *Biotropica*, **14**, pp. 40–9.

Pijl, L. van der (1982). *Principles of dispersal in higher plants*. Berlin: Springer-Verlag.

Proctor, M. & Yeo, P. (1973). *The pollination of flowers*. London: Collins.

Real, L. (ed.) (1983). *Pollination biology*. London: Academic Press.

Redhead, J. F. (1960). A study of mycorrhizal associations in some trees of Western Nigeria. Thesis submitted for the Diploma in Forestry. Oxford: Imperial Forestry Institute.

Redhead, J. F. (1980). Mycorrhiza in natural tropical forests. In *Tropical mycorrhiza research*, ed. P. Mikola, pp. 127–42. Oxford: Oxford University Press.

Room, P. M. (1973). Ecology of the mistletoe *Tapinanthus bangwensis* growing on cocoa in Ghana. *Journal of Ecology*, **61**, pp. 729–42.

Schemske, D. N. (1981). Floral convergence and pollinator-sharing in two bee-pollinated tropical herbs. *Ecology*, **62**, pp. 946–54.

Schnell, R., Cusset, G. & Quenum, M. (1963). Contribution à l'étude des glandes extraflorales chez quelques groupes de plantes tropicales. *Revue générale de botanique*, **70**, pp. 269–342.

Schnell, R. & Grout de Beaufort, F. (1966). Contribution à l'étude des plantes à myrmecodomaties de l'Afrique intertropicale. *Mémoire de l'Institut fondamentale d'Afrique noire*, **75**.

Snow, D. W. (1981). Tropical frugivorous birds and their food plants; a world survey. *Biotropica*, **13**, pp. 1–14.

Starr, M. P. (1975). A generalised scheme for classifying organismic associations. *Symposia of the Society for Experimental Biology*, **29**, pp. 1–20.

Steentoft, M. (1988). *Tropical plant biology*. New Delhi: Wiley Eastern, in press.

Vareschi, V. (1980). *Vegetationsökologie der Tropen*. Stuttgart: Verlag Eugen Ulmer.

Visser, J. (1981). *South African parasitic flowering plants*. Capetown: Juta.

Vogel, S. (1973). Fungus/gnat flowers and fungus mimesis. In *Pollination and dispersal*, ed. N. B. M. Brantjes & H. F. Linskens, pp. 13–18. Nijmegen: University of Nijmegen.

Wiens, D. (1978). Mimicry in plants. *Evolutionary Biology*, **11**, pp. 365–403.

Williams, C. N. (1960). *Sopubia ramosa*, a perennating parasite on the roots of *Imperata cylindrica*. *Journal of the West African Science Association*, **6**, pp. 137–41.

Woods, F. W. & Beck, K. (1964). Interspecific transfer of ^{45}Ca and ^{32}P by root systems. *Ecology*, **45**, pp. 886–9.

Yeo, P. F. (1973). Floral allurements for pollinating insects. In *Insect/plant relationships*, ed. H. F. van Emden. *Symposia of the Royal Entomological Society*, **6**, pp. 51–7.

2

Vegetation in West Africa

In the revised map and account of the vegetation of Africa prepared by White (1983), a dozen main vegetation types are recognised in West Africa, strand vegetation being excluded. The account is extensive, treating the whole of Africa. What is required for present purposes is a regional and more intensive one, providing a framework within which the families described in Chapters 3 to 40 may be appreciated.

Some kinds of vegetation occurring only in restricted or remote areas will be omitted. These areas include the Lake Chad and Niger basin swamps, mountain and plateau types, the coastal forest-savanna mosaic of the Accra Plains and the sahel.

Apart from strand vegetation, six zones will be described under names rather more traditional than, but equivalent to, those employed by White (1983):

Strand vegetation	Lowland forest
Mangrove swamp	Forest-savanna mosaic
Swamp forest	Northern guinea savanna
Sudan savanna	

Vegetation is described in terms of the habit of its most obvious components, and a minimum of half a dozen such terms, or combinations of terms, is needed.

Forest is composed of trees at least 10 m high, with touching crowns, which often support lianes and, in wetter areas, epiphytes. The undergrowth is of shrubs, and both this and the ground flora of perennial herbs vary in density according to the density of the tree canopy, which may be partially deciduous, though more often it is evergreen in West Africa. Forest trees succumb to fire as a rule.

Woodland is a less dense kind of vegetation than forest. The crowns often do not touch; indeed White (1983) puts the lower limit of canopy cover at 40% by area. It is perhaps unfortunate that the American term savanna (from the Spanish *sabana*), should have been imported to described non-forest areas of West Africa. Apart from the sahel, which White (1983) describes as being composed of 'wooded grassland and deciduous bushland', our so-called 'savanna' is a woodland–secondary grassland mosaic. The term savanna is, however, so well established in West Africa, that, with the provision that it does not mean grassland, it may still be used.

The ground cover under woodland is largely of grasses, but is accompanied by a rich flora of sedges, herbs, suffrutices and fire-hardy woody plants.

Grassland in West Africa occurs in three main forms: as wooded grassland in the sahel, where there is under 500 mm a year of rain and woody cover amounts to 10–40%; as edaphic grassland in flood plains in savanna, which become waterlogged in the rains, parched in the dry season, and carry a characteristic flora; and as secondary grassland throughout the savanna, and in some localities in forest, where man's activities and those of his herds have had a marked effect. In both edaphic and secondary grasslands, grasses up to 3 m high predominate, accompanied by a large flora of herbs, and with the cover provided by woody species amounting to at most 10% by area.

Bushland and thicket depends on White's definition of a bush as a woody plant intermediate in habit between a shrub and a tree, often 3–7 m high, and multiple-stemmed. When bushes grow closely enough together, thicket is formed, and this includes plants with a krummholz [with arched rooting branches] style of growth. Bushland occurs where there is 40% cover by bushes, and, when cover is more dense than this, scrub forest is formed.

The distinction between bush (3–7 m high) and shrub (up to 2 m high), the two multiple-stemmed forms, is not made in taxonomic works, and so the term bush has to be restricted in what follows to use only in connection with the terms bushland and thicket. Shrubs are otherwise regarded as being up to 5 m high.

On the vegetation map (Fig. 2.1), strand vegetation cannot be shown, but occurs on the coastline away from the influence of major rivers. Mangrove swamp and (freshwater) swamp forest are of local occurrence and only the main areas can be shown on the scale adopted. It is very important to remember to regard each zone as a mosaic, each area of the mosaic being made up of numerous communities.

An attempt is made to describe forest and woodland biomes from four aspects: the persistent (long-established) flora, the disturbed (regrowth) flora, the riverain flora and inselbergs.

Strand vegetation

This is the form of coastal vegetation alternative to mangrove swamp, being found away from the influence of rivers, on sandy shores and sand bars. The shore may be backed by lagoons or cliffs, the latter bearing any of a number of plant communities.

The sand of the shore is porous and well-drained, and the saltwater table lies at some depth, so that fresh-to-brackish water (from rainfall and by drainage) above it is available to plants. There is a nearly continuous on-shore salt-laden wind and the surface of the sand is somewhat unstable, especially so in areas of low rainfall. It is also liable to become very hot by day. In these circumstances, it is not surprising that most species show some xeromorphic features, and often halophytic ones as well.

Fig. 2.1. The vegetation zones of West Africa. Guin. Biss., Guinea-Bissaū.
(Adapted version of the map 'Northwestern Africa', *The Vegetation of Africa*,
© UNESCO 1981.)

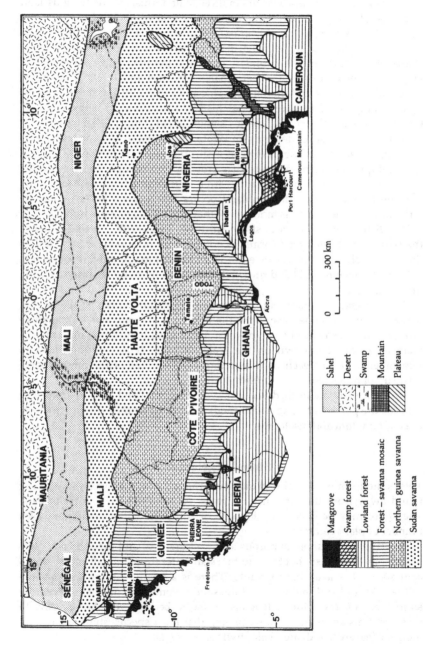

Pioneer zone

A large number of inland weed species flourish on the shore, even near high water level in the pioneer zone, where the sand is being colonised. Here, herbs and undershrubs occur as individuals and do not form continuous cover, although most species have rhizomes, or creeping stems which root above ground. Inland weeds likely to be seen include *Phyllanthus* spp., *Portulaca* spp. and *Schwenckia americana*. These and the species marked with an asterisk in the table below also occur higher up the shore, and there are few species solely characteristic of the pioneer zone.

Alternanthera maritima (Amaranthaceae)	*Ipomoea pes-caprae* (Convolvulaceae)
Canavalia rosea (Papilionaceae)	*I. stolonifera* (Convolvulaceae)
Commelina erecta subsp. *maritima* (Commelinaceae)	*Philoxerus vermicularis* (Amaranthaceae)
Diodia spp. (Rubiaceae)	*Scaevola plumieri*
Euphorbia glaucophylla (Euphorbiaceae)	*Sesuvium portulacastrum*

Scaevola plumieri (Goodeniaceae) is a small shrub up to 90 cm high, and can develop a krummholz habit. It has broad fleshy leaves leaving prominent scars on the thick branches. Short axillary cymes of white flowers are produced, often in the rains, the corollas with all 5 lobes drawn over to one side of the flower and the style and cupular stigma standing on the other. The fruit is a black drupe formed from an inferior ovary.

Sesuvium portulacastrum (Ficoidaceae or Aizoaceae) is a fleshy herb, rooting at the nodes, with narrow ± obovate, opposite leaves and solitary, axillary, apetalous flowers, the calyx pink or purple and 5-lobed, each lobe horned and with a papery margin. The 3-celled, 3-styled ovary has numerous seeds on axile placentae. The fruit is a circumcissile capsule. *Glinus lotoides* (sometimes placed in the Molluginaceae) (also in Gambia) is an annual weed also found in dry river beds, differing in having stellately hairy leaves and tiny white flowers in clusters. The sepals are free, round up to 9 forked staminodes, numerous stamens and a 5-celled ovary. The fruit is a 5-valved capsule, each seed having a long white strophiole.

Pioneer zone sedges/grasses

Sedges	Grasses
*Remirea maritima**	*Sporobolus virginicus**
	*Stenotaphrum secundatum**
	*Vetiveria fulvibarbis**

The pioneer zone is perhaps better characterised by the absence of certain grass species, common at higher levels on the shore, such as *Sporobolus robustus*, *Brachiaria falcifera* (probably *B. jubata*) and *Paspalum vaginatum*.

Main strand zone

Above the pioneer zone, a main strand zone is recognised, where these grasses occur together with a great number of weed species and the species marked with an asterisk in the table above. In addition, there are likely to be maritime species (lists, p. 25) and some distinctive geophytes, such as *Sansevieria liberica*, with blotched sword-shaped leaves (also on inselbergs inland), the bulbous *Urginea indica* and:

Fig. 2.2. *Sesuvium portulacastrum*. A. Rooting and flowering shoot × 1. B. Flower in longitudinal section × 4. C. Operculum of fruit × 8. D. Ovary in cross-section × 8. (From Jeffrey, 1961, Fig. 7 *pro parte*.) (British Crown copyright. Reproduced with permission of the Controller, Her (Britannic) Majesty's Stationery Office & the Trustees, Royal Botanic Gardens, Kew. © 1961.)

Crinum ornatum (Amaryllidaceae) (now *C. zeylanicum*), also found by streams in savanna. The strap-shaped leaves arise from the bulb and have rough edges. The inflorescence is cymose, but umbel-like, with papery bracts, on a scape. Each flower is 3-part, with an inferior ovary and long, pink-striped, white tepal lobes forming a slender perianth tube below. There is no false corona, a feature shared with blood lilies (*Haemanthus*, now *Scadoxus*, spp.; Friis & Nordal, 1976). *C. jagus*, in swampy ground in forest, has leaves approximately 3 times as broad and greenish-white flowers, while *C. ornatum*, the only other species with flowers this colour, is found in moist savanna areas and produces narrow glaucous leaves (with rough edges) after flowering (Nordal & Wahlstrøm, 1980). The third genus of the family, *Pancratium*, has flowers with a false corona formed from united filaments.

In addition to the species already indicated, several more coastal and/or inland species of sedges and grasses occur.

Main strand zone sedges/grasses

Sedges	Grasses
Eleocharis spp.	*Andropogon gayanus*
Fimbristylis spp.	*Cynodon dactylon*
Kyllinga peruviana	*Dactyloctenium aegyptiacum*
Pycreus polystachyos	*Heteropogon contortus*
Scleria naumanniana	*Schizachyrium pulchellum*

At the landward boundary of the main strand zone, truly strand species disappear and, quite often, the ground is cultivated or herds of swine are kept. Hedges of the introduced prickly pear (*Opuntia vulgaris*), sisal (*Agave sisalana*) and Mauritius hemp (*Furcraea gigantea*, now *F. foetida*) are commonly planted. Otherwise some form of coastal savanna thicket or grassland exists, the degree of man's interference and the kind and amount of water supply influencing species composition. This is likely to be a mixture of species with half a dozen different kinds of geographical distribution.

A. Maritime species
Eugenia calycina, p. 54   *Terminalia scutifera*
                                                (Combretaceae)

B. Maritime species extending inland
Chrysobalanus orbicularis   *Hydrocotyle bonariense*, p. 63
Crotalaria falcata   *Ipomoea mauritiana*
    (Papilionaceae)   (Convolvulaceae)
                Stylosanthes erecta (Papilionaceae)

C. Inland species, including riverain ones, extending to the coast
Cynometra vogeli   *Flacourtia flavescens*, p.37
    (Caesalpiniaceae)   *Ixora laxiflora* (Rubiaceae)
Eugenia spp., p. 54   *Napoleonaea vogelii*, p.45

| *Oxyanthus racemosa* | *Phoenix reclinata* (Palmae) |
| (Rubiaceae) | *Smeathmannia pubescens*, p. 32 |

| *D. Species of semi-arid conditions* | |
| *Calotropis procera* (Asclepiadaceae) | *Parinari macrophylla*, p. 46 |

In addition to mangrove shrubs (p. 29), the mangrove climber *Tetracera alnifolia* and, in Sierra Leone, two other maritime climbers, *Habropetalum dawei* and *Stigmaphyllon ovatum*, occur.

Fig. 2.3. *Chrysobalanus icaco* subsp. *icaco*. A. Flowering shoot × ⅔. B. Another leaf type × ⅔. C. Floral diagram. D. Half flower × 7. E. Fruit × ⅔. (A, B, D–F from Letouzey & White, 1978, Fig. 19 *pro parte*; C adapted from Eichler, 1875–8.)

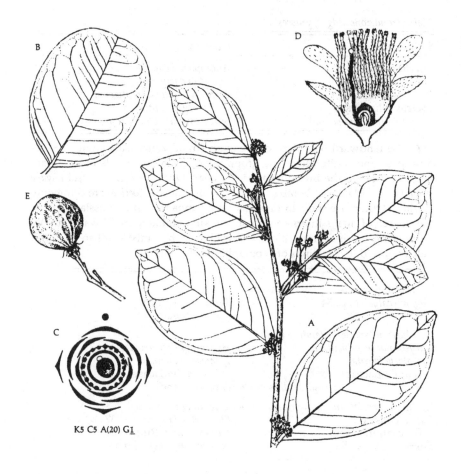

K5 C5 A(20) G<u>1</u>

Chrysobalanus orbicularis (White, 1976) (now *C. icaco* subsp. *icaco* including *C. ellipticus*) (Rosaceae, now Chrysobalanaceae) is a maritime shrub or small tree (also in Gambia) with suborbicular to widely elliptical evergreen leaves and the intrapetiolar stipules of the genus. At most times of the year there are axillary cymes of small, white 5-part flowers, each with a whorl of 15–20 fertile stamens joined basally and projecting from the edge of a receptacular cup, in the base of which is a single 1-celled carpel with a basal style. The fruit is a red-to-black drupe, *c.* 2.5 cm long, the stone opening by 5–8 valves.

Habropetalum dawei (Dioncophyllaceae) has 'leaves' with paired hooks at the tips, and white flowers, the petals of which fall early. The ovary opens soon after fertilisation, exposing seeds which develop an encircling wing several centimetres in diameter.

Stigmaphyllon ovatum (Malpighiaceae) is a slender twiner with opposite leaves, occurring up to the borders mangrove swamp. It has 5-part flowers with a glandular calyx and yellow clawed petals. These are followed by samaras.

Tetracera alnifolia (Dilleniaceae) (Caballé, 1982) (also in Gambia) is a twiner up to 15 m, the leaves deep green and glossy above with many pairs of nerves in grooves, the underside paler green with the nerves forming ribs. Large, apparently terminal, panicles of small 4-part flowers with pink calyces and white petals round numerous stamens and *c.* 4 free carpels are followed by tiny red follicles with a few seeds each. The arilloid is split and red.

In open areas of coastal scrub and in grassland several geophytes may be expected. These include *Haemanthus multiflorus*, which apparently has a bulb, though this is formed by a swollen rosette of leaf bases on the end of the rhizome. Another is *Anthericum warneckii*, and a third is:

Tacca leontopetaloides (Taccaceae) (also in Gambia). The tuber gives rise to 1 or 2 very large pinnately divided and lobed (decompound) leaves *c.* 1 m high and a peduncle 1–2 m high with a cluster of yellowish 3-part flowers with leaf-like outer bracts and thread-like purple inner ones. The fruit is a berry formed from an inferior ovary and containing arilloid seeds.

Mangrove swamp

Red mangroves (*Rhizophora* spp.) predominate here, and their appearance, standing on arching stilt roots, is unique. Although most extensively developed along the Nigerian coastline and along that of Sierra Leone, all West

African countries possess areas of this kind of vegetation. It exists at river-mouths and up river to the point to which salt water is conveyed at high tides (in creeks) and round lagoons.

Land building is going on in such areas. The mud or silt brought down by the river falls to the bottom when the flow of river water is checked on meeting the rising tide from the sea. All mangrove areas are regularly flooded by seawater, making them temporarily much more saline. The lowest-lying areas, the most recently formed fringes of mud or silt at the water's edge, are flooded twice daily, while earlier-formed, higher ground is flooded only twice monthly by the extra-high spring tides.

Seawater contains on average 35 g mineral salts (chiefly sodium chloride) per litre. The salt content of river (fresh) water is, by contrast, measured in parts per million (one part per million being 0.001 g per litre). Salt concentrations in brackish water lie between these two extremes, nearly as salt as seawater at high tides in the dry season, but only one-third as salt as seawater at low tides in the rains, when vast quantities of fresh water pour down the rivers.

Drainage, aeration and mineral concentrations are unique. The soil is a poorly aerated waterlogged mud or silt with fluctuating water levels and salinities, and low oxygen levels. These all make for low levels of root function (poor absorption and growth), with fluctuating water and high salt levels causing difficulties in obtaining enough water. Under these special edaphic conditions it is not surprising that mangrove areas support a limited number of species, and that these species show a number of special features which can be interpreted as adaptations to their special situation.

Red mangroves are often felled and seldom occur as other than shrubs and small trees, though they can become over 40 m high when undisturbed. All three species of *Rhizophora* have numbers of stout, arching stilt roots from the main trunk. Both trunk and stilt roots bear numerous lenticels, and, through these, gas exchange to the benefit of the rooting system can take place. When a stilt root reaches the substrate, it branches to form a raft of roots, firm enough to walk over. This solves another of the mangrove's problems – that of stability in unstable mud or silt washed over and through by large volumes of water.

The third problem, that of obtaining enough water against the greater salt concentration frequently prevailing outside the plant, is seen in the xeromorphic form of the leaves, which are simple, thick and leathery and of modest size – features likely to cut down on water loss by transpiration.

Vivipary is another interesting feature of red mangroves, shared with some other mangrove species. The seeds germinate precociously in the one-seeded berries, the hypocotyl protruding while the fruit hangs on the tree. No radicle is produced but adventitious roots arise from the hypocotyl whether it lodges vertically in mud on dropping from the tree or heels over and lodges horizontally after floating away in the water. Establishment takes place fastest at low tide and is said to occur between two high tides. Such rapid growth on top of precocious germination reflects the need for timing, if establishment is to be successful under such special conditions.

The fruit and seed habits of some mangrove species are described by Jackson (1964).

> *Rhizophora racemosa* (Rhizophoraceae) is the pioneer species appearing at the water's edge on newly deposited mud, but it is also the first species to disappear under human interference. It has many-flowered inflorescences of small 4-part flowers with thick sepals and whitish petals. Flower buds thick and blunt, hypocotyls 30–65 cm long. *R. harrisonii* (also in Gambia) forms thickets on rather higher and drier ground than *R. racemosa* and is distinguished from it by its fewer-flowered inflorescences with slim, ovate flower buds followed by few fruits. Hypocotyls *c.* 20 cm long. This species is possibly a hybrid between *R. racemosa* and *R. mangle.* The latter is a shrub, *c.* 5 m high, occupying the highest and driest ground, well aerated by the burrows of land crabs. The inflorescences are 2-flowered, flower buds slim, elliptical and very pointed, and the hypocotyls are from 15 to 20 cm in length.

Many kinds of vegetation can develop around a mangrove area according to local conditions. Other mangrove species occurring on higher ground (less frequently flooded) are notably *Avicennia africana* (see below), *Laguncularia racemosa* and *Conocarpus erectus.* Both *Avicennia* and *Laguncularia* have upward growing breathing roots (pneumorhizae), which almost carpet the ground around them, those of *Laguncularia* being white-tipped, where the inner tissue is exposed. *Conocarpus* is frequently wreathed with the parasite *Cassytha filiformis.* Both *Avicennia* and *Laguncularia* also show precocious germination of the seed. All three species have xeromorphic leaf characteristics.

> *Avicennia africana* (now *A. germinans*) (Avicenniaceae) (also in Gambia), black mangrove, is a grey-barked shrub or small tree with opposite leaves with glands excreting salt. In the dry season, axillary stalked clusters of white flowers with a yellow centre appear. The flowers are 4-part, 1 petal being adaxial. The pistil is 1-celled with free central placentation, with 4 apical ovules, and forms a 2-valved capsule with 1 seed, which soon produces a hairy radicle.

A number of semi-mangrove species, on the borders of the swamp, belong to the Papilionaceae. They have one-seeded pods or lomenta, which float, and remain viable while doing so for several months. A thicket of such species tends to develop, together with:

> *Flagellaria guineensis*   *Ochna multiflora* (Ochnaceae)
> *Hibiscus tiliaceus* (Malvaceae)   *Raphia* spp. (Palmae)

> *Flagellaria guineensis* (Flagellariaceae) is a herbaceous monocotyledonous climber with leaftip tendrils, occurring near rivers in swamp forest and

the wetter parts of lowland forest, also being somewhat salt tolerant. The aerial stems are unbranched and end in panicles of small, yellow-to-white, 3-part flowers, followed by small red drupes.

Cashew nut (*Anacardium occidentale*) is frequently cultivated on the borders of mangrove areas. Intensive farming in these areas leads to the formation of forest-savanna mosaic.

Closed lagoons (seldom in direct connection with the sea) are common in the low rainfall area of the eastern part of Ghana. They are not only brackish, but salinities vary widely between wet and dry seasons. The water-pollinated herb *Ruppia maritima* occurs, and a selection of the woody plants mentioned above (not red mangroves) occurs on the banks. If the lagoon dries out, grasses and sedges appear, together with *Philoxerus vermicularis* and *Sesuvium portulacastrum*.

Sometimes saline herbaceous swamp develops, and one of the most characteristic plants here is:

> *Xyris anceps* (Xyridaceae), a tufted, grass-like monocotyledon, with flattened peduncles *c.* 40 cm high, bearing globular inflorescences, *c.* 1 cm ∅, with prominent, dry bracts. The flowers are minute, 3-part and yellowish, followed by a 3-valved capsule with parietal placentae. *X. straminea*, on wet flushes on inselbergs, has tiny ± globular inflorescences with semi-transparent bracts, carried on a cylindrical peduncle.

Mangrove swamp sedges/grasses

Sedges	Grasses
Eleocharis geniculata	*Panicum strictissimum*
E. mutata	*Paratheria prostrata*
Kyllinga peruviana	*Paspalum vaginatum*
K. robusta	*Sacciolepis africana*
Scirpus litoralis	*Sporobolus virginicus*

The sedges also grow in salt water, as does *Paratheria prostrata*, while *Panicum strictissimum* favours less saline areas, for example round the high water springs level or above. *Sporobolus robustus* replaces *S. virginicus* at even higher levels (less saline situations).

Swamp forest

This includes both large and small areas of still-standing or flowing water, the high (fresh-)water table influencing the flora decisively. Continuous areas of freshwater swamp forest occur mainly in Nigeria, Côte d'Ivoire and Sierra Leone, while in Ghana there are only smaller, rather species-poor patches within the lowland forest area.

The soil in swamp is an unstable mud or silt, which is poorly aerated and poor in minerals. Many of the principle species show helophytic adaptations, such as stilt roots and pneumorhizae, just as mangrove species do.

There are three main habitats within freshwater swamp forest: open water, whether still or flowing; the exposed edges of the forest around these waters, where more light is available; and the interior of the forest, admitting less light and being extremely difficult to move around in.

In open water which is still, the commonest plant is probably water lettuce (*Pistia stratiotes*), though individuals of the tiny duckweeds outnumber them. Hornwort (*Ceratophyllum demersum*) is also present, but, as it is totally submerged, its frequency is more difficult to estimate. None of the free-floating plants can maintain station in a current; all are washed downstream and perish on meeting brackish water.

Water lettuce can spread so fast that floating islands of vegetation are formed, and in these non-aquatic plants can root. Even woody species such as *Alchornea cordifolia* can do so. The islands tend to be broken up and swept away in the rains.

Aquatic and swamp species (in or on water) of sedges/grasses

Sedges	Grasses
Cyperus articulatus	*Acroceras* spp.
C. nudicaulis	*Leersia hexandra*
Eleocharis dulcis	*Panicum subalbidum*
E. naumanniana	*P. strictissimum*
Scirpus cubensis	*Sacciolepis* spp.

In addition, the grasses *Echinochloa* spp. and *Vossia cuspidata* may both form mats on water and grow out from the banks. Among the sedges, *Eleocharis naumanniana* is restricted to forest areas, while the remainder can occur further north into the sudan zone. The grasses tend, with the exception of *Panicum subalbidum*, to have a more southerly distribution.

Swamp and riverbank species of sedges/grasses

Sedges	Grasses
Eleocharis acutangula	*Coix lacryma-jobi*
E. dulcis	*Panicum repens*
Fiurena umbellata	*Pennisetum purpureum*
Rhynchospora perrieri	*Phragmites karka*
Scleria depressa	*Saccharum spontaneum* var. *aegyptiacum*
S. melanomphala	

Again, the sedges occur in the sudan zone also, the grasses tending to have a more southern distribution. *Saccharum spontaneum* var. *aegyptiacum* has quite a restricted distribution: Upper Volta–Ghana–Nigeria.

In shallow water, a variety of bottom-attached plants are found, and in flowing water, only such species can maintain station. The moss-like members of the family Podostemonaceae are found in fast-flowing water, attached to rock.

Round these areas of water, *Raphia* and rattan palms, together with other climbers and shrubs of the Papilionaceae and Sapindaceae occur, together with:

> *Sphenoclea zeylancia* (Sphenocleaceae), an erect, much-branched, weedy herb with spongy stems and narrow leaves. There are terminal narrow, but prominent, spikes of closely packed 5-part white flowers, each about 6 mm Ø. The 2-celled capsule, with numerous seeds on axile placentae, opens transversely. Even plants only a few centimetres high (and unbranched) are likely to have at any rate 1 spike. The species tolerates some salt and is found in creeks as well as in swampy places inland; also in Gambia.

Other trees and shrubs include members of the Annonaceae, Euphorbiaceae and Guttiferae and:

Cynometra spp.	*Pandanus candelabrum*
(Caesalpiniaceae)	*Smeathmannia pubescens*
Napoleonaea vogelii, p. 45	*Syzygium owariense*, pp. 53
Nauclea pobeguinii (Rubiaceae)	*Ximenia americana*, p. 50

> *Pandanus candelabrum* (Pandanaceae), generally known as screwpine, is a coastal and inland swamp tree up to 10 m high, often growing in groups. The trunk is thorny and bears prominent stilt roots. The few branches bear crowded, spirally arranged strap-shaped leaves, sharply toothed along the edges and midrib. ♂ plants produce long, thick furry spikes of flowers each with approximately a dozen joined stamens, while ♀ plants produce thick whitish spikes of flowers, each of a superior pistil. These fuse in the fruit to form an infructescence of yellowish-green angular drupes, equivalent to a mini-pineapple. Generally found by rivers inland, this species can also occur by water on some inselbergs and near the coast (see also Huynh, 1987).

> *Smeathmannia pubescens* (Passifloraceae) is an under-storey shrub or small tree with stipulate leaves with glandular teeth round the blade. From the late rains onwards, single, white, 5-part scented flowers, 6 cm Ø appear in the axils of the leaves. Each has a small corona between the petals and numerous stamens, and a hairy 1-celled ovary with parietal placentae and free styles in the centre. The fruit is a brittle capsule, round which the calyx persists. The species is found by rivers as far north as in fringing forest, and also in secondary forest; in both kinds of places relatively open conditions exist.

In the interior of the swamp, an irregular kind of forest develops, with patches of standing water surrounded by rattans, straggling shrubs such as *Leea guineensis*, and clumps of large aroids and marantaceous herbs.

> *Leea guineensis* (Ampelidaceae or Leeaceae) has ribbed stems pinched at the nodes. The large, compound bipinnate leaves have serrate leaflets, the stalk of each of which is also pinched. Large, spreading leaf-opposed cymes of small yellowish-red flowers appear in the rains, followed by 5–8-lobed, waxy red berries, *c.* 6 mm ∅. These later turn black. The species also appears in forest regrowth. Other genera of Ampelidaceae are *Cissus*,

Fig. 2.4. *Leea guineensis*. A. Flowering shoot × ⅓. B. Flower × 8. C. Staminal tube (joined to base of petals) in side view × 8. D fruit × 2. (From Descoings, 1972, Pl. 48 *pro parte*.)

of similar habit to *Leea* but with simple or digitately compound leaves, and leaf-opposed tendrils, a species of which will be found in most habitats, including on inselbergs. *Ampelocissus* has simple leaves (except for *A. multistriata* in savanna) and inflorescence-opposed tendrils.

One of the most characteristic (and tallest) trees is *Mitragyna ciliata*, but the canopy is, in general, no more than 30 m high, and is composed of members of the Euphorbiaceae and:

Alstonia boonei	*Haplormosia monophylla*
(Apocynaceae)	(Papilionaceae)
Berlinia spp.	*Parinari robusta*, p. 46
(Caesalpiniaceae)	*Xylopia* spp. (Annonaceae)

The under-storey, up to about 10 m high, again contains euphorbiaceous species together with *Raphias*, rattan palms, *Carapa procera*, *Chrysobalanus icaco* and *Rothmannia megalostigma*.

On drier ground, conditions more resemble those of moist lowland forest, and species such as *Lophira alata*, and, to the west of Nigeria, *Tarrietia utilis*, appear.

Lowland forest

In West Africa this occurs in an interrupted belt, more or less parallel to the coastline. The eastern section, continuous with the great forests of the Congo basin, covers the southern part of Nigeria. The western part covers western Ghana and stretches westwards to just within the eastern border of Sierra Leone. The eastern section contains many of the same species as the Congo forests, but is less rich in species, and the western section is even poorer.

West African lowland forest is, nevertheless, one of the great tropical rain forests of the world. Very little of it has never been disturbed by man, and, in addition, edaphic conditions are far from uniform throughout its area. The forest must, therefore, be thought of as a complex of mosaics, partly determined by man's activity, and representing a stage in recovery from that disturbance, and partly determined by soil conditions such as aeration and nutrient supply.

The forest zone receives at least 1200 mm of rainfall a year, but this is highly seasonal in distribution, so that, near the coast, the growing season lasts at least nine months (with two rainfall peaks), while further inland it may approach seven months only. Western Liberian forests also have a growing period of this length but a single-peak rainy season and very high rainfall.

In order to simplify presentation, older (or mature, or closed, or high) forest will be described first, followed by a description of secondary (more recently disturbed) forest.

High forest

This is probably best thought of as secondary forest of sufficient age to be very similar to primary forest. Although the tree crowns touch, the very tallest-growing species are termed emergents, since in West Africa their crowns never form a continuous layer or storey. They are at least 35 m high. Under them, the upper-storey crowns lie at a height of 20–35 m. Below this is the under-storey (5–20 m). The shrub layer is regarded as occupying a zone from 2 to 5 m, and beneath the shrubs can be a herb layer, 2(+) m high. Saplings belong successively to each storey until mature height is reached.

The wetter parts of lowland forest – rainfall over 1600 mm a year – form a coastal belt (north to lat c. 6° N) in Liberia and W. Ghana, where such forest is known as moist evergreen forest. It reappears in the southern part of W. Nigeria, forms the entire forest of E. Nigeria and continues through Cameroun into the Congo basin. Some areas near the coast qualify for inclusion in the wet evergreen forest type.

Between latitudes c. 6° and 8° N, a drier type of forest occurs (rainfall 1200–1600 mm a year). This is found in W. Nigeria, the E. Ghana–Togo triangle (see Fig. 2.1) and from W. Ghana into Côte d'Ivoire. It is also composed of evergreen forest, but with a greater number of species which lose at least a proportion of their leaves, especially in the dry season. This kind of forest may be described as semi-deciduous or, better, as 'peripheral semi-evergreen' (White, 1983). Leaf shedding is most obvious among emergent and upper-storey species and separate lists are given for these storeys in the two kinds of forest.

White (1983) describes the remaining inland forest, which extends from Côte d'Ivoire to just within the Sierra Leone border, as a mosaic of the wet and moist forest types mentioned above.

The three kinds of forest grade into each other, unless a sharp boundary is brought about by, for example, an edaphic discontinuity, and other intermediate kinds of forest certainly are also to be found. Nine types and subtypes have been distinguished in Ghana alone.

Emergent and upper-storey species of wet or moist evergreen lowland forest

Afrormosia elata (Papilionaceae)	*Lophira alata* (Ochnaceae)
Anopyxis klaineana	*Parinari excelsa*, p. 46
Klainedoxa gabonensis	*Xylopia quintasii* (Annonaceae)

The families Bombaceae, Caesalpiniaceae, Guttiferae, Meliaceae, Mimosaceae, Rubiaceae, Sapotaceae and Sterculiaceae are well represented, the Meliaceae particularly so in Nigeria.

Anopyxis klaineana (Rhizophoraceae) is an evergreen emergent, lacking buttresses, with leathery leaves in pairs, or whorls of 3–4, with linear interpetiolar stipules. Small, yellowish 5-part flowers are produced in stiffly stalked axillary clusters most of the year, K(5) velvety, C5 A(10) forming a tube *Entangdrophragma* style (p. 173). The fruit is a grey, 5-valved woody capsule with winged seeds, standing on an enlarged, persistent calyx.

Fig. 2.5. *Klainedoxa gabonensis*. A. Flowering shoot × $\frac{1}{2}$. B. Flower × 5. C. Fruit × $\frac{1}{2}$. D. Fruit in cross-section × $\frac{1}{2}$. E. *Irvingia gabonensis* fruit × $\frac{1}{2}$. (From Voorhoeve, 1965, Fig. 68.)

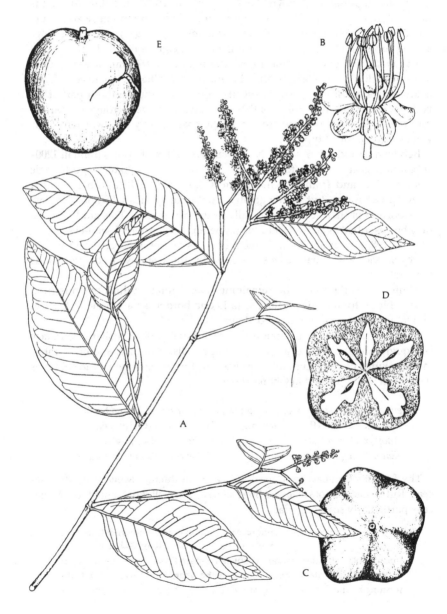

Klainedoxa gabonensis var. *oblongifolia* (Irvingiaceae) is also an evergreen forest emergent, but with wide-spreading buttresses. Trunks of young trees may have root spines. In any case, the stipules, 10–20 cm long, encircle the terminal buds as in *Rhizophora*. Terminal panicles of small, pink, 5-part flowers appear in the dry season, each with an orange disc between the 10 stamens and 5-sided ovary. This becomes a black 5-lobed, 5-stoned drupe up to 7.5 cm Ø, looking rather like an outsized tomato.

Emergent and upper-storey species of the drier parts of lowland forest (peripheral semi-evergreen forest) occur particularly in the following families: Bombaceae, Combretaceae, Meliaceae, Mimosaceae, Moraceae, Sterculiaceae, Ulmaceae.

Lower-storey species of lowland forest

Although species lists for the lower storeys of wetter and drier forests are rather different, the same families are involved in both. Just as there are more deciduous species among emergent and upper-storey species of drier forest, so are there also among lower-storey species. Families well represented include the Annonaceae, Apocynaceae, Bignoniaceae, Caesalpiniaceae, Euphorbiaceae, Guttiferae, Moraceae, Rubiaceae, Sapindaceae, and Sterculiaceae; also:

Anthocleista spp., p. 60	*Oncoba spinosa*
Aptandra zenkeri, p. 50	*Diospyros* spp.
Caloncoba gilgiana	*Fagara* spp., p. 45

Caloncoba (Flacourtiaceae) is a genus of forest shrubs and trees, having acuminate leaves with petioles at least 2.5 cm long, and stipules that fall early. When young, leaves are scaly or hairy. The large white flowers (2.5–10 cm Ø) with numerous stamens are followed by large woody fruits. The most widespread species is *C. gilgiana*, flowering while nearly leafless in the dry season. Clusters of 2–4 large long-stalked flowers with smooth ovaries are followed by smooth woody fruits. *Oncoba spinosa* differs in its shorter petioles, less than 2.5 cm long. Both genera have 1-celled fruits with parietal placentae. Only *O. spinosa* occurs in Gambia.

Flacourtia flavescens, in fringing forest and by rocky hills in savanna and by rivers down to the coast, is a small tree with branch spines which are long and straight, like those of *Oncoba spinosa*, and leaves which are also similar, but rounded rather than acuminate. Trees are ♂ or ♀, the flowers apetalous, K(4–5), petaloid, imbricate. ♂ flowers have an extrastaminal disc, which is glandular, and numerous free stamens. ♀ flowers have a pistil, of 3–5 joined carpels with the same number of styles, on a disc. Ovary incompletely 3–5-celled with numerous ovules, forming a red or purple berry-like drupe (Niger plum).

Diospyros (Ebenaceae), ebonies (timber) and persimmons (fruit) – a genus of dioecious trees and shrubs, with about 40 species in West Africa, about half of these being confined to the eastern end of the area. Most are lowland forest species, though *D. tricolor* is found in coastal thicket. A few species extend into savanna along rivers, including *D. monbuttensis*, (Côte d'Ivoire–Zaïre), the only species with stem spines. The flowers in the genus are whitish, small, K(3–7) C(3–7), with contorted corolla lobes (compare Chapter 22, Apocynaceae) and either ♂ or ♀. In ♂ flowers there are 2–1000 stamens, usually apiculate and often joined in groups; in ♀ flowers the pistil is superior, composed of 2–several joined carpels with at least as many stigmas. There are 2 pendulous ovules in each cell, or 1 per cell where there is a septum dividing the cell into 2. The calyx persists round the base of the berry. *D. mespiliformis* is decidedly the most wide-spread species, occurring from drier forest areas northwards into riparian woodland (also in Gambia). It has a bole fluted at the base, and the bark is black and fissured in squares. The leaves are elliptical-obovate, and at least acute, if not acuminate, with many pairs of lateral veins. Flowers are 4–5-part, with a prominent tube and small corolla lobes, paired anthers in the clusters of ♂ flowers, and stigmas in the solitary ♀ flowers, being included in that tube. The fruit is yellow and globular, c. 2.5 cm ∅, hairy only when young, standing on a large persistent calyx with 4–5 wavy-edged lobes. *D. sanza-minika* is common in Ghana and may there be confused with *D. mespiliformis*. However, it lacks flutes to the trunk, has longitudinally fissured bark, and has larger lanceolate leaves. The berry is cylindrical, up to 5 cm tall, standing on a plate-like calyx with small irregular lobes.

The shrub layer of high forest is evergreen and sparse, and the Rubiaceae are particularly well-represented. The 5 m tall herbs of the Marantaceae are ecologically part of the shrub layer also. The families commonly represented include the Annonaceae, Apocynaceae, Combretaceae, Compositae, Euphorbiaceae, Moraceae, Sapindaceae and Solanaceae.

The lianes of lowland forest are in general under-collected, and may make up as much as 40% of the species present. The Apocynaceae and Papilionaceae are well represented, together with many of the following genera:

Afrobrunnichia	*Neuropeltis* (Convolvulaceae)
Artabotrys (Annonaceae)	*Pararistolochia*
Combretum (Combretaceae)	*Salacia*, p. 60
Entada (Mimosaceae)	*Strychnos*, p. 60
Hippocratea, p. 59	*Uncaria* (Rubiaceae)
Hugonia	*Uvaria* (Annonaceae)

Afrobrunnichia erecta (Polygonaceae) (Cremers, 1974), in the same family as the garden corallita, climbs in the lower canopy by means of axillary tendrils. In the dry season clusters of apetalous flowers are produced, each with a green perianth tube with 5 pink lobes. Each pedicel is 2-winged, and this structure enlarges up to 7 cm long as the nut, enclosed in the perianth tube, matures. Ptyxis is revolute, as in other members of the family.

Hugonia planchonii (Linaceae) (Badré, 1971) belongs to a genus of low-climbing shrubs producing recurved hooks on the stems, and with finely divided more or less persistent stipules. Young growth is hairy. Flowers are bright yellow, short-lived and rather malvaceous-looking (p. 118), but with only 10 stamens (with 2-celled anthers) and pistil with 10 pendulous ovules and 5 styles.

Pararistolochia spp. (Aristolochiaceae). The 12 species are forest twiners of mostly limited distribution, *P. flos-avis* being one of the more widespread. It has twining petioles and ovate-elliptic leaf blades, but the spectacular flowers, produced on old wood, are more commonly seen. There is a large, curved and constricted petaloid perianth tube, purple-spotted and up to 8 cm long, with 3 purple lobes of similar length. The gynostemium stands above an inferior ovary. The fruit is pepo-like, 6-sided and *c.* 3 cm Ø.

Forest herbs
Growing in deep shade, these tend to be scattered, but in clearings, where there is more light, very dense communities may develop; broad-leaved grasses appear in shade, sedges more often in clearings. Families well represented include the Acanthaceae, Araceae, Commelinaceae, Marantaceae, Rubiaceae and Zingiberaceae.

Forest species (also near water) of sedges/grasses

Sedges	Grasses
Cyperus diffusus	*Beckeropsis uniseta*
C. fertilis	*Centotheca lappacea*
Hypolytrum heteromorphum	*Commelinidium gabunense*
H. purpurascens	*Cyrtococcum chaetophoron*
Mapania spp.	*Megastachya mucronata*
Scleria boivinii	*Oplismenus* spp.
S. verrucosa	*Pseudechinolaena polystachya*

The bambusoid grasses (genera 2–8 in the Gramineae of the *Flora of West Tropical Africa*) are, with the exception of *Oxytenanthera abyssinica*, forest species with rather wide leaves (3–7 cm). The remaining genera of forest grasses have rather narrower leaves (1.5–3.5 cm) with the exception of those of *Setaria megaphylla*, which are up to 10 cm wide. This is also the only non-bambusoid species with cross-veins in the leaf, and has a pseudopetiole as well. *Hypolytrum* and *Mapania* are genera particularly prominent in the wetter parts of forest. *M. amplivaginatum* and *M. macrantha* in Nigeria are replaced by *M. baldwinii* and *M. coriandrum* in Ghana and further west, accompanying *Tarrietia utilis* very often.

Secondary forest
Whenever high forest is disturbed, by felling, farming or the fall of decayed trees, the canopy is broken and more light enters, altering conditions for growth. A new growth of colonising species soon appears, partly composed of secondary forest species – which only survive in strong-light conditions, and soon disappear again – and partly of high forest species, seed being supplied by the surrounding forest. These species survive in the shade of the secondary forest species, grow through them, and persist.

Secondary growth on fallow land in forest has been well investigated. When only very short fallow periods are allowed, say two years, secondary growth is species poor, and the succession is curtailed. If the fallow period is prolonged, the plot first becomes covered with an impenetrable thicket of shrubs, saplings and climbers, which form a canopy 7–8 m above ground. Penetrating and standing above this canopy are the trees preserved during clearance. These include large trees which are difficult to remove (list A below, together with *Lophira alata*, Caesalpiniaceae, Mimosaceae, Sapotaceae) and planted and preserved trees such as oil palms and *Irvingia gabonensis*.

> *Irvingia gabonensis* (Irvingiaceae) (African mango) is an evergreen tree up to 35 m tall, with dark green shining foliage. The stipules are of the same kind as in *Klainedoxa* (p. 37), but only 1.5 cm long. The flowers are also similar, but greenish and scented, and produced in axillary racemes. The ovary is only 2-celled, and becomes a pale yellow ellipsoid drupe with 1 stone.

Later, faster growing tree species form a new canopy at a higher level (high bush stage) and later still, slower growing but shade-tolerant high forest species penetrate both these canopies and eventually a high forest canopy with emergents is formed.

Invading species are mostly persistent, but among shrubs, small climbers and herbs the proportion of seral species may be up to one-fifth of the species present early in the succession, while among taller tree species (over 8 m at maturity), there may be no seral species. Concerning list C, below, three-quarters of the persistent species can germinate in at least light shade, but the remaining one-quarter, needing light for germination, is dependent on the occurrence of gaps in the forest canopy if the species are to regenerate (Swaine & Hall, 1983). Such species are marked with an asterisk in the lists below.

Lianes are nearly all persistent and, of these, all can germinate in shade except *Adenia lobata*. However, there is some doubt as to the importance of germination in colonisation by lianes. Peñalosa (1984) has shown invasion from surrounding forest by rapidly growing orthotropic shoots which root. Two African species, *Entada scelerata* and *Tetracera alnifolia* also do this (Caballé, 1982).

A. *High forest species, mature height over 30 m, all species persistent*

*Alstonia boonei**	*Entandrophragma angolense*
(Apocynaceae)	(Meliaceae)
Anopyxis klaineana, p. 35	*Nauclea diderrichii** (Rubiaceae)
*Ceiba pentandra**	*Ricinodendron heudelotii**
(Bombaceae)	(Euphorbiaceae)
Celtis zenkeri	*Terminalia ivorensis**
(Ulmaceae)	(Combretaceae)
*Chlorophora excelsa**	*Triplochiton scleroxylon*
(Moraceae)	(Sterculiaceae)

B. *Under-storey species, mature height 8–30 m*

*Cussonia bancoensis**, p. 49	*Hymenostegia afzelii*
Funtumia africana	(Caesalpiniaceae)
(Apocynaceae)	*Macaranga* spp.
*Harungana madagascariensis**	(Euphorbiaceae)

Monodora myristica (Annonaceae)   *Trichilia monadelpha*
Musanga cecropioides*   (Sterculiaceae)
   (Moraceae)   *Vernonia conferta*
Trema guineensis* (Ulmaceae)   (Compositae)

C. *Small trees and shrubs, mature height up to 8 m*
Alchornea floribunda*   *Ixora laxiflora*
   (Euphorbiaceae)   (Rubiaceae)
Bridelia micrantha*   *Olax gambecola*, p. 50
   (Euphorbiaceae)   *Pavetta owariensis*
Clerodendrum capitatum*   (Rubiaceae)
   (Verbenaceae)   *Psychotria* spp. (Rubiaceae)
Coffea rupestris (Rubiaceae)   *Rauvolfia vomitoria**
Hymenodictyon floribundum*   (Apocynaceae)
   (Caesalpiniaceae)

D. *Lianes*
Combretum spp.*   *Rhaphidophora africana*
   (Combretaceae)   (Araceae)
Hippocratea spp., p. 59   *Salacia* spp., p. 59
Landolphia spp.   *Strophanthus preussii*
   (Apocynaceae)   (Apocynaceae)
Neuropeltis acuminata   *Strychnos* spp., p. 60
   (Convolvulaceae)

E. *Smaller climbers*
The Cucurbitaceae and Rubiaceae are well represented together with:
Baissea spp. (Apocynaceae)   *Pararistolochia macrocarpa*, p. 38
Cissus spp., p. 33   *Parquetina nigrescens**
Clerodendrum umbellatum*   *Secamone afzelii*
   (Verbenaceae)   (Asclepiadaceae)
Mikania cordata* (Compositae)   *Urera* spp., p. 43

F. *Herbs*
A large number of families is represented, both by forest species (germinating
in shade) and by species needing light for germination (more than half the
species present as a rule). Both annual and perennial species are present,
also sedges and grasses.

Harungana madagascariensis (Hypericaceae) (also in Gambia) is a shrub
or small tree up to 10 m high, all young growth being covered by reddish
stellate hairs. The leaves are opposite and covered with black glands.
Small, white, scented, 5-part flowers are produced in terminal corymb-like
cymes at various times of the year. The flowers are also dotted with black
glands and have 5 bundles of stamens, G(5) topped by 5 persistent styles,
and, eventually, numerous tiny black drupes, with a crusty skin and 2–5
stones, develop.

Parquetina nigrescens (Periplocaceae) is a common twiner of regrowth
forest, becoming woody. The leaves are opposite, leathery and glossy
above, whitish below, with apiculate tips. Leaf-opposed cymes of small
greenish flowers appear in the rains, each flower 5-part, the corolla tubular
with lobes *c.* 1 cm long, green outside like the calyx but red or purple

and velvety inside. The relationship between the 5 stamens with hairy anthers, and the 2 carpels, joined only by their styles, is complex (compare

Fig. 2.6. *Harungana madagascariensis*. A. Flowering shoot × ½. B, C. Extremes of leaf shape × ½. D. Flower × 8. E. Part of inflorescence × 1. F. Fruit (flesh removed) with stones exposed and one cut open to show seed × 8. (From Milne-Redhead, 1953, Fig. 5 *pro parte*.) (British Crown copyright. Reproduced with permission of the Controller, Her (Britannic) Majesty's Stationery Office & the Trustees, Royal Botanic Gardens, Kew. © 1953.)

Asclepiadaceae). The pollen is in tetrads, the fruit a pair of follicles up to 45 cm long, containing seeds with a tuft of silky hairs on each. In the same family, *Tacazzea apiculata* is a similar twiner of inselbergs and forest patches in forest-savanna mosaic northwards. The leaves are broad, up to 10 cm long, with a tiny apicule, and woolly below. Masses of small, yellowish-white flowers are produced, each with a similarly short corolla tube but glabrous anthers. The follicles are very different from those of *Parquetina*, hairy and only 5 cm long, almost conical in fact.

Urera spp. (Urticaceae) climb by means of adventitious roots and have stinging hairs. *U. oblongifolia* and *U. robusta* occur in secondary growth, the former and *U. rigida* also as facultative epiphytes.

Where fallow periods are reduced to two to four years, the number of species present is also reduced, and only thicket is formed before the onset of the next farming period. Weed species, of both woody and herbaceous kinds, predominate, accompanied by oil palms, *Combretum* spp. and *Trema guineensis*. *Rauvolfia vomitoria* is one of the last species to disappear when this regime is prolonged, only grasses, especially *Imperata cylindrica*, remaining. A selection of the species of lists E and F, above, is commonly typical of a short fallow plot together with:

Alchornea cordifolia	*Icacina trichantha*, p. 58
(Euphorbiaceae)	*Mallotus oppositifolius*
Byrosocarpus coccineus	(Euphorbiaceae)
Clausena anisata	*Microglossa volubilis* (Compositae)
Cnestis ferruginea	*Rauvolfia vomitoria*
Ficus exasperata (Moraceae)	(Apocynaceae)
Glyphaea brevis, p. 50	*Solanum torvum* (Solanaceae)
Hoslundia opposita	*Vernonia amygdalina*
(Labiatae)	(Compositae)

Byrsocarpus coccineus (Connaraceae) is a low deciduous shrub or climber with compound imparipinnate leaves with *c*. 5 pairs of small, rounded leaflets appearing with or after the flowers in the dry season. The flowers are small, white and 5-part, carried in short axillary racemes. The flowers are heterostylous and self-incompatible but there is a discrepancy in the description of filament length in East and West Africa (Baker, 1962). Only one of the 5 free carpels ripens, forming a red, fleshy follicle, about 1.5 cm long, containing an apparently yellow seed. The yellow layer is the sarcotesta, which can be scraped off, revealing the black 'seed'. *Cnestis ferruginea* is another member of the same family commonly found in similar situations; also in Gambia. It flowers after the new leaves have appeared, however, and the sepals are longer than the petals. Velvety red follicles (2–5) are formed per flower, each with a black seed with a basal yellow arilloid.

Clausena anisata (Rutaceae) is a deciduous shrub or small tree, regularly encountered in forest regrowth. It has gland-dotted, aromatic, imparipinnate compound leaves and axillary inflorescences of 4-part (A8) flowers

Fig. 2.7. *Clausena anisata*. A. Flowering branch × $\frac{2}{3}$. B. Fruiting branch × $\frac{2}{3}$. C. Underside of leaflet × 4. D. Flower × 6. One anther is missing. Eight stamens may be expected. The sepals are orthogonal, the petals diagonal. E. Pistil × 6. F. Ovary in cross-section × 12. G. Stamen × 8. (From Mendonça, 1963, Fig. 31, with permission from the Editorial Board of *Flora Zambesiaca*.)

with G(4), followed by black, shining ellipsoidal berries less than 1 cm long. These are usually 2-locular with a brown seed in each compartment. *Fagara* (now *Zanthoxylum*) in the same family, is a genus of under-storey forest trees and shrubs with similar leaves but has prickles, generally broad-based on all stems and stalks. The minute 4–5-part ♂ or ♀ flowers are carried in terminal panicle-like inflorescences (which are also prickly in *F. macrophylla*, now *Z. gilletii*). There are only 4–5 stamens or staminodes per flower and, by abortion, a single carpel, which forms a glandular follicle with a single blue-black seed. The flowers of both genera resemble in plan those of the Sapindaceae and Anacardiaceae, though with a stamen or staminode opposite each sepal, the single carpel making *Zanthoxylum* flowers .l., perhaps obliquely so.

Forest regrowth sedges/grasses

Sedges	Grasses
Bulbostylis spp.	*Andropogon gayanus*
Cyperus spp.	*Digitaria* spp.
Fimbristylis spp.	*Eragrostis* spp.
Kyllinga squamulata	*Imperata cylindrica*
Lipocarpha chinensis	*Loudetia arundinacea*
Mariscus squarrosus	*Pennisetum purpureum*
Pycreus pumilus	*Rottboellia exaltata*

Riverain forest

This occurs on well-drained riverbanks, though most species can withstand temporarily swampy conditions. In permanently swampy areas, the species are those of freshwater swamp forest. Woody species of the families Annonaceae, Caesalpiniaceae, Euphorbiaceae, Mimosaceae, Moraceae, Papilionaceae and Rubiaceae are prominent. In Sierra Leone *Parkia bicolor* is characteristic, but generally the following occur:

Cola laurifolia (Sterculiaceae)	*Napoleonaea vogelii*
Combretum cuspidatum	*Ochna multiflora* (Ochnaceae)
(Combretaceae)	*Olax gambecola*, p. 50
Hippocratea macrophylla, p. 60	*Parinaria congensis*
Manilkara obovata (Sapotaceae)	*Smeathmannia pubescens*, p. 32

Syzygium rowlandii, p. 54

Napoleonaea vogelii (Lecythidaceae) is an evergreen tree up to 15 m with marginal glands on the leaves. In the dry season, complex flowers *c.* 4 cm ∅ appear in the axils of leaves or leaf scars (or even on trunks). They are 5-part, with a velvety calyx, the lobes tipped with glands, round a ribbed, reflexed, yellowish-red-to-purple corolla. Within is a double false corona of staminodes tipped with purple round 10 united stamens and, in the centre, is a 5-celled inferior ovary with a thick short style. The fruit is a rough yellow berry with a thin skin and 5 cells, each with 2 seeds in a jelly-like pulp. The calyx stands on top of the fruit, which is *c*. 6 cm ∅. *N. imperialis* is an under-storey forest species with 2 glands at the base of the leaf acumen as well as a pair at the base of the blade. The calyx is glabrous, in contrast to that of *N. vogelii*. *N. imperialis* is a Nigerian (–Angola) species, occurring also in forest outliers and relic forest in forest-savanna mosaic, while *N. vogelii* occurs from coastal areas up river into high forest, Sierra Leone–S. Nigeria.

Parinari congensis (Chrysobalanaceae, formerly Rosaceae) belongs to a genus of evergreen trees and shrubs in the same family as *Chrysobalanus* (p. 27). *Parinari* spp. have short-petioled leaves with a pair of glands on the petioles. Stipules are mostly deciduous. That the cavities behind the stomata are filled with hairs is a useful, though microscopic character. Terminal, branched leafy inflorescences of white 5-part flowers are produced mainly in the dry season. These differ from *Chrysobalanus* flowers in being .l. with a single 2-celled carpel standing on 1 side of the receptacle cup, on the rim of which are 7(–10) stamens and *c*. 6 staminodes or teeth. Large deciduous bracts and bracteoles are a feature of the young inflorescence. The fruit is a rough-skinned drupe up to 5 cm long, the stone opening by 2 basal germination pores. *P. congensis* is a riverain forest and forest outlier species and, in forest, *P. excelsa* (also in Gambia) needs to be distinguished. *P. congensis* has leaves which are white and cottony beneath with petiolar glands on the upper half of the petiole, while *P. excelsa* has new leaves covered with a felt of golden brown hairs beneath and gland at the middle or on the lower half of the petiole. In Gambia and Sierra Leone, *P.* (now *Neocarya*) *macrophylla* (White, 1976) is found in sandy coastal areas, reappearing in N. Nigeria in drier savanna. The foliage is pale green with hairs on the upper side of the midrib only. The flowers are in stout little branched inflorescences with a few small bracts and bracteoles, each flower with 12–17 fertile stamens, an inflated receptacle tube and obtuse calyx lobes. *P. curatellifolia*, a widespread savanna species, also has pale green foliage, though the leaves are white beneath and glossy above, and the inflorescences are loose, the flowers with 7–8 stamens each.

Maranthes (including the former *Parinari kerstingii*, *P. polyandra*, *P. glabra* and *P. robusta* among others; White, 1976), differs from *Parinari* in having intrapetiolar stipules, no hair-filled leaf cavities, but glands at the base of the leaf (which has only up to 10 pairs of lateral nerves), and large, robust corymb-like leafless inflorescences. Each flower has 40–60 stamens and nectar is produced by the tissue lining the tubular receptacle. There are 1–3 carpels, each 2-celled and persisting independently in the fruit.

Inselbergs

Inselbergs (of the bornhardt type) usually have large areas of exposed rock. On this, the sedge *Afrotrilepis pilosa* forms mats and, with age, its erect stems carry many species of epiphytic orchids. The mats are colonised by another monocotyledon, *Cyanotis lanata* and also by such woody species as:

Cissus spp., p. 33   Euphorbia spp. (Euphorbiaceae)
Hildegardia barteri   Hymenodictyon floribundum
(Sterculiaceae)   (Euphorbiaceae)

Seasonally damp areas (flushes) develop temporary communities in the rains, and include a number of characteristic sedges, and herbs, such as:

Burmannia latialata, p. 54   Phyllanthus nigericus
Drosera indica   (Euphorbiaceae)
Eriocaulon nigericum   Utricularia spp.
                    Xyris straminea, p. 30

Drosera indica (Droseraceae) is a delicate little plant with a rosette of insect-trapping leaves bearing sticky glands; also in Gambia.

Eriocaulon nigericum (Eriocaulaceae) belongs to a family of inconspicuous monocotyledonous herbs with rosettes of narrow grass-like leaves and small heads of ♂ and ♀ flowers on leafless glabrous peduncles. The heads are c. 4 mm Ø, with pale involucral bracts and 3-part flowers.
Mesanthemum prescottianum, mainly seen on inselbergs in forest-savanna mosaic, is larger, up to 45 cm tall, softly hairy, with short hairs on the peduncle, and heads with several series of prominent white bracts round small ♂ and ♀ flowers.

Utricularia spp. (Lentibulariaceae) have leaves bearing bladder-like traps into which microscopic organisms are drawn. U. subulata is a common terrestrial species in flushes, but a few species are epiphytes or twiners. Another 7 species are aquatics.

Surrounding forest may continue up the flanks of the inselberg to some height in gullies and ravines, where the soil may be deeper. Woodland and scrub, with the structure of regrowth, covers areas of shallow soil, usually surrounding forest areas.

Tree species of woodland and scrub
   Afzelia africana   Hildegardia barteri
    (Caesalpiniaceae)   (Sterculiaceae)
   Albizia ferruginea (Mimosaceae) Napoleonaea imperialis, p. 45
   Elaeophorbia drupifera   Trema guineensis
    (Euphorbiaceae)   (Ulmaceae)

Shrubs and woody climbers
The Apocynaceae, Ochnaceae and Rubiaceae are well represented, together with other species of both high forest and regrowth forest, including *Hippocratea welwitschii* and *Salacia* spp. in the former group, and *Acridocarpus smeathmannii* in the latter.

The broad-leaved life plant, *Bryophyllum pinnatum* may form pure stands on the edge of woodland and scrub patches.

Inselberg sedges/grasses

Sedges	Grasses
Cyperus cuspidatus	*Andropogon tectorum*
C. haspan	*Loudetia arundinacea*
Fimbristylis alboviridis	*Panicum griffonii*
F. bisumbellata	*P. lindleyanum*
Mariscus dubius	*Sporobolus paniculatus*

The sedge *Microdracoides squamosus* is confined to inselbergs (in Cameroun, the extreme east of S. Nigeria, Guinée and Sierra Leone).

Forest-savanna mosaic

This is essentially a mixture of forest or woodland with secondary grassland. The greater part of the area, from Sénégal to west of the Niger, is a mosaic of lowland forest and secondary grassland. To the east and northeast of the Niger, in Nigeria, there is a mosaic of lowland forest and *Isoberlinia* woodland with secondary grassland (White, 1983). This latter kind of mosaic, formerly known as southern guinea savanna, is little developed outside Nigeria.

The whole area of inland forest-savanna mosaic, which has a dry season of four to five (plus) months, is considered to have been covered originally by forest, but it has been so altered by farming, firing, etc. that large parts of it now support only woodland at most. Forest species, which are thin-barked, survive only on such sites as escape frequent burning, including, for example, those by rivers and on rocky hills and plateaux, unsuitable for farming or grazing. In the southern part of the forest-savanna mosaic, such fire-tender species form patches of relic forest, while in the northern part, these species form only closed woodland.

Apart from such patches, and from areas of secondary grassland, whether these are determined by man's interference or by edaphic factors, open woodland (the crowns not touching) is the usual form of vegetation, composed of woody species also found in the northern guinea savanna (sudanian woodland) to the north.

There is a ground cover of flat-leaved grasses up to 4.5 m high, with scattered shrubs and herbs. There are at most two tree storeys composed mostly of deciduous species, always with thick fire-resistant bark. These species readily sprout from damaged stumps and may show seedling features such as crypto-geal germination, as in some Combretaceae, and the early formation of a root crown, from which new shoots can arise at need (*Butyrospermum paradoxum*) (Jackson, 1968, 1974). The tallest tree species do not exceed 15 m as a rule.

Open woodland

Sudanian tree species
The Caesalpiniaceae, Combretaceae, Euphorbiaceae and Mimosaceae are well represented, together with:

Butyrospermum paradoxum
(Sapotaceae)
Crossopteryx febrifuga
(Rubiaceae)
Cussonia barteri
Diospyros mespiliformis, p. 37
Lannea spp. (Anacardiaceae)
Lophira lanceolata (Ochnaceae)
Parinari curatellifolia, p. 46

Pterocarpus erinaceus
(Papilionaceae)
Stereospermum kunthianum
(Bignoniaceae)
Syzygium guineense var.
macrocarpum, p. 53
Trichilia emetica (Meliaceae)
Vitex doniana
(Verbenaceae)

Cussonia barteri (now *C. kirkii* var. *kirkii*) (Araliaceae) is a twisted deciduous tree up to 10 m high, with thick bark and a rounded crown. Young growth is covered with curly hairs. The leaves are stipulate and digitately compound, with 7–11 leaflets each up to 25 cm long, and a petiole up to 85 cm long. The leaves are produced in clusters at the ends of the thick branchlets. Clusters of long, thick, hairy spikes appear at the end of the dry season, each bearing small, greenish, 5-part flowers. The ovary is inferior, and develops into a tiny 1–2-stoned drupe. *C. bancoensis* is a forest species of more restricted distribution (Côte d'Ivoire–S. Nigeria) with smaller leaflets but longer petioles.

Shrubs and climbers
The Rubiaceae are represented, together with

Annona senegalensis (Annonaceae)
Cochlospermum planchoni
Combretum spp. (Combretaceae)
Grewia spp.

Maytenus senegalensis
Ochna spp. (Ochnaceae)
Psorospermum spp.
Strychnos spinosa, p. 60

Ximenia americana

Cochlospermum planchoni (Cochlospermaceae) is a shrub up to 2.5 m high, also found to the north in the northern guinea zone. It has digitately lobed leaves, and, towards the end of the rains, large, yellow, 5-part flowers appear at the tips of the shoots. The numerous stamens, opening by pores, and the ovary with 3–5 parietal placentae are characteristic. The fruit is a 3–5-valved capsule containing numerous seeds. The outer hairy testa is easily separated from the inner shining one. *C. tinctorium* is a savanna herb of ironstone areas, flowering near the ground after fires. The leafy shoots appear later, from a woody rootstock. *C. planchoni* has a massive xylopodium, buried at least 30 cm deep, from which new shoots arise.

Grewia spp. (Tiliaceae) have stipulate leaves, the blades 3–7-nerved at the base and stellately hairy, at least on 1 surface. Small 5-part flowers with a valvate calyx, free smaller petals each with a glandular patch on the inner face, and numerous all fertile stamens round a 2–4-celled ovary, are produced in cymes. The fruit is a drupe or consists of 1–2 drupelets (mericarps). *G. carpinifolia* is a climbing shrub in forest with broad-based buds and lobed drupes (Hall & Siaw, 1980). *Glyphaea brevis* is a tree of regrowth forest differing from *Grewia* mainly in flower and fruit characters. The flowers are 4-part with petals as large as the sepals, round numerous stamens joined basally in groups. There is no androgynophore, and the external stamens are sterile, the fertile anthers long, and with apical pores. The ovary is 8–10-celled, with numerous ovules on axile placentae, the fruit being woody and spindle-shaped, indehiscent.

Maytenus senegalensis (Celastraceae) (also in Gambia) is a shrub or small tree, up to 5 m high, with pale, leathery fine-toothed leaves, (ovate)-obovate-(orbicular) in shape, though cuneate at the base. In the axil of each leaf is either a spine, or a branched inflorescence of small, whitish, 5-part flowers each with a prominent intrastaminal disc and a 3-celled ovary. The fruit is a widely-opening, 3-valved, red capsule, with 6 seeds, each with a thin cupular arilloid.

Psorospermum spp. (Hypericaceae). Two species, *P. corymbiferum* and *P. febrifugum*, are savanna shrubs up to 4.5 m high, with (sub)-opposite gland-dotted leaves. Small, yellowish 5-part flowers like those of *Harungana* (p. 41), are produced in short, branched terminal inflorescences in the dry season. The fruits are red-to-purple, few-seeded berries. In *P. corymbiferum* (also in Gambia) the leaves are alternate-subopposite, in *P. febrifugum* they are opposite.

Ximenia americana (Olacaceae) is a shrub or small tree with zig-zag branches bearing stiff axillary spines and pale, elliptical leaves up to *c.* 5 cm long. Axillary racemose clusters of small, white, scented, hairy flowers are produced in the dry season, each 4-part with a 1-celled ovary. The placenta is a projection from the ovary floor, carrying 4 pendulous ovules from its tip. The fruit is a yellow drupe with green flesh. Like some other members of the family, *Ximenia* is a root parasite. When growing near the coast, it often has rather fleshy leaves and is very spiny, while inland it may be unarmed and have thin leaves.

Olax gambecola, a shrub up to *c.* 1.5 m high, is seen in riverain forest to the south, but is also prominent in forest outliers. It lacks spines and has narrow racemes of glabrous flowers with only (3–)4 stamens (and 5–6 staminodes) instead of the 8 stamens of *Ximenia*. The drupes are globular and red.

Aptandra zenkeri is a dioecious shrub or small tree, the ♀ possessing fruits with a remarkably expanded, red waxy calyx up to 8 cm ∅ round the black drupe.

Herbs
The monocotyledonous families are well represented, for example by *Curculigo pilosa*, together with the Compositae, Papilionaceae and Rubiaceae. Gregarious flowering often takes place after fires. *Cienfuegosia heteroclada* and *Cochlospermum tinctorium* (p. 49) are suffrutices also flowering after fires. 'Weed' genera are commonly present.

> *Curculigo pilosa* (Hypoxidaceae) (also in Gambia) is a hairy grass-like herb of damp places, with a stem tuber. In the late dry season, solitary yellow flowers appear, each 6-part with a petaloid perianth. The very long perianth tube is fused with the style, and the pedicel is very short, the inferior ovary being hidden in the basal rosette of plicate leaves.

Species of perennial grasses of well-drained woodland are numerous, sedges much less so. *Cyperus tenuiculmis* is an example of a perennial sedge in such a situation. It becomes dormant by the beginning of the dry season.

Perennial grasses of open woodland

Andropogon gayanus	*H. subplumosa*
A. schirensis	*Loudetia phragmitoides*
Brachiaria brizantha	*Monocymbium ceresiiforme*
B. jubata	*Panicum* spp.
Ctenium newtonii	*Paspalum orbiculare*
Hyparrhenia cyanescens	*Pennisetum purpureum*
H. rufa	*Schizachyrium sanguineum*

Isoberlinia doka is particularly common in forest-savanna mosaic in Nigeria east and northeast of the Niger, and although it occurs elsewhere, particularly in Ghana, it is nowhere as frequent. On hills it occurs with *Monotes kerstingii*.

> *Monotes kerstingii* (Dipterocarpaceae), the only member of its family in West Africa, is a tree up to 15 m high, with resinous wood and hairy branchlets on which the stipules leave circular scars. The leaves are notched at the base and apex, woolly beneath, and with very regular pinnate venation. There is a pit-like gland on the upper side of the base of the midrib. Short softly woolly cymes of greenish-yellow flowers appear in the rains, each flower being 5-part, with hairy sepals and petals, numerous stamens and a 3-celled ovary. This forms a distinctive fruit, a nut, surrounded by 5 pink or red wings formed from the enlarged persistent calyx.

Savanna regrowth
This takes place beneath scattered preserved and planted trees, which form the upper, very discontinuous, canopy. The Caesalpiniaceae and Sapotaceae are present, together with:

Adansonia digitata	*Mangifera indica*
(Bombacaceae)	(Anacardiaceae)

Parkia clappertoniana   Treculia africana
(Mimosaceae)   (Moraceae)
Vitex doniana (Verbenaceae)

Grasses, herbs (including weeds) and sucker shoots of woody species com-
bine to form scrub, eventually with rather little undergrowth. In the southern
part of the zone, thickets with fire-tender drier forest species and oil palms
are common, though some protection from fire has to be available. In the
northern area of forest-savanna mosaic, the oil palm is confined to forest
outliers.

The sedges of disturbed ground are likely to include those given for disturbed ground in forest (p. 45). The grasses included a number of 'weed'
species, as well as Imperata cylindrica.

Annual and perennial grasses of disturbed ground
    Adropogon pseudapricus   Loudetia hordeiformis
    Cynodon dactylon   Pennisetum spp.
    Dactyloctenium aegyptiacum   Schizachyrium exile
    Eragrostis spp.   Sporobolus pyramidalis

Secondary grassland

This is maintained by repeated disturbance and firing, and therefore tends
to be a mosaic of associations reflecting the history of each patch. *Imperata
cylindrica* survives best in these situations.

Relic forest

This occurs in patches from which fire is excluded in the forest-savanna
mosaic, in the southern part of it in Nigeria. The trees are smaller in stature
than they are when growing in lowland forest, but a great many species
may be expected. Forest grasses (p. 39) may be found. Oil palms, Meliaceae
and Sterculiaceae are common, also *Celtis zenkeri* and *Chlorophora excelsa*.

In the northern part of the forest-savanna mosaic, the equivalent of relic
forest is closed woodland with a canopy of fire-tolerant savanna trees mixed
with fire-tender forest species, over thin undergrowth of mainly evergreen
forest shrubs, and savanna herbs and grasses.

Closed woodland trees: forest species
    Albizia zygia (Mimosaceae)   Malacantha alnifolia (Sapotaceae)
    Cola millenii   Phyllanthus discoideus
        (Sterculiaceae)   (Euphorbiaceae)
    Holarrhena floribunda   Sterculia tragacantha
        (Apocynaceae)   (Sterculiaceae)

Closed woodland trees: sudanian species
Crossopteryx febrifuga
  (Rubiaceae)
Entada spp. (Mimosaceae)
Hymenocardia acida
  (Euphorbiaceae)

Lannea spp.
  (Anacardiaceae)
Maranthes polyandra, p. 46
Pterocarpus erinaceus
  (Papilionaceae)

Forest outliers
These are strips of lowland forest confined to river valleys. They are not necessarily continuous with the northern forest border, but show the same species as both the mature and secondary parts of drier forest, together with many of the riverain species of that zone. The structure of an outlier is much like that of lowland forest, though the storeys are generally lower.

Depending on how intensively farming is carried out, the outlier–savanna boundary may be very sharp, draped with climbers, or there may be an intermediate belt of transition woodland, composed of fire-tender *Anogeissus leiocarpus* over a thicket of forest species.

Tree species
Each of a large number of families is likely to be represented by two to three species: Annonaceae, Bombacaceae, Sterculiaceae, Euphorbiaceae, Caesalpiniaceae, Meliaceae, Moraceae, Sapindaceae, Bignoniaceae and Rubiaceae. Other species include:

Aubrevillea kerstingii (Mimosaceae)
Celtis spp. (Ulmaceae)
Diospyros spp., p. 37
Elaeis guineensis (Palmae)
Lonchocarpus sericeus
  (Papilionaceae)

Manilkara multinervis (Sapotaceae)
Maranthes and Parinari spp., p. 46
Pterocarpus santalinoides
  (Papilionaceae)
Raphia hookeri (Palmae)
Syzygium guineense var. guineense

Syzygium guineense (Myrtaceae) is a fire-tolerant shrub or tree, up to 15 m high, with opposite gland-dotted leaves and terminal corymb-like inflorescences of white, 4-part flowers with numerous stamens and an inferior ovary. The calyx is merely a 4-lobed rim round the receptacle, in which the ovary is buried. The petals fall off like a cap in 1 piece when the flower opens. The receptacle is cupular, with the stamens on the edge, and the cup is lined with an intrastaminal disc. The fruit is a berry, with numerous seeds on axile placentae, and a persistent calyx. All 4 West African species are ecologically important, and all have leathery leaves with close-set lateral nerves and marginal nerve as well. *S. guineense* exists in 3 varieties, var. *guineense* in forest outliers and fringing forest, often with galled flowers and inedible greyish fruits; var. *macrocarpum* is a fire-tolerant shrub or tree of open woodland savanna, with edible purple fruits up to 2 cm Ø (both also in Gambia); var. *littorale* is the coastal form with rather smaller but otherwise similar fruits to var. *macrocarpum*. *S. owariense* is a freshwater swamp forest tree with ovate leaves, the other species having obovate ones. In *S. rowlandii* (riverain forest) young shoot and inflorescence branches are 4-sided. This is a useful character by which to separate this species and *S. guineense* var. *guineense*. *Eugenia* spp. differ in their axillary flowers, with distinct sepals, the 4 white petals providing the flower colour.

Beneath this tree cover, forest shrubs and climbers are found, together with forbs from the families Commelinaceae, Marantaceae and Zingiberaceae, together with *Crinum jagus*.

In wide swampy valleys there may be a grassy flood plain between the forest outlier and the nearest woodland. In northern parts of the zone in Ghana and Nigeria, these areas are occupied by savanna bamboo (*Oxytenanthera abyssinica*), other grasses, *Raphia sudanica* (replacing *R. hookeri* of the south), and *Borassus* palm, together with *Mitragyna inermis*, *Pseudocedrela kotschyi*, *Terminalia glaucescens* and *T. macroptera*.

Inselbergs

Such areas are often surrounded at the base by relic forest in the southern part of the zone, by closed woodland in the northern part. Bare rock, sedge and grass mats, pools, boulder falls, cave mouths and ravines all support special communities.

The pioneer species on bare rock is the arborescent sedge *Afrotrilepis pilosa*. In the north and northeast of Sierra Leone, this is replaced by a much larger sedge of similar habit, *Microdracoides squamosa*. Neither carries the epiphytic orchid flora typical of these two species when growing on lowland forest inselbergs. A thin layer of soil accumulates under sedge mats, and these are colonised by grasses, other sedges and herbs, the commoner ones being:

Aloe schweinfurthii　　　　　*Dipcadi taccazzeanum*
(Liliaceae)　　　　　　　　　　(Liliaceae)
Bryophyllum pinnatum, p. 48　*Ilysanthes* spp.
Cyanotis lanata (Commelinaceae)　*Pancratium trianthemum*, p. 25

Ilysanthes spp. (Scrophulariaceae). *I. gracilis* and *I. schweinfurthii* are inconspicuous erect herbs, probably annuals, up to 10 cm high, with opposite leaves 5–15 mm long, many with a solitary flower in its axil. The flowers are tiny, 2-lipped and pale, K(5), 5-toothed, C(4), with 2 stamens and 2 staminodes at the mouth of the corolla tube. Each anther has 2 divergent cells. G(2), forming a narrow septilicidal capsule 5–10 mm long.

Seepage from the edge of sedge and grass mats allows the development of associations of unusual species such as insectivorous plants (p. 47), together with:

Burmannia latialata　　　　　*Mesanthemum prescottianum*, p. 47
Eriocaulon nigericum, p. 47　　*Neurotheca loeselioides*
　　　　Xyris straminea, p. 30

Burmannia latialata (Burmanniaceae) (now *B. madagascariensis* and including *B. liberica* and *B. welwitschii*) is a slender annual herb up to 30 cm high with a basal rosette of leaves and a scape of purple, yellow or blue flowers with a persistent 3-winged perianth tube and 3+3 white lobes. There

Fig. 2.8. *Burmannia latialata*. A. Plant in flower × 1. B. Flower × 5. C. Flower, the centre in longitudinal section × 5. D. Anther × 20. (From Geerinck, 1970, Fig. 1 *pro parte*.)

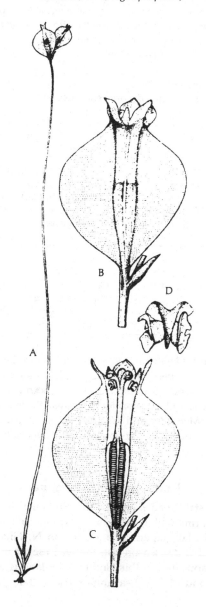

are 3 stamens and a 3-celled gynaecium with axile placentae. In other species, the perianth tube lacks wings. These are saprophytes of a yellow/ white colour.

Neurotheca loeselioides (Gentianaceae) is similar in habit to *Ilysanthes*, though 4–5 times taller, up to 45 cm high. It has opposite leaves up to 2.5 cm long, many of them with solitary bright blue or purple flowers in their axils. The flowers are up to 1 cm long, and conspicuous. Each is 4-part with a ribbed calyx tube, a tubular corolla with 4 contorted free lobes, the stamens being inserted in the tube, and a pistil of G(2), 1-celled with parietal placentae. The fruit is a capsule.

On boulder falls, deeper soil exists in pockets between boulders. In it, geophytes, climbers and shrubs become established.

Geophytes, climbers and shrubs

Geophytes	Climbers
Aloe schweinfurthii (Liliaceae)	*Ampelocissus* spp., p. 34
Amorphophallus dracontioides	*Cissus* spp., p. 33
(Araceae)	*Ipomoea* spp. (Convolvulaceae)
Anchomanes difformis (Araceae)	*Sarcostemma viminale*
Haemanthus multiflorus, p. 25	(Asclepiadaceae)
Sansevieria liberica, p. 24	*Taccazzea apiculata* p. 43

Shrubs	
Cochlospermum planchoni, p. 49	*Ochna membranacea* (Ochnaceae)
Euphorbia kamerunica	*Pavetta saxicola* (Rubiaceae)
(Euphorbiaceae)	(now *P. gardeniifolia* var. *gardeniifolia*;
Ficus abutilifolia (Moraceae)	Bridson, 1977)

Ficus abutilifolia, rather less common here than on inselbergs of the northern guinea zone, eventually becomes a small tree. Other tree species include *Diospyros mespiliformis* and *Manilkara multinervis* (both also in forest outliers) and *Monodora tenuifolia* and *Napoleonaea imperialis*, all four species also occurring in relic forest. *Monodora* and *Napoleonaea* are especially prominent in ravines and sheltered hollows. On boulder falls *Markhamia tomentosa* is common in Nigeria.

Northern guinea savanna

This is the longer established name for the 'Sudanian *Isoberlinia* woodland' of White (1983). It forms a slim, more or less wedge-shaped zone, the broad end in Guinée and Mali, tapering to a point in N. Nigeria, and not well investigated of late. The dry season lasts for five to seven months.

In structure and appearance, this kind of woodland is very similar to that of the open woodland of forest-savanna mosaic, though the grasses may come into leaf rather later and do not grow so tall, perhaps to 2.5 m. Oil palms are absent and there are no patches of relic forest, or even closed woodland of forest species, apart from in fringing forest or on inselbergs.

Open woodland savanna is usually dominated by *Isoberlinia* spp., *Daniellia* and *Lophira* becoming rarer than in forest-savanna mosaic, while *Monotes* is much more common, dominating dry eroded slopes.

Open woodland

Tree species
Isoberlinia doka and other Caesalpiniaceae are characteristic, together with:

Afrormosia laxiflora	*Lannea* spp.
(Papilionaceae)	(Anacardiaceae)
Bridelia ferruginea	*Monotes kerstingii*, p. 51
(Euphorbiaceae)	*Ochna afzelii*
Butyrospermum paradoxum	(Ochnaceae)
(Sapotaceae)	*Sterculia setigera*
Diospyros mespiliformis,	(Sterculiaceae)
p. 37	*Uapaca togoensis* (Euphorbiaceae)

In addition, the southern and northern parts of the zone differ in the frequencies of a number of other species. In the southern part, the following species are more frequent than in the northern part:

Antiaris africana (Moraceae)	*Parinari curatellifolia*, p. 46
Crossopteryx febrifuga	*Parkia clappertoniana*
(Rubiaceae)	(Mimosaceae)
Daniellia oliveri	*Psorospermum corymbiferum*,
(Caesalpiniaceae)	p. 50
Hymenocardia acida	*Pterocarpus erinaceus*
(Euphorbiaceae)	(Papilionaceae)
Lophira lanceolata	*Stereospermum kunthianum*
(Ochnaceae)	(Bignoniaceae)
Terminalia laxiflora (Combretaceae)	

Two further woody species are particularly characteristic of the southern part of the zone, though only from Ghana eastwards: *Faurea speciosa* and *Protea elliottii*.

> *Faurea speciosa* (*Proteaceae*) is a tree up to 12 m high, with densely hairy twigs bearing shortly petiolate, acute elliptical-to-lanceolate leaves. The leaves are woolly beneath, shining above, with numerous pairs of prominent veins. In the dry season, robust, terminal, reddish, erect bottlebrush spikes of apetalous flowers are produced. The long calyx tube splits basally, releasing the style to one side, then the tip of the calyx splits into 4 lobes and bends as whole to the opposite side. The 4 anthers, attached by very short filaments to the upper part of the calyx tube, are thus exposed. The 1-celled ovary ripens into a nut with a long persistent

style, and covered in golden brown hairs. The calyx abscisses circularly round the fruit, and drops off. The spikes are thus highly distinctive whether flowering or fruiting.

Protea elliottii (now *P. madiensis*, and including all West African species; Chisumpa & Brummitt, 1987) differs in being a shrub with blunt-tipped leaves, which are sessile with faint nerves; the blade sometimes stands vertically. At most times of the year, strong-smelling flower heads may be found. These are *c.* 20 cm ∅, with numerous stiff bracts, pink on the outside and white on the inside, round the edge. They surround numerous tiny flowers each with a white petaloid calyx, which splits into 2 lobes.

In the northern part of the zone, the following species are more frequent than in the southern part:

Cussonia barteri, p. 49   *Lannea microcarpa* (Anacardiaceae)
Entada africana (Mimosaceae)   *Terminalia avicennioides*
Isoberlinia tomentosa   (Combretaceae)
(Caesalpiniaceae)   *Ximenia americana*, p. 50

In addition, *Burkea africana* and *Piliostigma thonningii*, which are at their southern limit, and *Acacia* spp., notably *A. albida*, are found.

Eroded areas generally develop a flora composed predominantly of *Dichrostachys glomerata*, together with species such as *Ficus abutilifolia* and *Stereospermum kunthianum*, and some open woodland savanna species such as *Isoberlinia* spp., *Lannea* spp. and *Ximenia americana*. *Sclerocarya birrea*, a riparian species further north, and *Combretum lamprocarpum* may be seen.

Savanna regrowth

In flood plains, vegetation structure and composition is much the same as in forest-savanna mosaic (p. 54), but in savanna regrowth different species occur from those encountered further south. Grazing is a more important factor, keeping the grass short – fires occurring early in the dry season are unlikely to be so fierce. The preserved and planted farm trees are the same species as in forest-savanna mosaic (although *Daniellia* is less common) and these stand over scrub woodland of sucker shoots (of *Isoberlinia* spp., *Piliostigma* spp. and *Terminalia avicennioides* principally) shrubs and climbers (*Combretum* spp.) and shoots of species such as *Icacina senegalensis*, which has deeply buried tubers.

The grasses form a mosaic of associations depending both on soils and on the degree and kind of interference suffered. The species found are likely as a rule to include those listed for forest-savanna mosaic (p. 52).

Icacina senegalensis (Icacinaceae) is particularly common in farmed savanna areas of Ghana. It dies down each dry season, and puts up new woody shoots at the beginning of the rains. The shoots bear ovate to obovate leaves up to 15 cm long and very prominently reticulate. Lax terminal inflorescences of small, yellowish white, 5-part flowers, the petals each with a belt of hairs on the inner face. The 1-celled, 1-ouled superior ovary becomes a red berry, about 3 cm long. This species also occurs in Gambia, but is not common in Nigeria. *I. mannii* (Sierra Leone – Nigeria) (Cremers, 1973) and *I. trichantha* (Nigeria) are climbing shrubs in forest, and have longer leaves.

Fringing forest

This is also known as gallery forest. It is confined to stream valleys and is an interesting mixture of open woodland species and species from the open woodland in forest-savanna mosaic (list A below), together with species from closed and transition-type woodlands and forest outliers of the forest-savanna mosaic (list B below).

A. *Species also associated with inselbergs, i.e. surviving where some protection is afforded*

Anogeisus leiocarpus
  (Combretaceae)
Diospyros mespiliformis, p. 37

Lophira lanceolata (Ochnaceae)
Pterocarpus erinaceus
  (Papilionaceae)
Stereospermum kunthianum (Bignoniaceae)

B. *Forest-savanna mosaic species from forest outliers or closed or transition woodland*

Albizia zygia (Mimosaceae)
Cola millenii (Sterculiaceae)
Dialium guineense
  (Caesalpiniaceae)
Ficus exasperata (Moraceae)

Holarrhena floribunda (Apocynaceae)
Hoslundia oppositifolia (Labiatae)
Malacantha alnifolia (Sapotaceae)
Maranthes kerstingii, p. 46
Pachystela brevipes (Sapotaceae)

Other fringing forest and inselberg species are mentioned below, and a fourth group, species of fringing forest and also of eroded areas, includes *Pseudocedrela kotschyi* and *Tamarindus indica*.

The species mentioned in fringing forest so far are mainly tall trees, *c.* 15–18 m in height and mostly deciduous. Under these is an evergreen under-storey, described as tunnel-like, of smaller trees, shrubs and climbers, some of them forest species. The families Annonaceae, Sapindaceae, Apocynaceae and Rubiaceae are well represented, together with:

Alchornea cordifolia
  (Euphorbiaceae)
Caloncoba gilgiana, p. 37
Entada abyssinica (Mimosaceae)
Garcinia spp. (Guttiferae)

Harungana madagascariensis, p. 41
Hippocratea africana
Maytenus senegalensis, p. 50
Opilia celtidifolia
Strychnos spp., p. 60

Hippocratea (Celastraceae or Hippocrateaceae) (Hallé, 1962, 1984) is a forest genus of twining shrubs and lianes, *H.* (now *Loesneriella*) *africana* being commonly present in fringing forest (also in Gambia). The leaves are paired and leathery, and in their axils, much-branched cymes of tiny greenish or white, 5-part flowers appear. There are only 3 stamens, extrorse and opening by a gaping transverse split. Each flower has an intrastaminal disc round the pistil of 3 joined carpels with a single style. The flat capsule contains seeds attached by their basal wing. *H. africana* has triangular green petals and orange anthers, and the flowers are scented. The capsule is 2–3 times as long as broad and the seed wing is much longer than the seed itself. *H.* (now *Cuervea*) *macrophylla* is a riverain forest species, with very large leaves, up to 23 cm long × 7 cm wide, capsules scarcely longer than broad and the seed wing shorter than the seed body. The first internode of each plagiotropic branch twines (Cremers, 1973). *H.* (now *Simirestis*) *welwitschii* is a forest climber particularly seen on lowland forest inselbergs. It has narrow capsules *c.* 2 cm wide, but otherwise resembles *H. africana* in fruit and seed characters. *Salacia* spp. are common forest climbers, differing from *Hippocratea* spp. in having berries and, almost always, alternate leaves, scale leaves subtending the branches. Identification of species is very difficult, but a key to vegetative characters exists for the 13 Ghanese species (out of a total of *c.* 40 species) (Hall & Lock, 1975).

Strychnos (Loganiaceae or Strychnaceae) are either shrubs or small trees of savanna, from forest-savanna mosaic northwards, or forest lianes, with paired or single branch hooks (the hardened bases of tendrils) (Cremers, 1973). *S. spinosa*, though more a sudan zone species, first becomes prominent in the northern guinea zone. It is a small-leaved shrub or tree, with interpetiolar ridges between opposite leaves, each of which has one or two pairs of basal nerves which continue well towards the leaf tip. Short terminal cymes of 5-part gamopetalous flowers, the 5 stamens alternating with the corolla lobes, are followed by large, yellow berries the size of a grapefruit, containing yellow-brown, acid, edible pulp. *S. innocua* is rather less common, non-spiny and with rather larger leaves, up to 10 cm long. The flowers are yellowish, some of the clusters axillary. The fruit is lemon-sized, with a woody shell, containing sticky, sweet orange pulp. *Anthocleista* spp. differ in being under-storey forest trees with frangipani-like branching and very large leaves (up to 1.5 m long × 0.5 m wide) at the ends of the branches. These are armed with paired (or a single forked) spine(s) in place of axillary branches. The inflorescence is a massive, erect terminal panicle of robust white flowers each about 2.5 cm long, gamopetalous and *c.* 10–16-part, followed by leathery berries. The genus is sometimes placed in the Potaliaceae. *A. vogelii* is a tree of wetter forest areas. About one-third of the length of the flower bud is contributed by the 2+2 calyx, the tip remaining rounded and the bud closed. *A. djalonensis* is a tree of drier forests and savanna, the flowers with a very small calyx and tapering corolla in the bud, which opens widely revealing the contorted corolla lobes and central staminal column (12–13 part). The more or less sessile leaves, the blade tapering away to the base, are distinctive.

Opilia celtidifolia (Opiliaceae) (also in Gambia) is a liane with oblong-obovate leaves, pointed and tapering at both ends, up to 13 cm long. Before expansion, the inflorescences resemble small cones of overlapping bracts. They later expand to hanging racemes *c.* 5 cm long, the bracts drop off and tiny, scented, greenish, 5-part flowers in clusters along the axis are exposed. The 5 stamens stand opposite the petals, and the 5 lobes of the disc alternate with the stamens. The pistil is 1-celled and forms a drupe about 2 cm long. The species is a root parasite.

Near the stream in swampy ground, clumps of two palms, *Raphia sudanica* and *Phoenix reclinata* are common, together with *Adina microcephala* in places, though this species is more common in the same situations in sudan savanna. *Dichrostachys glomerata* often forms thickets.

If undisturbed, fringing forest is bordered by a belt of transition woodland or closed woodland savanna with such species as:

Cussonia barteri, p. 49	*Terminalia glaucescens*
Nauclea latifolia	(Combretaceae)
(Rubiaceae)	*Vitex doniana* (Verbenaceae)

With fairly intensive farming, the forest strip is likely to be abruptly edged with climbers, and if farming is very intensive, the fringing forest itself is likely to be composed of only a reduced number of species out of those mentioned in the lists above.

Ironstone areas

In ironstone areas a great many open woodland species will be found, but in addition the following are likely to be present:

Combretum binderianum	*Dichrostachys glomerata*
(Combretaceae)	(Mimosaceae)
C. glutinosum	*Gardenia erubescens*
(Combretaceae)	(Rubiaceae)
Crossopteryx febrifuga	*Psorospermum corymbiferum*,
(Rubiaceae)	p. 50
Vitex simplicifolia (Verbenaceae)	

In shallow ironstone soils grasses with *Cochlospermum tinctorium*, *Combretum nigricans* var. *elliottii* and the small pyrophytic *C. sericeum* may be found.

Inselbergs

Many otherwise common species are lacking from these features, but *Bombax costatum* is virtually characteristic, although this species only appears occasionally on dry upper slopes and plateau edges. The shrub *Byrsocarpus coccineus* is similarly characteristic.

Another group of species also occur as fringing forest and riverside species:

Albizia zygia (Mimosaceae)	*Grewia mollis*, p. 50
Anogeissus leiocarpa	*Holarrhena floribunda* (Apocynaceae)
(Combretaceae)	*Hoslundia opposita* (Labiatae)
Carissa edulis (Apocynaceae)	*Khaya senegalensis* (Meliaceae)
Diospyros mespiliformis, p. 37	*Pachystela brevipes* (Sapotaceae)
Ficus abutifolia (Moraceae)	*Steganotaenia araliacea*
F. exasperata (Moraceae)	*Stereospermum kunthianum*
Flacourtia flavescens p. 37	(Bignoniaceae)

Steganotaenia araliacea (Umbelliferae) is a soft-wooded tree up to 12 m
high with compound imparipinnate leaves with up to 4 pairs of toothed

Fig. 2.9. *Steganotaenia araliacea*. A. Leaves × ⅓. B. Infructescence × ⅓. C.
Flower in longitudinal section × 6. D. Floral diagram of an umbelliferous
species. E. Fruit in cross-section × 6. (A–C and E from Burger, 1967,
Fig. 51.2; D from Eichler, 1875–8.)

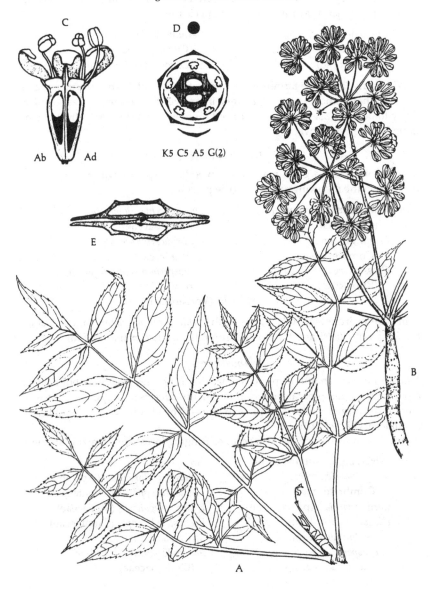

K5 C5 A5 G(2)

leaflets, teeth ± marginal bristles. In the dry season, compound umbels of tiny, white flowers with inferior 2-celled ovaries with 2 styles appear. The winged fruit splits into 2 mericarps, which hang from a common stalk. This species also occurs in well-watered valleys. *Hydrocotyle bonariense* is a creeping herb of sandy shores, with peltate leaves.

Some woodland species are also present, but in smaller numbers than is the rule elsewhere, for example:

Annona senegalensis	*Entada africana* (Mimosaceae)
(Annonaceae)	*Terminalia avicennoides*
Cussonia barteri, p. 49	(Combretaceae)

Climber tangles of *Acacia ataxacantha* are common. The first species to appear on bare rock is usually the sedge, *Afrotrilepis pilosa*, though occasionally *Vellozia schnitzleinia* (now *Xerophyta*; Smith & Ayensu, 1974) replaces it in Nigeria. Other communities follow in sequence, and some unusual plants, such as bladderworts (*Utricularia* spp.) may be expected if there is sufficient moisture.

The woodland found round the base of an inselberg is sometimes confusingly referred to as fringing woodland. It is closed woodland savanna with a rather high proportion of southern savanna species such as:

Daniellia oliveri	*Pterocarpus erinaceus*
(Caesalpiniaceae)	(Papilionaceae)
Parinari curatellifolia, p. 46	*Vitex doniana* (Verbenaceae)

On dry ridges and slopes, for example on inselbergs, and bordering erosion gullies, particularly in the southern part of the zone, *Monotes kerstingii* and *Protea elliottii* are characteristic. They are accompanied by *Lophira lanceolata* and *Securidaca longepedunculata* and, quite often, *Faurea speciosa*, as well as other open woodland species (both general and southern).

> *Securidaca longepedunculata* (Polygalaceae) is a shrub or small tree (also in Gambia) with small, blunt-tipped oblong-lanceolate greyish leaves with a few pairs of faint nerves. At various times of the year, simple racemes of purple .l. flowers are produced, each with 3 petals and 8 stamens round the superior ovary. This develops into a samara with a wing about 5 cm long and up to 1.5 cm wide, sometimes accompanied by a second, very small, wing.

Sudan savanna

This is the 'Sudanian undifferentiated woodland' of White (1983) and is considerably drier than the preceding zone, with under 1000 mm of rain per year and a dry season lasting seven to nine months. Open woodland, not more than 12 m high, mostly of deciduous species, covers the area. About half the species concerned have small leaves or compound leaves with small leaflets. About a quarter of the tree species are also thorny.

Open woodland

Tree species

The Anacardiaceae, Combretaceae and Mimosaceae, together with:

Adansonia digitata (Bombacaceae)
Annona senegalensis
  (Annonaceae)
Anogeissus leiocarpus
  (Combretaceae)
Balanites aegyptiaca
Boscia spp.
Boswellia dalzielii
Butyrospermum paradoxum
  (Sapotaceae)
Commiphora spp.
Diospyros mespiliformis, p. 37
Ziziphus spp.

Gardenia sokotensis (Rubiaceae)
Hyphaene thebaica
  (Palmae)
Lannea spp. (Anacardiaceae)
Lonchocarpus laxiflorus
  (Papilionaceae)
Maytenus senegalensis, p. 50
Piliostigma reticulata
  (Caesalpiniaceae)
Sterculia setigera (Sterculiaceae)
Stereospermum kunthianum
  (Bignoniaceae)
Strychnos spinosa, p. 60

Shrubs

Rubiaceae, together with:

Grewia mollis, p. 50
Guiera senegalensis
  (Combretaceae)

Securinega virosa
  (Euphorbiaceae)
Ximenia americana, p. 50

Climbers

Acacia ataxacantha (Mimosaceae)   *Capparis corymbosa*
                Combretum spp. (Combretaceae)

Balanites aegyptiaca (Zygophyllaceae or Balanitaceae). Although this deep-rooting species is also found in forest-savanna mosaic, it is probably of greater potential importance in the sudan and sahel zones. Its fruits have been shown to yield charcoal (from the woody endocarp), alcohol and precursors of steroid drugs (from the treated pulp), edible seeds and seed oil, and cattle cake. It is a Ghanian or N. Nigerian tree up to 9 m high, with green branches and simple axillary spines up to 7 cm long. Each is in the axil of a leaf of 2 leaflets. In the early rains, clusters of small, 5-part, greenish flowers appear between the leaf and the spine above it. The superior ovary of each flower can develop into an edible drupe, though few actually seem to do so. This species also occurs in forest-savanna mosaic.

Boscia spp. (Capparidaceae, now Capparaceae) are deciduous trees up to 12 m high, and confined to the driest savanna and the sahel (Sénégal– N. Nigeria). *B. salicifolia* is the tallest species, with drooping branches bearing long, narrow dull and leathery leaves with several pairs of indistinct veins. In the early dry season, dense leafless and rather woolly axillary racemes of flowers each with 4 green sepals appear. There is a cupular nectary round *c.* 12 stamens and a stalked ovary of 2 joined carpels. The ovary is 1-celled, with an ovule on each of its 2 parietal placentae. The berry is *c.* 1 cm ∅, and has a smooth shining shell. *B. angustifolia* is a smaller tree with smaller leaves, about one-third as long (5 cm), rather compact inflorescences and rough berries. *B. senegalensis* is a shrub with much larger (9 cm long × 4 cm broad) clearly veined leaves, leafy racemes and larger warted berries.

Capparis corymbosa is one of several species of scrambling climbers with stipular spines. It has hairy branchlets and oblong leaves up to 5 cm long. Small, white petaloid flowers appear in terminal corymbs on short shoots at the beginning of the rains. Numerous stamens surround a stalked ovary, the stalk lengthening under the yellow berry which later develops.

Fig. 2.10. *Balanites aegyptiaca.* A. Shoot with immature fruit × $\frac{1}{2}$. B. Flowering shoot × $\frac{1}{2}$. C. Flower × 5. D. Centre of flower in longitudinal section × 5. E. Fruit × 13. F. Floral diagram. (From Burger, 1967, Figs. 25.1 and 25.3.)

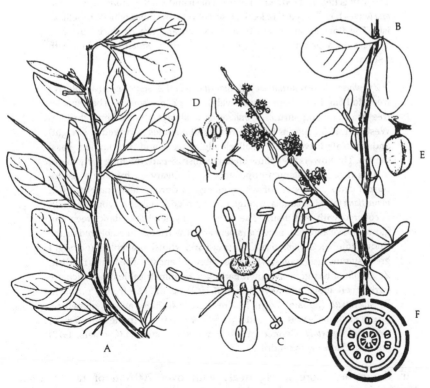

K5 C5 A10 G(5)

Boswellia dalzielii (Burseraceae) is found in the more easterly drier savannas (Côte d'Ivoire–northern Nigeria). It is a deciduous tree up to 12 m high, with greyish-yellow papery bark peeling off in smallish sheets. The wood is resinous and strongly scented in most members of the family, providing incense and fumigants when burnt, e.g. *Boswellia* spp. yield the frankincense (and *Commiphora* spp. the myrrh) of the Bible. The compound imparipinnate leaves have opposite or subalternate serrate leaflets are carried on thick branchlets, terminal clusters of racemes of small, white, scented flowers being produced in the dry season after the leaves fall. The flowers are 5-part (10 stamens) with an intrastaminal disc and 3-celled ovary. The fruit is a 3-valved capsule, the valves falling off to release the winged seed in each cell (see also Geerling, 1984). In *Commiphora* spp., the branches frequently form spines. *C. africana* and *C. pedunculata* are both widespread species (Sénégal–Sudan at least). The flowers are small, 4-part (8 stamens) with a cupular hypanthium and intrastaminal disc. Some of the flowers are unisexual. The fruit is a 1–2-stoned drupe. *C. africana* has 3-foliolate leaves on reddish branch spines from which the bark later peels off in strips. The flowers are also reddish, produced in ± sessile axillary clusters in the dry season while the tree is leafless. The fruit is tiny and red, with resinous flesh and a white stone. *C. pedunculata* leaves have 2–4 pairs of leaflets, sometimes on spines which are yellowish. The flowers are greenish or yellowish and are produced in hairy clusters on a peduncle in the rains. The fruit is small and black with a sharp tip.

Ziziphus spp. (Rhamnaceae). *Z. mauritiana* and *Z. spina-christi* are always trees (to 12 and 18 m high, respectively) while *Z. abyssinica* can be a smaller tree or scrambling shrub. *Z. mucronata* is always a scrambling shrub in West Africa. All species have paired stipular spines (1 curved, 1 straight) and petiolate leaves with small glandular teeth and 3 prominent basal nerves. The flowers are yellowish or greenish, 5-part, opening flatly round a large disc, each stamen opposite a petal. Ovary ÷, forming a red-to-purple drupe, the stalk of which elongates during development. In *Z. mauritiana* (also in Gambia, though not recorded for Ghana), young growth is covered with pink hairs. The leaves are symmetrical, dark green and glossy above and with grey hairs beneath. The fruit is an edible red drupe (Indian jujube). *Z. spina-christi* has grey bark, whitish branchlets and bright green symmetrical leaves, which are glabrous beneath. The fruit is reddish. *Z. abyssinica* has grey-to-brown bark, but asymmetrical leaves which are dark green above and have brown hairs beneath. The fruit is purplish. *Z. mucronata* (buffalo thorn) has reddish branchlets and asymmetrical nearly glabrous leaves, and bitter purple-brown fruits. In the same family, *Gouania longipetala* (Cremers, 1974) and *Ventilago africana* (Cremers, 1973) are forest lianes with tendrils.

In the rather more moist areas, with over 700 mm of rain a year, *Afzelia africana*, *Burkea africana*, several members of the Combretaceae

(*Combretum nigricans, Pteleopsis suberosa, Terminalia laxiflora*) and *Afrormosia laxiflora* are present.

Fig. 2.11. *Ziziphus* spp. A–E. *Z. abyssinica*. A. Flowering and fruiting shoot × 1. B. Leaf × $\frac{2}{3}$. C. Flower × 6. D. Flower after petals and stamens have fallen × 6. E. Part of infructescence × 1. F–H. Leaves, all × $\frac{2}{3}$. F. *Z. mauritiana*. G. *Z. spina-christi*. H. *Z. mucronata*. (From Johnston, 1972, Figs. 8 and 9 *pro parte*.)

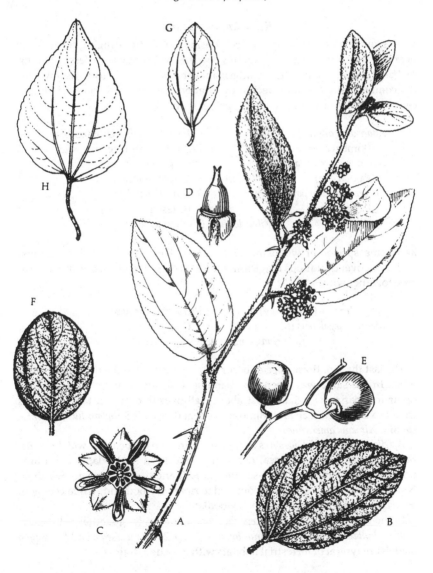

Savanna regrowth

The preserved trees include some of those also preserved in northern guinea savanna, such as *Adansonia* and *Ceiba*, *Blighia sapida*, *Butyrospermum* and *Tamarindus*. *Acacia albida* and *Balanites aegyptiaca* are more frequent, but *Parkia clappertoniana* is prominent only in the very south of the zone. Under the trees is a thin scrub of grasses, weeds and shrubs, such as *Cassia singuineana* and *Piliostigma reticulata*, *Combretum micranthum* and *Guiera senegalensis*, and *Dichrostachys glomerata*.

Riparian woodland

This is quite different in species composition from the fringing forest seen near streams further south. The only lowland forest species present belong to *Afzelia* and *Diospyros*. The savanna mahogany, *Khaya senegalensis*, is however the commonest emergent species. In addition, the Anacardiaceae, Bignoniaceae and Palmae are represented, together with:

Adansonia digitata	*Celtis integrifolia* (Ulmaceae)
(Bombacaceae)	*Ficus* spp. (Moraceae)
Adina microcephala	*Pterocarpus erinaceus*
(Rubiaceae)	(Papilionaceae)
Butyrospermum paradoxum	*Tamarindus indica*
(Sapotaceae)	(Caesalpiniaceae)
Vitex doniana (Verbenaceae)	

and some species of woodland (list, p. 64). These species are deciduous, and stand over deciduous tangles and thicket composed of species of Mimosaceae together with:

Ficus glumosa (Moraceae)	*Pterocarpus erinaceus*
Khaya senegalensis (Meliaceae)	(Papilionaceae)
Tamarindus indica (Caesalpiniaceae)	

The last three of these species occur in the same way in the northern guinea zone. In the same way open woodland species of the sudan zone can also occur in riparian woodland (list above), all over thicket and tangle species from the Combretaceae and Mimosaceae together with *Strophanthus sarmentosus* and *Ximenia americana*.

Afzelia africana and *Diospyros mespiliformis* are two of the most widely distributed species in West Africa, occurring from the drier lowland forest area northwards throughout forest-savanna mosaic and the northern guinea zone. North of this, *Afzelia africana* is confined to riparian woodland, while *Diospyros mespiliformis* is still present in open woodland.

In the river bed, *Adina microcephala* is sometimes seen, together with *Acacia albida*, *Phyllanthus beillii* and *Glinus lotoides* (see p. 23). *Acacia seyal* and *Mitragyna inermis* are typical of terrestrial habitats with impeded drainage.

Rocky hills

In the Sudan zone, such hills are seldom subject to heavy grazing or fierce fires. The vegetation consists of large trees standing over thickets with deciduous climber tangles and sparse grass. The tree species *Boscia salicifolia* and *Isoberlinia tomentosa*, and the shrub *Gardenia sokotensis* are confined to these sites. As in the zone to the south, some riparian species also occur on inselbergs. These include:

Bauhinia rufescens   *Commiphora pedunculata*, p. 66
(Caesalpiniaceae)   *Leptadenia hastata*
Capparis corymbosa, p. 65   (Asclepiadaceae)
Combretum micranthum   *Xeromphis nilotica* (Rubiaceae)
(Combretaceae)   *Ximenia americana*, p. 50
                    Ziziphus spp., p. 66

Bibliography

Further references may be found in White, 1983, pp. 271–324.

Badré, F. (1971). *Hugonia* africains et leurs fruits. *Adansonia*, sér. 2, **11**, pp. 95–106.

Baker, H. G. (1962). Heterostyly in the Connaraceae with special reference to *Byrsocarpus coccineus*. *Botanical Gazette*, **123**, pp. 206–11.

Berghen, C. van den (1979). La végétation des sables maritimes de la Basse Casamance meridional (Sénégal). *Bulletin du Jardin botanique national de Belgique*, **49**, pp. 185–238.

Bernhard-Reversat, F., Huttel, C. & Lemée, G. (1978). Structure and functioning of evergreen rain forest ecosystems of the Ivory Coast. In *Tropical forest ecosystems*, pp. 557–74. Paris: UNESCO/UNEP/FAO.

Bridson, D. M. (1977). Studies in *Pavetta* (Rubiaceae, sub-family Cinchonoideae) for part 2 of *Flora of tropical East Africa: Rubiaceae*. *Kew Bulletin*, **32**, pp. 609–52.

Burger, W. C. (1967). *Families of flowering plants in Ethiopia.* Stillwater, Oklahoma: Oklahoma State University Press.

Caballé, G. (1982). Caracteristiques de croissance et multiplication végétative en forêt dense du Gabon de la liane à eau *Tetracera alnifolia* Willd. (Dilleniaceae). *Adansonia*, sér. 2, **19**, pp. 467–76.

Chisumpa, S. M. & Brummitt, R. K. (1987). Taxonomic notes on tropical African species of *Protea*. *Kew Bulletin*, **42**, pp. 813–51.

Cremers, G. (1973). Architecture de quelques lianes d'Afrique tropicale. *Candollea*, **28**, pp. 249–80.

Cremers, G. (1974). Architecture de quelques lianes d'Afrique tropicale 2. *Candollea*, **29**, pp. 57–110.

Descoings, B. (1972). *Flore du Cameroun: Vitacées, Leéacées.* Paris: Muséum national d'histoire naturelle.

Eichler, A. W. (1875–8). *Blüthen-Diagramme.* Leipzig: Engelmann.

Ewusie, J. Y. (1974). Delimitation of ecological zones in strand vegetation at Elmina, Ghana. *Ghana Journal of Science*, **14**, pp. 59–67.

Friis, I. & Nordal, I. (1976). Studies on the genus *Haemanthus* (Amaryllidaceae) 4. *Norwegian Journal of Botany*, **23**, pp. 63–77.

Geerinck, D. (1970). *Flore du Congo etc.: Burmanniaceae.* Meise: Jardin botanique national de Belgique.

Geerling, C. (1984). *Boswellia odorata* Hutch. and *B. occidentalis* Engl. (Burseraceae), syn nov. *Bulletin du Muséum national d'histoire naturelle,* Paris, sér. 4, 6, sect. B, *Adansonia* pp. 233–5.

Hall, J. B. & Lock, J. M. (1975). Use of vegetative characters in the identification of species of *Salacia* (Celastraceae). *Boissiera,* **24**, pp. 331–8.

Hall, J. B. & Siaw, D. E. K. A. (1980). The varieties of *Grewia carpinifolia* Juss. (Tiliaceae). *Adansonia,* sér. 2, **20**, pp. 339–47.

Hall, J. B. & Swaine, M. D. (1981). *Distribution and ecology of vascular plants in a tropical rain forest: forest vegetation in Ghana.* The Hague: Junk.

Hallé, N. (1962). Monographie des Hippocrateacées d'Afrique occidentale. *Mémoires de l'Institut fondamentale d'Afrique noire,* sér. A, **64**.

Hallé, N. (1984). Révision des Hippocrateae (Celastraceae). *Bulletin du Muséum national d'histoire naturelle,* Paris, sér. 4, 6, sect. B, *Adansonia,* pp. 179–88.

Hopkins, B. (1970). Vegetation of the Olokemeji Forest Reserve VII. *Journal of Ecology,* **58**, pp. 795–825.

Huynh, K.-L. (1987). Étude des *Pandanus* (Pandanaceae) d'Afrique occidentale, 5. *Candollea,* **42**, pp. 129–46.

Jackson, G. (1964). Notes on West African vegetation I. *Journal of the West African Science Association,* **9**, pp. 98–110.

Jackson, G. (1968). Notes on West African vegetation III. *Journal of the West African Science Association,* **13**, pp. 215–22.

Jackson, G. (1974). Cryptogeal germination and other seedling adaptations to the burning of vegetation in savanna regions. *New Phytologist,* **73**, pp. 771–80.

Jeffrey, C. (1961). *Flora of tropical East Africa: Aizoaceae.* London: Crown Agents.

John, D. M. (1986). The inland waters of tropical West Africa. *Ergebnisse der Limnologie/Advances in Limnology,* **23**.

Johnston, M. C. (1972). *Flora of tropical East Africa: Rhamnaceae.* London: Crown Agents.

Lawson, G. W. ed. (1986). *Plant ecology in West Africa. Systems and processes.* Chichester: John Wiley.

Letouzey, R. & White, F. (1978). *Flore du Cameroun: Chrysobalanacées.* Paris: Muséum national d'histoire naturelle.

Markham, R. H. & Babbedge, A. J. (1979). Soil and vegetation catenas on the forest–savanna boundary in Ghana. *Biotropica,* **11**, pp. 224–34.

Menaut, J. C. & César, J. (1982). The structure and dynamics of a West African savanna. In *Ecology of tropical savannas,* ed. B. J. Huntley & B. H. Walker, *Ecological Studies* 42, pp. 80–101. Berlin: Springer-Verlag.

Mendonça, F. A. (1963). *Flora zambesiaca: Rutaceae,* vol. 2, part 2. London: Crown Agents.

Milne-Redhead, (1953). *Flora of tropical East Africa: Hypericaceae.* London: Crown Agents.

Nordal, I. & Wahlstrøm, R. (1980). A study of the genus *Crinum* (Amaryllidaceae) in Cameroun. *Adansonia,* sér. 2, **20**, pp. 179–98.

Peñalosa, J. (1984). Morphological specialisation and attachment success in two twining lianes. *American Journal of Botany,* **69**, pp. 1043–5.

Smith, L. B. & Ayensu, E. S. (1974). Classification of Old world Velloziaceae. *Kew Bulletin,* **29**, pp. 181–205.

Swaine, M. D. & Hall, J. B. (1983). Early succession on cleared forest land in Ghana. *Journal of Ecology*, **71**, pp. 601–27.

Tomlinson, P. B. (1986). *The botany of mangroves*. Cambridge: Cambridge University Press.

Voorhoeve, A. G. (1965). *Liberian high forest trees*. Wageningen: Centrum voor Landbouwpublikaties en Landbouwdokumentatie.

White, F. (1976). The taxonomy, ecology and chorology of African Chrysobalanaceae (excluding *Acioa*). *Bulletin du Jardin botanique national de Belgique*, **46**, pp. 308–50.

White, F. (1983). *The vegetation of Africa*. Paris: UNESCO.

3

Annonaceae – soursop family

A pantropical but especially Old World family of woody plants, found in West Africa mainly as shrubs, lower-storey trees and lianes in forest. Only four species are recorded for Gambia, *Hexalobus monopetalus* and *Annona senegalensis* in savanna, *A. glabra* in coastal swamps, and *Xylopia aethiopica* in forest-savanna mosaic.

Members of the family may be recognised by their two-ranked (alternate), conduplicate, simple, existipulate, entire leaves, aromatic wood and fragrant three-part, often downward-facing flowers, with many spirally arranged stamens and carpels. The fruits are also distinctive and contain large arilloid seeds, which are mottled or streaked in section (ruminate endosperm).

Artabotrys spp. are forest lianes with thick flat peduncles formed as hooks (Cremers, 1973), *A. velutinus*, a yellow-flowered species, being fairly widespread. The other *c.* 10 species in the genus are very locally distributed, as is common in many other genera in the family.

Flowers ± ⊕ and ♂. Sepals 3, petals 3+3 (*Uvariopsis* ♂ or ♀ K2 C4; *Enantia* C3, the petals opposite the sepals; *Hexalobus*, *Isolona* C(6), *Monodora* C(3+3)). Stamens numerous, with a thickened connective, tightly packed in a spiral on the convex receptacle. G∞ free and spirally arranged (joined and in 1 whorl only in *Isolona* and *Monodora*); ovules 1 to several per carpel, attached to the adaxial carpel margin where the carpel is free, or basal (parietal only in *Isolona* and *Monodora*; Deroin, 1985).

Pollination *Annona* is reported to be pollinated by beetles, but only hand-pollination seems to ensure adequate fruit-set. It is probably disadvantageous that the flower is so markedly protogynous that the styles fall off before the anthers open. The continued interest of adequate numbers of beetles is then very important. Stigmatic exudate is produced, but there are no nectaries.

Fruit Either aggregate, composed of 1 to several-seeded berries (red in *Artabotrys velutinus*, black in *Enantia* and the introduced *Cananga odorata*), or (dehiscent) follicles or (indehiscent) monocarps, or the carpels joined in a fleshy mass with the receptacle (*Annona*), or joined to form a hard-rinded 'berry' (*Monodora*). In the fleshy fruits such as some *Artabotrys* spp. and *Monodora*, the flesh is pulpa, while in other species of *Artabotrys* the flesh is formed by the arilloid. Situations similar to that in *Artabotrys* exist in *Xylopia*.

Fig. 3.1. A–D. *Monodora tenuifolia*. A. Floral diagram. B. Flower × 1. C. Flower in longitudinal section × 6, the outer petals shortened. D. Stamen × 27. E. *M. myristica* fruit × $\frac{1}{3}$, part of pericarp cut away. F. *Annona squamosa* fruit × $\frac{1}{3}$; a few carpels removed. *Isolona* has a more knobbly, but similarly multilocular fruit. G. *Uvaria* sp. flower × $1\frac{1}{4}$; one outer petal removed. H. *U. chamae* fruit × $\frac{5}{8}$.

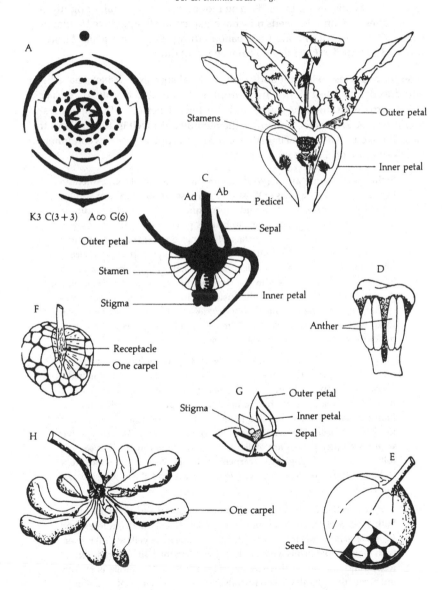

Dispersal The sweet aromatic flesh of the fleshy fruits is attractive to a variety of animals; e.g. *Annona* spp. attract bats. Where the follicles open, as in *Xylopia*, and a strong colour contrast is presented (between the carpel lining, pulpa or sarcotesta and seed), the possibility of bird dispersal arises. *Enantia polycarpa, Polyalthia* spp. (now *Greenwayodendron*) and some *Hexalobus* spp. are reported to be dispersed by birds. The monocarps are generally brightly coloured, often red, and the seeds may have a spicy smell, attractive to primates. *Uvaria ovata* seeds germinate from baboon dung, *Brieya fasciculata, Monodora* spp. and *Pachypodanthium staudtii* from elephant dung.

Economic species yield fruit and seeds. Several species of *Annona* have been introduced into West Africa from tropical America, but only *A. squamosa* (sweetsop), with knobbly fruits, and the hybrid between this species and cherimoya (*A. cherimola*), known as the atemoya, do well in the lowland tropics. The seeds of *Monodora myristica* and *Xylopia* spp. are used as condiments, and even those of *Annona* are peppery.

> **Field recognition** *Annona* (West Indian name) is largely an American genus, with only 3 species in West Africa. These are small trees or shrubs, of which *A. senegalensis* is found in savanna, while *A. glabra* occurs in coastal swamps (*A. glauca* seems to take its place in Ghana). The flowers are *Uvaria*-like, with valvate petals, the inner ones smaller and thinner than the outer ones. The fruit is a yellow-netted syncarp. *A. senegalensis* (including *A. arenaria* as spp. *oulotricha*) has hairy shoots bearing bluish leaves, and yellow flowers in the dry season; the fruit is *c.* 5 cm long. *A. glabra* has white flowers, the inside of the inner petals being conspicuously red. The fruit has rather dry flesh and is *c.* 12 cm long × 8 cm Ø. In America, these fruits are said to be eaten by alligators and iguanas, hence the common name – alligator apple.
>
> *Cleistopholis* (Gr. closed and scale, the inner petals folding over the stamens and carpels). Of this small West African genus, *C. patens* is widespread as an upper-storey riverside tree from swamp forest northwards into forest outliers. It has horizontal branches, pale-grey bark and a fragrant, moist slash 'looking like palm oil and tasting salty'. The bark is used for cordage. The small, greenish-yellow flowers (*Uvaria*-type but with long outer petals) are axillary, on leafy shoots in the dry season, followed by knobbly, globular carpels (each with 2 seeds) on short, thick stalks, in the rains. Each carpel becomes black and fleshy outside and hard inside.
>
> *Hexalobus* (Gr. 6-lobed) is a small genus of tropical and southern Africa, both West African species being widespread, *H. crispiflorus* in forest and *H. monopetalus* in savanna. Both produce conspicuous yellow C(6) flowers with long narrow petals which, while covered in the bud, are transversely folded and are wrinkled when they first appear. The corolla is slightly imbricate (though often taken for valvate) and the inner petals are slightly longer than the outer ones. The fruit is a collection of rather fleshy red

monocarps. *H. crispiflorus* is a forest tree up to 25 m high, producing, in the dry season, small axillary clusters of flowers with petals up to 8 cm long, followed by woody thick monocarps 8 cm long × 4 cm broad. *H.*

Fig. 3.2. *Annona senegalensis* var. *senegalensis*. A. Flowering branch × $\frac{2}{3}$. B. Fruiting branch × $\frac{2}{3}$. C–D. Stamen, front and side view × 12. E. Receptacle and carpels × 2. F. Fruit × $\frac{2}{3}$. G. Tangential section of fruit × $\frac{2}{3}$. H. Seed in longitudinal section × 2. (From Verdcourt, 1971, Fig. 27 *pro parte*.) (British Crown copyright. Reproduced with permission of the Controller, Her (Britannic) Majesty's Stationery Office & the Trustees, Royal Botanic Gardens, Kew. © 1971.)

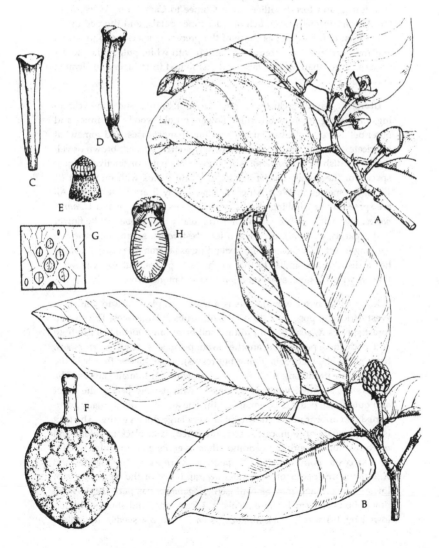

monopetalus is a shrub or small tree and has much smaller flowers (petals up to 3 cm long) mainly in the rains. The monocarps are smaller, only 2 cm thick and the walls contain juice.

Monodora (Gr. single gift, the solitary flowers?). This is a small tropical African forest genus of large shrubs and small trees, flowering while leafless in the dry season. The flowers are conspicuous, in shades of white, green and yellow, blotched with red or purple. The fruit is a green or brown hard-shelled 'berry', 7–15 cm ∅, hanging down beneath the foliage. In it is sticky brown pulp in which the seeds are embedded. *M. tenuifolia* is a handsome tree of forest margins and forest-savanna mosaic, especially relic forest, and forest outliers from Guinée to Cameroun. *M. myristica* (an African nutmeg) has much broader inner petals, and the peduncles are at least twice as long as those of the preceding species (up to 20 cm) and the fruits twice as large (10–15 cm ∅) with white pulp. It is a forest species, larger than *M. tenuifolia*, and distributed from Liberia to Uganda and Angola.

Uvaria (L. grape, the cluster forming the fruit). Shrubs, sometimes climbing, producing green or brownish flowers on old wood. The inner and outer petals are ± equal and up to 2 cm long in most cases, and somewhat imbricate. The monocarps are usually only shortly stalked, brown or yellow-to-reddish in colour, and carry the stellate hairs distinctive of most species of the genus. *Uvaria* is an Old World genus with one-tenth (*c.* 15) of its species in West Africa. Of these only 3 are widespread. All are spreading or climbing shrubs or small trees, *U. chamae* and *U. annonoides*, both with a calyx which splits irregularly into 3 lobes as the flower bud expands. *U. chamae* occurs in drier forest, forest-savanna mosaic and fringing forest, *U. annonoides* not extending as far from the coast. *U. afzelii* has a distribution similar to that of the last species, but has yellowish flowers with 3 distinct sepals and long-stalked monocarps.

Xylopia (Gr. bitter wood). This is a mainly tropical African genus, in West Africa composed of *c.* 10 species of shrubs or large trees. The swamp forest species *X. staudtii* can produce stilt breathing roots up to 2 m high and both this species and another swamp forest species, *X. rubescens*, have plank-like stilt roots (flying buttresses). Most species flower in the dry season, producing long flower buds with very small calyces. The flowers are scented, white or yellowish, with narrow petals, the inner and outer ones being ± of equal length. The anthers are internally partitioned and appear netted; they contain very large pollen grains. The fruit is an aggregate of monocarps, sometimes red, later black in colour and brightly coloured within. In addition there is either a fleshy pulpa within, or a sarcosta, both coloured. *X. quintasii*, *X. rubescens* (in Nigeria) and *X. staudtii* are species of freshwater swamp forest or the wetter parts of lowland forest, and may be confused. *X. quintasii* has particularly small flower buds and small white petals, *c.* 1.5 cm long, and short styles followed by 3–5 sessile thick monocarps each with 2–4 seeds, the arilloid being laciniate. *X. rubescens* has long flower buds and outer petals, *c.* 3.5 cm, the inner petals being only approximately one-sixth as long. The monocarps are in groups of 6–12, thick, twisted and/or beaded, with up to 8 seeds with a net-like arilloid each. *X. staudtii* has ± globular flower buds in small clusters and yellow flowers which are followed by groups of 3–5 black stalked, curved and beaded monocarps lined with red, each seed with a yellow, cupular arilloid round the black seed. Like *X. rubescens*, the rest of the widespread genera have long, pointed flower buds, though like *X. quintasii* the inner and outer petals are ± the same length. *X. acutiflora* is a forest shrub or small tree with solitary sessile white or yellowish flowers with petals 4.5 cm long. These are followed by 5–10 red monocarps, which later blacken. Each is stalked, rather stout and obliquely beaded, containing 'frilly' flesh and 7–9 seeds with sarcotestas. The remaining 2 widespread species are in drier areas. *X. aethiopica* occurs in lowland forest northwards into forest-savanna mosaic (in forest outliers and round relic forest). The flowers are white with rusty hairs, solitary or in small clusters. Numerous narrow sessile monocarps – resembling children's fingers – each contain 4–9 seeds. The monocarps are red, becoming black, and the arilloid is papery and yellow. *X. parviflora* extends from lowland forest into fringing forest in the northern guinea zone. It has flowers which are solitary or in small clusters, red or purple at the base and yellow at the tip of the 5 cm long petals. There are 3–5 stout red sessile carpels with a red to pink lining, and green seeds (6–9) attached to bright pink 'frilly' pulpa.

Bibliography

Cremers, G. (1973). Architecture de quelques lianes d'Afrique tropicale. *Candollea*, **28**, pp. 249–80.

Deroin, T. (1985). Contribution à la morphologie comparée du gynécée des Annonaceae–Monodoroideae. *Bulletin du Muséum national d'histoire naturelle*, Paris, sér. 4, 7, sect. B, *Adansonia*, pp. 167–76.

Verdcourt, B. (1971). *Flora of tropical East Africa: Annonaceae*. London: Crown Agents.

4

Amaranthaceae – amaranth family

Annual or perennial herbs, creeping or with taproots, mostly weeds of disturbed ground. Many species are pantropical or almost cosmopolitan in distribution, often by introduction.

Members of the family may be recognised (in 10/14 genera) by their opposite leaves, which are also simple and exstipulate, and condensed, erect, leafless racemose inflorescences (often spikes or clusters) with dry bracts and bracteoles, of minute apetalous flowers, each with a dry, often coloured calyx. The fruit is a utricle (an inflated achene, but see Chapter 39, Cyperaceae), which may, however, dehisce.

Introduced cultivated ornamentals include bachelor's button (*Gomphrena globosa*), cockscomb (*Celosia argentea* f. *cristata*), love-lies-bleeding (*Amaranthus* sp., see below) and *Alternanthera* (now *tenella* var.) *bettzickiana*.

There are very few climbers in the family and *Sericostachys scandens*, in the forests of southern Nigeria and Cameroun, is the only ± woody one. It has opposite leaves and large inflorescences, each unit of which is composed of one fertile and two (to three) sterile flowers. These latter develop white feathery hairs as the fruit ripens, and the infructescences are very striking.

Two more species are distinctive by both habit and habitat. *Centrostachys aquatica*, the only species in its genus and the only aquatic species in the family, has occasionally been collected along the southern border of the Sahara, to where its North African distribution extends. It has grooved floating stems rooting at the lower nodes and bearing opposite leaves, and flower spikes up to 30 cm long.

Along the West African coast, growing on the upper shore, is *Philoxerus vermicularis*, the only West African species of a small genus of coastal herbs extending to the West Indies, tropical Americas and Australia. It is woody-rooted, with creeping stems rooting at the nodes. The bracts and sepals (and thus the spikes) are white.

Flowers ⊥ ⊕ ♂ (♂ and ♀ in *Amaranthus*). K5 or (5), dry. C0. Fertile flowers: A(4–5) (1–2 in *Nothosaerva*, 3 or 5 in *Alternanthera*), the stamens opposite the sepals, anthers 2-celled (1-celled in *Alternanthera*, *Gomphrena* and *Philoxerus*); the edge of the staminal tube may bear pseudostaminodes in the form of teeth or fringed tongues (*Aerva*, *Centrostachys* etc.) and the tube itself is longer than the pistil in *Gomphrena*. G(2–3), 1-celled with a single basal ovule (several in *Celosia*) suspended on a long funicle; stigmas 1–3; the embryo is curled

Amaranthaceae   79

Fig. 4.1. A–B. *Celosia sp.* A floral diagram. B. Half flower × 8. C–E. *Amaranthus spinosus*. C. Flowering shoot × $\frac{2}{3}$. D. Stamens × 16. E. Pistil × 10. F–G. *Sericostachys scandens*. F. Part of infructescence × $\frac{2}{3}$. G. Cymule of two sterile flowers and a single fertile one × 4. (A from Eichler, 1875–8; B from Baillon, 1886, Fig. 229; C–G from Cavaco, 1974, Pls. 3 and 4 *pro parte*.)

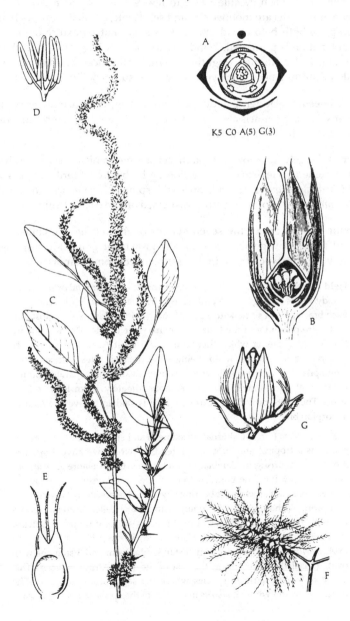

K5 C0 A(5) G(3)

round a mass of perisperm. Sterile flowers: 2–3 accompany 1–2 fertile flowers in each cymule in some species; bracts (*Pupalia*) or calyces (*Cyathula*) with hooks, see also *Sericostachys*, above.

Pollination has been investigated only in the grain amaranths, which are wind pollinated. Each cymule has 1 to few ♀ flowers and many ♂'s, and the grain amaranths are monoecious and self-fertile, pistillate flowers becoming receptive both before and after anthesis, so that a mixture of self- and cross-pollination is promoted. Wind pollination may be prevalent in the rest of the family also, though brightly coloured inflorescences may be visited by pollen-collecting bees. Such visits may not result in pollination however.

Fruit A 1-seeded bladder like 'utricle', indehiscent, circumcissile or opening irregularly, or a circumcissile several-seeded capsule (pyxidium) in *Celosia*; surrounded by the persistent calyx.

Dispersal Largely unknown, though censer mechanisms seem probable. Seeds of some species survive ingestion by herbivores. Sterile flowers with hooks (*Cyathula, Pupalia*) or fertile ones with spines (*Achyranthes, Alternanthera* sp., *Amaranthus* sp.) form burr fruits dispersed on the coats of animals.

Economic species Six of the seven species of *Amaranthus* are cultivated or gathered as a leaf vegetable, but none is grown for grain. Most species in the family are crop weeds and, in that sense, of economic importance.

> **Field recognition** *Achyranthes* (Gr. chaff and flower). Annual or perennial woody herbs with green-to-pink spiny terminal spikes, up to 2 m high. Each bracteole is a spine with a pair of chaffy wings, fringed pseudostaminodes are present and the fruit is a circumcissile utricle. This is a mainly Old World genus of which there are only 3 West African species, only *A. aspera*, a pantropical weed, being widespread in West Africa. It has long spikes, in which the flowers become reflexed and separated on the older basal part, above ovate leaves approximately 10 cm long × 5 cm wide. The roots provide an extract which is said to relieve the pain of a scorpion bite.
>
> *Alternanthera* (Gr. the alternating stamens and staminodes). Like *Achyranthes*, is a tropical and subtropical genus of which we have 4 species, 3 of them occurring in Gambia. Of these, only the seashore *A. maritima* is widespread. It is a perennial with thick leaves and fleshy stems rooting at the nodes, the erect stems bearing silvery-white sessile axillary clusters of flowers. Each flower has 5 stamens with linear 1-celled anthers alternating with laciniate staminodes, and the fruits are tiny rounded utricles. *A. repens* (now *A. pungens*), an inland weed, has sepals of 2 lengths, as in *A. maritima*, but there are only 3 staminodal teeth and 3 stamens with ovate (1-celled) anthers as in the rest of the West African species. This is the only West African species where the sepal has a spiny tip. The utricle is flat and winged, also as in the rest of the species in West Africa.

Amaranthus (Gr. unfading bracts, calyx) (amaranth). Annuals up to 1.5 m high, with alternate leaves and ♂ and ♀ flowers, the former without pseudostaminodes. Only 2 of the 7 West African species are not cultivated. These are *A. spinosus* (also in Gambia) and *A. graezicans*, the former with

Fig. 4.2. A–C. *Philoxerus vermicularis*. A. shoot × $\frac{2}{3}$. B. Flower × 16. C. Pistil × 16. D–G. *Alternanthera maritima*. D. Shoots × $\frac{2}{3}$. E. Paired (short) bracteoles and three sepals from adaxial side × 16. F. Bracteoles and calyx from abaxial side; the two abaxial sepals are strongly veined. G. Stamens, pseudostaminodes and pistil × 12. (From Cavaco, 1974, Pl. 8 *pro parte*.)

paired axillary spines and spiny-tipped sepals to the ♀ flowers, the latter
lacking spines and with only 3 sepals. The fruits are utricles, circumcissile
in these 2 species, opening irregularly in 2 other species. The inflorescences
are erect spikes except in the cultivar of *A. hybridus* subsp. *cruentus*, love-
lies-bleeding, where they are long, red and pendant. *Cruentus* means
'blood red'.

Celosia (Gr. burned, the flame-like shape of the inflorescence or its dry
appearance or colour?). Mostly annual herbs, tropical and temperate in
distribution, up to 2 m high, erect or straggling and with alternate leaves.
The *Celosia* spp. lack pseudostaminodes and have several-seeded circumcissile capsules. The silvery white-to-pink spikes of the introduced and
cultivated *C. argentea*, the sepals of which are up to 1 cm long, are striking,
the fasciated red or yellow spikes of f. *cristata* even more so. Most West
African species have minute yellowish or brownish flowers in interrupted
spikes, though *C. trigyna* (including *C. laxa*) (also in Gambia), with 2–3
stigmas, has white flowers with red stamens.

Cyathula (Gr. little cup) is a small genus with a disjunct distribution
(Africa, Madagascar, Ceylon, China and Malaysia), all but 1 of the 6 West
African species having hooks on the sterile flowers of the cymule. Two
species (Cameroun) are climbing herbs, the remaining, more or less widespread species, are weeds and forest herbs, with slender spikes of green
flowers with pseudostaminodes and burr fruits containing the utricle.
C. prostrata (also in Gambia) is a stout much-branched herb, up to 90 cm
high, the flowers often galled. In the fruiting stage, the hooked sepal
spines are nearly as long again as the sepal itself, and the burr is reflexed
as in *Achyranthes*.

Pandiaka (L. from Pandiaki, place in N. Nigeria) is a small genus of
tropical and southern Africa with only 2 species in West Africa (Gambia–S.
Nigeria), both herbs up to 90 cm high with terminal head-like clusters
of flowers (with pseudostaminodes) subtended by leafy bracts. *P. involucrata*, found on inselbergs and dry slopes in savanna, is a hairy plant with
small white clusters of flowers on paired opposite peduncles, subtended
by leafy ovate bracts 1 cm long. The leaves are also ± ovate. *P. heudelotii*
(now *P. angustifolia*; Townsend, 1980) has linear leaves and bracts and
reddish inflorescences. It is only slightly hairy.

Pupalia (Indian name). This is another small genus of tropical Africa
(with Madagascar and India) of which *P. lappacea*, is common and widespread in West Africa (also in Gambia). It is a branched erect, sometimes
woody, weed with terminal spikes of cymules, the sterile flowers of which
bear branched hooks on their bracts. The sepals are green and the fertile
flowers have yellow or red stamens. The utricle is circumcissile and hidden
among the bracts etc. of the burr fruit. *Digera alternifolia* is now included
as *P. micrantha*.

Bibliography

Baillon, H. E. (1886). *Histoire des plantes*, vol. 9. Paris: Librairie Hachette.
Cavaco, A. (1974). *Flore du Cameroun: Amaranthacées*. Paris: Muséum national d'histoire naturelle.
Eichler, A. W. (1875–8). *Blüthen-Diagramme*. Leipzig: Engelmann.
Townsend, C. C. (1975). The genus *Celosia* (subgenus *Celosia*) in tropical Africa. *Hooker's Icones Plantarum*, **38**, pp. 1–123.
Townsend, C. C. (1980). *Achyranthes, Achyropsis* and *Pandiaka. Kew Bulletin*, **34**, pp. 423–33.

5

Cucurbitaceae – gourd family

A family of trailing and climbing annual and tuberous perennial herbs in the warmer parts of the world, in West Africa common in the drier parts, though not as weeds. Since the publication of the account of the family in the *Flora of West tropical Africa* in 1954, several new genera have been recognised (names given in parentheses in the following account). The family is still probably undercollected, ♀ flowers and fruits of dioecious species mostly being needed. As the fruits are difficult to preserve and transport, local studies on plants grown from seed would be appropriate.

The West African genera almost all have Old World affinities, and are on the western edge of their range.

Members of the family may be recognised by their exstipulate, simple, digitately 3–7-lobed leaves, each accompanied by a tendril. Small cymes of ♂ flowers and (often) solitary ♀ flowers are produced, both kinds of flowers shallow, with five fragile, veined yellow or white petals. In the ♂ flower, each anther has only two loculi – it opens by one slit – and the anthers are often complexly folded. In ♀ flowers, the ovary is inferior and one-celled, with three stigmas.

Introduced species are of economic interest, though many species can be grown as decorative climbers.

A few genera can be distinguished on vegetative characters, but usually flowers and fruits are needed. Entire leaves are seen in *Gerrardanthus* (and sometimes in *Momordica* (*Zehneria*) *capillacea*), pinnately divided ones in *Citrullus*, and digitately compound ones in *Telfairia* and *Momordica cissoides*. *T. occidentalis* the only West African species, has bifid tendrils and is probably the most woody and extensive climber in the family in West Africa; it occurs in forest, while *M. cissoides* has simple tendrils and occurs in savanna.

Several genera have extrafloral nectaries. *Lagenaria* (including the former *Adenopus*) and *Cayaponia* (also in Gambia) have a pair of glands at the top of the petiole or base of the leaf blade; other examples may show glands at nodes, on leaves, bracts or on the calyx.

Bracts, glandular or not, can be distinctive, occurring either singly at nodes ('stipular' bracts, e.g. in the dry savanna species *Ctenolepis cerasiformis* and *Trochomeria macrocarpa* (including *T. atacorensis*, *T. dalzielii*, and *T. macroura*), where they are circular and fringed. Probracts, in association with peduncles, occur in *Citrullus*, *Luffa* and *Telfairia*, large ones in most *Momordica* spp.

The pyrophytic habit is pronounced in species of *Coccinia* and *Momordica*, and in *Trochomeria macrocarpa*.

Flowers ⊹ ⊕ ♂ or ♀, on the same or separate plants, perianth 5-part. The receptacle (hypanthium) generally forms a shallow cup containing stamens in ♂ flowers, or with the style arising from its floor in ♀ flowers (seldom tubular, *Peponium, Trochomeria*). K5, small sometimes pinnately divided, e.g. *Rhaphidiocystis*. C5 in *Luffa* and *Cucurbita*, but mainly C(5), with up to 3 scales in *Momordica*; the fringed petals of *Telfairia* and *Trichosanthes*, and the linear ones of *Trochomeria*, are unusual. Stamens rarely 5 (*Luffa, Bambekea*) or 4 (*Gerrardanthus*, the adaxial one represented by a staminode, and the whole flower .l. with 2 large and 3 small reflexed petals); stamens apparently 3 as a rule, the adaxial one 2-locular ('single' stamen), the other 4-locular, the result of fusion in pairs (2 'double' stamens), a condition easily seen in, for example, *Citrullus, Cucumis* and *Cucurbita*; anthers and/or filaments may be fused, and the anthers may be straight, i.e. the condition commonest in angiosperms, pleated into 3 folds (triplicate) or irregularly folded (contorted, convoluted); ♀ flowers may have staminodes, either separate ones or united in a ring. G(3̅) 1-celled, the odd carpel adaxial, the placentae fleshy, growing into the middle of the ovary and out again, the placentation thus secondarily parietal; the receptacle and ovary are fused, the outer surface often warted, bristled etc.; ♂ flowers may have pistillodes.

Fig. 5.1. *Luffa aegyptiaca*. A. ♂ flower floral diagram, the central pistillode (= nectary) omitted. B. ♀ flower floral diagram. C. Half ♂ flower × 1. D. Half ♀ flower × ⅘. E. Diagrammatic cross-section of ovary. In species with only three stamens, the single stamen is adaxial, opposite the adaxial stigma, and the two double stamens occupy positions roughly opposite the two other stigmata.

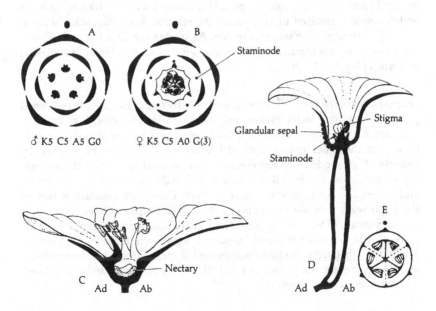

♂ K5 C5 A5 G0   ♀ K5 C5 A0 G(3̅)

Pollination is by insects, often bees, the existence of separate ♂ and ♀ flowers, and of large, sticky pollen grains, making the services of a pollinator necessary. Monoecious species maybe self- or cross-fertilised. The crop genera (*Citrullus, Cucumis, Cucurbita*) are monoecious, and have been best studied. ♀ flowers are fewer in number than ♂ ones, possibly because of the greater production of ♂ flowers in periods of greater day length and higher night temperature, possibly because, in dioecious wild species, fruits and seeds may be subject to attacks from herbivores, and, in any case, the formation of one to two fruits inhibits the further production of ♀ flowers in many species. About half the West African genera produce nectar in both ♂ and ♀ flowers, (in ♂s, the pistillode is glandular; in ♀s, the staminodes are glandular). A few species have no floral nectar production, and in a few more genera only the ♀ flower produces nectar. In about a quarter of the West African genera, production remains to be ascertained. ♀ flowers of dioecious species in particular need investigation. In *Anguria* and *Gurania* (Gilbert, 1975), pollinated by butterflies in Central America, ♀ flowers attract partly by deceit, bumps on the stigma resembling pollen grains, which are also collected.

Fruits are always crowned by the receptacle scar. Berries, usually small and red, are frequently produced, e.g. the 2-seeded berries of *Ctenolepis* and the 3-seeded ones of *Cayaponia*, though otherwise many seeds are the rule. The other kind of fruit which is common is the gourd (pepo), a 1-celled 'berry' formed from an inferior ovary, with a firm-to-hard rind. The outside may be bristly or warted, or smooth and mottled or striped. Most economic species have indehiscent pepos, but dehiscent pepos open at the apex by a lid (*Luffa*) or expel their seeds through the pedicel (*Oreosyce*). Pepos which open more widely are also referred to as capsules (*Momordica, Rhaphidiocystis, Telfairia*), but only *Gerrardanthus* has a dry capsule. All seeds have thick coats, the outermost one being contributed by the pulpa, sometimes as a fleshy (*Momordica*) or fibrous (*Telfairia*) layer.

Dispersal of berries (e.g. *Coccinia*) is probably by birds, and the sweet-fleshed pepos (melon and water melon) may be eaten by primates, e.g. baboons consume *Cucumis melo*. The red arilloids of *Momordica* are also attractive to birds, but the other pepos, with rather insipid flesh, may be gnawed by rodents – their seeds are at any rate full of oil. In southern Africa, the burrowing aardvark eats the fruit flesh of *Cucumis humifructus* for its water content, and buries the seeds in dung near its nest. Given the resistant nature of the pepo wall, it is not surprising that basal and apical dehiscence occur, through tissues of lesser resistance. There are no spectacularly explosive fruits in West Africa, though those of *Momordica* 'burst', a mild form of explosion. The bitter substances so often encountered in pepos deter herbivore attack, and pepos are generally slow to rot. If the seeds are adequately long-lived, active dispersal is not essential.

Economic species provide fruits grown and consumed locally, as fruits or vegetables, or the seeds or their oil may be consumed. Three annual species of *Cucurbita* have been introduced from America and naturalised. *C. maxima* and *C. moschata* (pumpkins and squashes) are eaten young, or used as fodder at any stage. *C. pepo* (marrow) is also best eaten young (courgettes). Young fruits of *Cucumis sativus* (cucumber, which probably originated in India and came to West Africa via Europe) and *Luffa aegyptiaca* (loofah, now *L. cylindrica* from tropical Asia) are eaten raw as salads. Parthenocarpically produced fruits have two advantages, not developing the bitter compounds of the fertilised fruit and not reducing the number of fruits developing later, as has been shown for cucumber, and could prove true for other vegetable species also. The sweet fruits *Cucumis melo* (melon) and *Citrullus lanatus* (formerly *Colocynthis citrullus*, water melon), and the slightly acid fruits of *Coccinia grandis*, are also eaten raw. *Telfairia occidentalis* seeds and (white) egusi (from *Cucumeropsis mannii*, including *C. edulis*) are eaten or the oil extracted, and the pepo flesh of the latter provides a vegetable, as does the flesh of certain varieties of bottle gourd, *Lagenaria siceraria*, and snake gourd, *Trichosanthes anguina*. Tubers are seldom eaten. The Asiatic annual wax gourd, *Benincasa hispida*, is worth introducing, since the fruit grows rapidly producing three to four crops a year, and can be eaten at any stage of growth. The mature fruit can attain 2 m in length, 1 m ∅ and weigh 35 kg. The neutral flavour means that it can be used to 'stretch' both sweet and savoury dishes and because of its waxy coat, it will keep up to a year without refrigeration. The flesh of non-sweet gourds is often tasteless when young, but indigestible, bitter or even poisonous, when mature. These conditions are improved by leaving the cut up flesh in salt, and throwing away the juice extracted, also before cooking.

Field recognition *Adenopus*, see *Lagenaria*.

Coccinia (L. scarlet, the colour of the ripe berry) is a tropical African genus of 30 species, 2 of the 5 West African species being widespread, *C. grandis* in savanna and *C. barteri* in forest regrowth. Both species had gamopetalous yellow flowers, the ♂ ones with 3 stamens joined in a column, the anthers triplicate. The berries are red ± green, ellipsoidal and 2–7 cm long. *C. grandis* is a smaller climber with bifid tendrils, *C. barteri* a tall climber with simple ones.

Lagenaria (L. bottle-shaped, the gourd), including the former *Adenopus*, is a small tropical African genus, 3 of the 4 West African species being widespread. All have white polypetalous flowers, opening at night. *L. siceraria*, bottle gourd, often cultivated, is a savanna plant, pantropical by introduction elsewhere, while *L. breviflora* (including *Adenopus ledermannii*) and *L. guineensis* are forest species, the latter in regrowth. All have a pair of glands at the apex of the petiole or base of the leaf blade (*L. siceraria*). This species has bifid tendrils, and solitary, narrowly funnel-shaped flowers, the ♂'s with 3 free stamens inserted in the receptacle

tube, the anthers triplicate and contorted. The remaining species are dioecious, with racemes of ♂ flowers and solitary ♀ ones, the receptacle tubes
3–5 cm long, at least twice as long as those of *L. siceraria*. *L. breviflora*
has simple or bifid tendrils and sepals without glands, while *L. guineensis*
has simple tendrils and a glandular calyx.

Melothria, see *Mukia* and *Zehneria*.

Momordica (L. to bite, the seeds may look bitten) – 45 Old World species
(now including *Dimorphochlamys mannii = M. cabraei*), of which 4 of the
8 West African species are widespread. ♂ flowers are solitary or in umbels,
♀ flowers solitary, both with a bract on the peduncle. The flowers are
small and polypetalous, with up to 3 scales within the corolla. The fruit
is a fleshy dehiscent pepo, like a capsule, reddish-yellow and warted,
opening to reveal 2 rows of red arilloid seeds on each valve. Stamens
2 or 3, free. *M. cissoides* is a dioecious perennial climber with compound
leaves and simple tendrils, occurring in forest. The flowers are white,
with a dark spot at the base of each petal and scales at the base of only
2 petals. The bract is kidney-shaped. In ♂ flowers there are 2 stamens,
in ♀ flowers 2 staminodes, and the receptacle is spiny. The fruit has
soft prickles. *M. foetida* (including *M. cordata*) is found as a weed and
in forest regrowth; its leaves smell unpleasant when bruised. It has simple
or bifid tendrils and is also a dioecious perennial, but the flowers have
3 scales, the ♂ ones 3 stamens (the anthers fused), the fruits rather longer
than those of *M. cissoides*. The monoecious species are *M. balsamina* (balsam
apple) in savanna, and *M. charantia* (also in Gambia) (balsam pear) in
forest. These are annuals with simple tendrils and solitary flowers. The
former species has shortly pedicillate flowers, the bract being near the
flower, the latter has long pedicillate flowers, the bract far from the flower.
Scale and stamen characters are as those of *M. foetida*. The fruits of *M.
balsamina* are broadly ellipsoid, up to 35 mm long, with rows of small
tubercles. Those of *M. charantia* are longer, over 35 mm, with tubercles
both scattered and in rows.

Mukia (East Indies name) – 4 Old World species, with only *M. maderaspatana* (formerly *Melothria maderaspatana*; also in Gambia) in Africa. This is
a weedy, monoecious perennial climber with a woody rootstock and simple tendrils, and rough stems and leaves. There are separate clusters of
tiny yellow ♂ and ♀ flowers, both with prominent discs. ♂ flowers have
3 free stamens with very short filaments and straight anthers, the fruits
are small red berries.

Physedra, now *Ruthalicia* (compounded of *Ruth* and *Alice*). Only two
species, both endemic to West Africa, are known. *R. eglandulosa* is a widespread, dioecious perennial climber in forest and forest-savanna mosaic,
with simple tendrils. It has large ± polypetalous, hairy yellow corollas
marked with purple, the petals 3 cm wide. ♂ flowers are in racemes,
each with 3 free stamens, ♀ flowers solitary, followed by 5 cm ∅ scarlet

Fig. 5.2. *Melothria maderaspatana*. A. Flowering shoot × $\frac{2}{3}$. B. ♂ inflorescence × 4. C. ♂ flower × 7. D. ♂ flower opened, half the adaxial single
stamen and sepal removed × 10. E. Double stamen × 20. F. ♀ inflorescence
with fruit × 4. G. ♀ Flower opened × 13. H. Seed × 4. (From Jeffrey,
1967, Fig. 19 *pro parte*.) (British Crown copyright. Reproduced with permission of the Controller, Her (Britannic) Majesty's Stationery Office &
the Trustees, Royal Botanic Gardens, Kew. © 1967.)

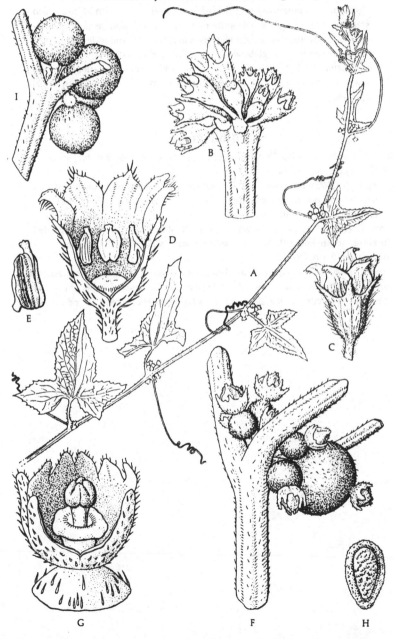

pepos with yellow spots. *R. longpipes* is a drier savanna species, with glabrous petals 5 mm wide.

Zehneria (J. Zehner, Austrian botanical artist), formerly *Melothria* spp., is another Old World genus (of *c*. 30 species), with only 6 species in West Africa, 3 of them widespread. These are all slender monoecious climbers with simple tendrils and small white flowers (yellowing with age), with prominent discs. ♂ flowers in clusters of 1–3, with 3 free 2-celled stamens with straight sessile anthers, ♀ flowers solitary, on slender pedicels, each glabrous or at most slightly hairy, with 3 staminodes. The fruits are small red berries, globose in *Z. capillacea*, ellipsoid or spindle-shaped in *Z. hallii* and *Z. thwaitesii*. The former has white seeds, the latter brownish ones. *Z. capillacea* and *Z. thwaitesii* are both swamp and riverain forest species.

Bibliography

Chakravarty, H. L. (1968). Cucurbitaceae of Ghana. *Bulletin de l'Institut fondamentale d'Afrique noire*, sér. A, **30**, pp. 400–68.

Gilbert, L. E. (1975). Ecological consequences of a coevolved mutualism between butterflies and plants. In *Coevolution of animals and plants*, ed. L. E. Gilbert & P. H. Raven, pp. 210–40. Austin: University of Texas.

Jeffrey, C. (1964). Key to the Cucurbitaceae of West tropical Africa, with a guide to localities or rare and little-known species. *Journal of the West African Science Association*, **9**, pp. 79–97.

Jeffrey, C. (1967). *Flora of tropical East Africa: Cucurbitaceae*. London: Crown Agents.

Okoli, B. E. & Onofeghara, F. A. (1984). Distribution and morphology of extrafloral nectaries in Cucurbitaceae. *Botanical Journal of the Linnean Society*, **89**, pp. 153–64.

6

Ochnaceae – ironwood family

Although there are only five genera in West Africa, three of them are important ecologically. These are trees and shrubs, *Ouratea* being a large genus (of *c*. 40 species in West Africa), containing, on a pantropical basis, half the species of the family (*c*. 300).

Members of the family may be recognised by their alternate, simple, stipulate leaves with prominent, numerous pinnate veins, shallow conspicuous white or yellow flowers (with 10 to numerous stamens) in racemose inflorescences, and by their fruits (see below).

Fleurydora felicis is endemic to West Africa and confined, so far as is known, to Guinée, where it grows as a yellow-flowered shrub or small tree among rocks. The only West African herb of the family, *Sauvagesia erecta*, is, however, widespread in wet places (also in Gambia), and has solitary, axillary white-to-pink flowers.

Flowers ± ⊕ ♂ 5-part. K5, coloured and/or persistent (caducous in *Fleurydora*). C5 (sometimes joined basally, *Ouratea*), contorted (valvate in *Sauvagesia*). A5 (*Fleurydora* and *Sauvagesia*), or 10 (*Ouratea*), alternating with the petals, or numerous (*Lophira* and *Ochna*); the filaments persistent, anthers opening by slits (terminal pores in *Lophira*, some *Ochna* spp., *Ouratea*, *Fleurydora*). G(2) (*Lophira*), (3) (*Sauvagesia*) or (5), uncommonly to (10), deeply lobed in *Ochna* and *Ouratea* with a gynobasic style; 1-celled in *Lophira* (with an intrusive basal placenta with several ovules) and in *Sauvagesia* with intrusive parietal placentae above, axile placentae below; 5–10-celled otherwise, very deeply lobed with a single basal ovule per cell in *Ochna* and *Ouratea* (several axile ovules in *Fleurydora*).

Pollination Mostly unknown, though species with large numbers of stamens are probably attractive to bees. *Lophira* spp. belong to this group, and, in addition, possess a nectary round the base of the ovary.

Fruit A capsule (*Fleurydora*, *Sauvagesia*) or samara with unequally enlarged sepals forming the wings in *Lophira*; *Ochna* and *Ouratea* are pseudoapocarpous, the pseudomonocarps drupoid and black, the sepals and central syncarpous part of the fruit fleshy and brightly coloured.

Dispersal *Ochna* and *Ouratea* are clearly adapted to dispersal by birds, *Lophira* to wind. A censer mechanism usually operates for capsules.

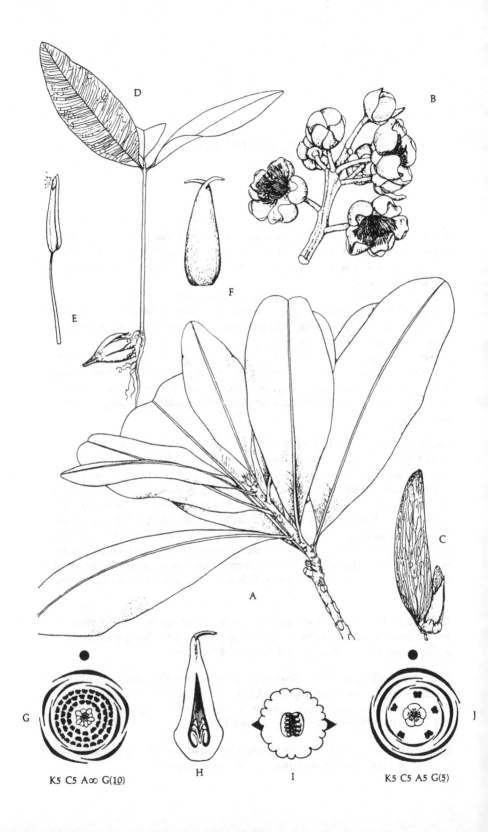

D

B

E

F

C

A

G

H

I

J

K5 C5 A∞ G(10)

K5 C5 A5 G(5)

Economic species Only timber (ironwood), from *Lophira alata*, is of importance.

Field recognition *Lophira* (Gr. crest, the enlarged calyx). The only 2 species are West African endemics, *L. alata* being a forest emergent, *L. lanceolata* (also in Gambia) a savanna species. Both species have strap-shaped leaves with close pinnate venation and wavy margins, the leaves being red when young. The flowers are white, in terminal panicles 30 cm long in the dry season, and these are followed by reddish samaras, the unequally developed sepals providing the wings around the nut. *L. alata* has leaves up to 4 times as long as broad, on a short petiole and clustered at the ends of the twigs. *L. lanceolata* has leaves up to 7 times as long as broad on petioles *c.* 7 cm long.

Ochna (Gr. name). Small trees and shrubs with deciduous stipules and oblanceolate ovate, finely toothed leaves, flowering late in the dry season, or early rains, before the new leaves appear. The fruits are characteristic, and, after they fall, a peg (the basal part of the jointed pedicel) is left on the fruiting branch, which is generally part of a short axillary raceme or cluster. Two yellow-flowered species of northern savannas are *O. rhizomatosa* and *O. schweinfurthiana*, a small shrub and a small tree, respectively. *O. afzelii* is a white-flowered tree, up to 20 m high, in southern savanna, especially forest margins and sites protected from fire. *O. multiflora* is a smaller tree (also in Gambia) of riverain forest and by lagoons, with racemes of very short-lived yellow flowers. All the above species have anthers opening by slits, while *O. membranacea* is a small forest tree up to 8 m high, with yellow flowers in short axillary panicles, the anthers opening by terminal pores. *Ochna* occurs in tropical and southern Africa, and although all the above species are widespread in West Africa, only *O. schweinfurthiana* extends through tropical Africa.

Ouratea (Guianese name). Most species are of local distribution. Most are shrubs, sometimes with persistent stipules, with finely serrate leaves up to 60 cm long. The flowers are yellow, and resemble those of *Ochna* except for the stamens (10 only, the filaments very short and the anthers always opening by pores). The fruit also resembles those of *Ochna*, and the pedicels are also jointed, the base persisting as a peg. Of the 4 widespread species (none in Gambia), *O. flava* is easily the most common. It is a forest shrub (also on inselbergs), with spiny-serrate leaves with caducous stipules and large, loose, erect terminal panicles of flowers, mostly

Fig. 6.1. A–F, H, I. *Lophira alata*. A. Leafy shoot × ⅓. B. Inflorescence × 1. C. Fruit × ⅓. D. Seedling × ⅓. E. Stamen × 5. F. Ovary × 3. H. Ovary in vertical section, diagrammatic. I. Ovary in cross-section, diagrammatic. G. *Ochna* sp. floral diagram. J. *Fleurydora felicis* floral diagram. (A–F from Voorhoeve, 1965, Fig. 55; G–J from Engler & Prantl, 1895, Figs. 70A, 74H, 74I. J adapted from Fig. 70C.)

in the dry season. In riverain forest and extending north into fringing forest is another shrub (or small tree), *O. glaberrima*, which also has rather distant secondary veins in the leaves. However the stipules are persistent and the flowers are in a loose raceme. The 2 widespread small forest trees *O. calophylla* and *O. myrioneura* (up to 15–17 m high) have leaves with the close set pinnate veins similar to those of *Ochna* and *Lophira*, the former with leathery leaves, the latter with thin leaves.

Farron (1968) has split *Ouratea* into 4 genera, maintaining that *Ouratea sensu strictu* occurs only in the New World. The 3 African genera are as follows: *Idertia* (spp. 8 and 9 in the *Flora of West tropical Africa*, 1954); *Rhabdophyllum* (spp. 27, 30–4, *loc. cit.*), and *Campylospermum* (probably all the remaining species, *loc. cit.*). For present purposes, the genus *Ouratea* is maintained in the sense used in the *Flora of West tropical Africa*.

Bibliography

Engler, A. & Prantl, K. (eds.) (1895). *Die natürliche Pflanzenfamilien: Ochnaceae*, 1st edn. Leipzig: Engelmann.

Farron, C, (1968). Contribution à la taxonomie des Ouratéeae d'Afrique. *Candollea*, **23**, pp. 177–228.

Keay, R. W. J. (1954). *Flora of West tropical Africa*. London: Crown Agents.

Voorhoeve, A. G. (1965). *Liberian high forest trees*. Wageningen: Centrum voor Landbouwpublikaties en Landbouwdokumentatie.

7

Combretaceae – afara family

This is a small family in West Africa, of nine genera and $c.$ 80 species, of entirely woody plants, trees, shrubs and lianes. *Guiera* is a monospecific genus endemic to West Africa, while *Conocarpus* and *Laguncularia*, of two species each, and *Anogeissus* (11 species), are each represented by a single species in West Africa. Only *Combretum* and *Terminalia* are represented by several to many species. All the above genera are, however, important ecologically, *Terminalia* also economically.

Members of the family may be recognised by their simple, exstipulate entire leaves and congested racemose inflorescences which may be axillary, extra-axillary or terminal. The flowers are small, actinomorphic and epigynous, and are followed by winged fruits. Both opposite (verticillate) and alternate leaf arrangements occur in the family. Ptyxis is usually conduplicate, in *Quisqualis* and *Terminalia* supervolute.

Introduced species include *Terminalia catappa* (Indian almond), a handsome tree with 'whorls' of branches and red senescent foliage. The fruit is a green fibrous drupe, which floats in water.

The indumentum (in all hairy species) is characteristically composed of non-glandular compartmented unicellular ('combretaceous') hairs, ± glandular multicellular stalked glands and ± glandular multicellular scales, these last being distinguishable by a hand lens. *Quisqualis*, for example, has stalked glands, *Guiera* scales, but *Combretum* spp. may have either, while *Pteleopsis*, *Terminalia* and *Conocarpus* have neither. The epidermis in general provides valuable characters for both taxonomy and identification (Exell & Stace, 1972), but the pits and pockets in the epidermis referred to by these authors as domatia do not appear to have anything to do with an ant mutualism.

Flowers ⚤ ⊕ ♀ (or ♀ and ♂ in *Pteleopsis* etc.), 4–5-part. The receptacle forms a cup or tube ('calyx' tube, hypanthium), to which the style adheres in *Quisqualis*, and from which the sepals (and petals, present in 5 genera) arise; the lower part of the hypanthium is free from the pistil only in *Strephonema*. Four-part flowers have orthogonal sepals, diagonal petals. A5+5 in *Quisqualis*, *Pteleopsis*, *Strephonema* etc., both A5+5 and 4+4 in *Combretum*, but A5 only in *Conocarpus*; often projecting from the flower and with versatile anthers. An epigynous disc is commonly present. $G(\overline{2-6})$ 1-celled with 2–6 pendulous ovules; only in *Strephonema* is the ovary $\frac{1}{2}$ to almost-superior; the style is sometimes attached to the lower part of the hypanthium (*Quisqualis*).

Fig. 7.1. Epidermal hairs. A. Compartmented (internally divided) hairs (*Strephonema* spp.). These lie parallel to the epidermal surface. B is a two-armed hair, only one arm being shown complete. C. Type characteristic of *Guiera*. D–F. Multicellular scale-like hairs (*Combretum* spp.). D. Small, simple type. E. Large complex type, all subdividing walls radial. F. Large complex type, radial and tangential subdividing walls equal. G–K. Glandular hairs. G–H. *Combretum* spp. I. *Conocarpus erectus*. J. *Laguncularia racemosa*, in epidermal pit. K. *Quisqualis indica*. In *Combretum* and *Quisqualis*, the stalk is sometimes thick and parenchymatous, longer than that of *Laguncularia*. (A–C and G–K adapted from Stace, 1965, Figs. 1, 2, 11, 31, 37, 43–5; D–F adapted from Exell & Stace, 1972, Fig. 4.)

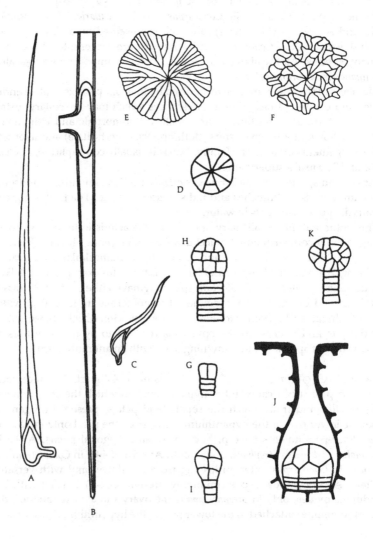

Fig. 7.2. A–D. *Combretum platypterum*. A. Flower in side view × 1¾. B. Flower in longitudinal section × 1¾. Many species of *Combretum* have a ring of hairs in the hypanthium at the base of the stamens, but in C. *platypterum*, hairs edge a series of pockets under the stamens. C. Floral diagram. This applies to all the petaloid genera except possibly *Laguncularia*, which, with *Terminalia* and *Conocarpus* (apetaloid genera), are shown by Engler (1964) to have the odd sepal adaxial. D. Fruit × ¾. E–F. *Anogeissus leiocarpus*. E. Fruiting head × ½. F. Flower × 7½. G–H. *Conocarpus erectus*. Fruit in G abaxial, and H side view × 6½. I. *Guiera senegalensis* fruiting head × ½. J–L. *Terminalia* spp. fruits × ¾. J. *T. scutifera*. K. *T. superba*. This is the only African sp. with a fruit wider than it is long. L. *T. ivorensis*.

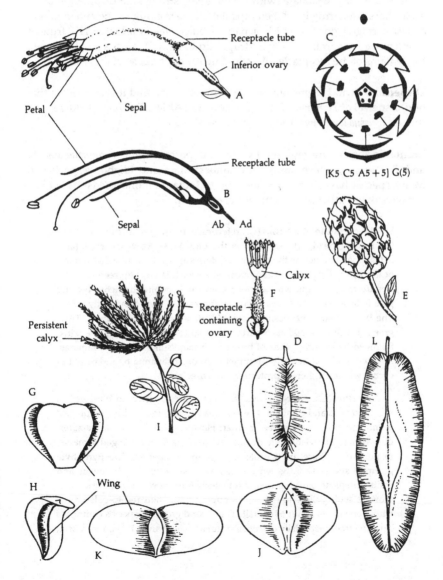

Pollination There is no recorded information for West Africa, but the considerable differences in shape and length of hypanthium suggest that a wide variety of insects is involved. *Combretum* with brush-type flowers are pollinated by humming birds in South America, but although the red and yellow colours are common, none of the West African species has that structure, and although *C. platypterum* suggests sunbird pollination, this has never been observed. *C. fruticosum* in South America is visited by seven species of primates by day, and by moths at night.

Fruit A 2–5 winged samara (with 1 seed) and a strong endocarp, but in *Quisqualis* there is a narrowly winged capsule, in *Guiera* a ridged one and in *Laguncularia* a ridged one to which a pair of bracteoles is adnate. The fruit of *Strephonema* is aberrant, being a 'drupe' up to 8 cm ∅, with a single 'stone', the remains of the receptacle tube standing round the base of the fruit.

Dispersal by wind seems obvious for winged fruits, and it is surprising that bird-dispersal of *Terminalia* fruits is recorded in Africa. Trochain (1940) reports that *Anogeissus leiocarpus* fruits are dispersed by ants.

Economic species are few, the two most important being *Terminalia superba* and *T. ivorensis*, which supply the timbers afara and idigbo, respectively. Many species have, however, local uses. Indian almond seeds are edible, but have to be extracted from the stony endocarp.

> **Field recognition** *Anogeissus* (Gr. referring to the scale-like fruits besetting the fruiting head). *A. leiocarpus* is the only West African species (also in Gambia), and lies on the extreme western edge of the area of distribution of the genus (tropical Africa–Southeast Asia). It is an evergreen savanna tree up to 27 m high, with an open crown of drooping branches bearing pale foliage. The leaves are alternate and each is tipped by a tiny spine. The bark is pale, the slash yellow with dark lines, exuding gum. Tiny, greenish-yellow apetalous 5-part (A5+5) flowers appear in the late rains, followed by fruiting heads of brown, *Conocarpus*-like 2-winged samaras. *A. leiocarpus* is particularly prominent on forest-savanna boundaries, but occurs wherever there is protection from fire.
>
> *Combretum* (L.). A large genus with *c.* 50 representatives in West Africa, mainly shrubs and climbers. The savanna species tend to show cryptogeal germination of some degree or other. Flowers are 4–5-part and petaliferous, either yellowish-white or red, followed by 4- or 5-winged samaras, respectively. There are 4 ecological groups of species. The pyrophyte group (4 species) can be represented by *C. sericeum*, which flowers and fruits conspicuously after fires in forest-savanna mosaic in the dry season. Flowers with yellow-to-red petals appear first, then red-to-pale-brown fruits, on shoots about 60 cm tall. The second group of species comprises savanna trees, especially of ironstone areas. There are *c.* 7 frequently seen

species, with small, yellowish-white 4-part flowers with cupular receptacles. *C. glutinosum* (Gambia–N. and S. Nigeria) is a gum-yielding species (*taramniya* in Hausa) especially of the sudan zone, with thick leaves, sticky above at least when young, white and densely hairy (not scaly) below. Further south, 2 species without gum, *C. molle* and *C. nigricans* var. *elliottii*, are commoner. Both have tiny fringed petals, but *C. molle* has leaves densely covered underneath with both silky hairs and scales, while *C. nigricans* var. *elliottii* has scarcely hairy leaves (no scales).

Savanna shrubs and climbers form the third group of species, mainly represented by *C. micranthum* (Gambia–N. Nigeria), the whole plant of which is covered with red scales. The commoner species in the fourth group, the forest shrubs and lianes, usually have red petals. *C. paniculatum* (also in Gambia) extends from the forest northwards into fringing forest and other protected situations, but in lowland forest it becomes a lofty liane, the petioles hardening into spines. *C. racemosum* (also in Gambia) also has red 4-part flowers, but the inflorescences are surrounded by white bracts, while *C. platypterum* has 5-part flowers, and its tubular receptacle is rather uncommon.

Conocarpus (Gr. cone-shaped fruit). *C. erectus*, the only species, is found as a mangrove shrub all along the west coast of Africa and that of eastern tropical America. It has alternate, lanceolate leaves narrowed to a short winged petiole bearing 2 glands. The dense ellipsoid heads of small apetalous flowers with 5 stamens otherwise resemble those of *Anogeissus*. The fruits are red, 2-winged and also *Anogeissus*-like, but without the beak (the remains of the receptacle tube). *Conocarpus* is frequently host to the parasite *Cassytha filiformis*.

Guiera (Senegalese name). *G. senegalensis*, a shrub with grey foliage of opposite leaves, each leaf covered underneath with black glands and distinctive hairs, is the only species, and is confined to northern tropical Africa (Gambia eastwards). Before opening, the dense heads of small, greenish-yellow petaliferous 5-part flowers are protected by 4 bracts. The fruits are covered by silky hairs and crowned by the persistent calyx.

Laguncularia (L. *lagena*, flask, the fruit shape). *L. racemosa* is a mangrove shrub or small tree, developing pneumorhizae with white tips. The leaves are opposite, with a pair of glands towards the top of each petiole, and the blades are thicker and fleshier in more saline conditions, and panicles of tiny, greenish (4–)5-part flowers with minute petals appear mainly in the rains. Bracts are deciduous, but a pair of bracteoles remains adnate to each receptacle tube. Fruits ribbed, with a persistent calyx.

Terminalia (L. *terminus*, referring to the tight spiral of leaves at the end of each branch). Trees with spikes of tiny, apetalous ♀ and ♂ flowers, greenish-white with a cupular receptacle, followed by 2-winged fruits. *T. scutifera* is a gnarled, deciduous seashore tree in Sierra Leone, flowering in the dry season. *T. superba* and *T. ivorensis* are deciduous emergents of drier lowland forest, the former with widely winged, stalkless fruits

at the end of the dry season, along with the new leaves. It is also prominent in forest regrowth as saplings. *T. ivorensis* has stalked, elongated fruits appearing in the rains and lasting on the tree for several months. Four savanna species also flower in the rains. These are *T. avicennioides* (also

Fig. 7.3. *Terminalia laxiflora*. A. Flowering shoot × $\frac{1}{2}$. B. Flower bud × 4. C. Opening flower × 5. D. Flower × $4\frac{1}{2}$. E. Flower in longitudinal section × $4\frac{1}{2}$. F. Fruits × $\frac{1}{2}$. (From Griffiths, 1959, Fig. 18.)

in Gambia), which is locally common on lighter soils and in more northerly savannas. It may appear in regrowth further south. Leaves and fruits are covered with grey-green felt (pink on young growth), and the fruits are widest across the bottom. *T. laxiflora* is more common to the south, especially in ironstone areas, and is distinguished by its stalked leaves and glabrous fruits, which are widest across the middle. Near streams and in flood plains with at least temporarily impeded drainage, *T. macroptera* and *T. glaucescens* are found. The former (also in Gambia) has large (8–13 cm × 3–4 cm) glabrous fruits and sessile, bright green leaves. The latter has hairy (stalked) leaves and flowers, followed by fruits, broadest across the bottom, covered with fine, silky hairs. They ripen from red through white to brown, with a fine grey or mauve bloom. The West African species are all illustrated by Griffiths (1959).

Bibliography

Engler, A. (1964). *Syllabus der Pflanzenfamilien*, 12th edn, vol. 2. Berlin: Gebruder Borntraeger.

Exell, A. W. & Stace, C. A. (1972). Patterns of distribution in the Combretaceae. In *Taxonomy, phytogeography and evolution*, ed. D. H. Valentine, pp. 307–23. London: Academic Press.

Griffiths, M. E. (1959). A revision of the African species of *Terminalia*. *Journal of the Linnean Society (Botany)*, **55**, pp. 818–907.

Jenik, J. (1970). Root systems of tropical trees. 5. The peg roots and pneumathodes of *Laguncularia racemosa* Gaertn. *Preslia, Praha*, **42**, pp. 105–13.

Stace, C. A. (1965). The significance of the leaf epidermis in the taxonomy of the Combretaceae. *Journal of the Linnean Society (Botany)*, **59**, pp. 229–52.

Stace, C. A. (1980). The significance of the leaf epidermis in the taxonomy of the Combretaceae: conclusions. *Botanical Journal of the Linnean Society*, **81**, pp. 327–39.

Trochain, J. (1940). Contribution à l'étude de la végétation du Sénégal. *Mémoires de l'Institut française de l'Afrique noire*, **2**.

8

Guttiferae (Clusiaceae) – butter tree family

Except for some species of the genus *Garcinia*, these are upper-storey, ever-green forest trees. Most are glabrous, the slash exuding yellow or orange resinous sap or latex. The family is a pantropical one, of about 600 species. No purely decorative species have been introduced. The Hypericaceae may be included as a subfamily.

Members of the family may be recognised by their pairs of exstipulate, simple entire leaves, which are streaked and spotted with resin glands, and carried on branches which appear to be whorled. The flowers are axillary, solitary or in cymes, red or white and conspicuous, with numerous stamens often in bundles. The fruits are large and berry like, and of economic importance.

Flowers $\div$ $\oplus$, $\vec{O}$, $\female$ and $\vec{O}$, often 5-part, though 4-part in *Calophyllum* etc., when the sepals are orthogonal and the petals diagonal. Anthers mostly united in bundles each opposite a petal and alternating with the stigmata, but free in *Garcinia* sp., united only basally in *Calophyllum* and *Mammea*, and united into a staminal tube in *Symphonia*, cf. Chapter 18, Meliaceae; present as stami-nodes in $\female$ flowers. An extrastaminal disc is usually present. G(2)–(5)–(8) with the same number of cells and axile placentation, the terminal style sometimes very short; 1-celled with 1 basal ovule in *Calophyllum* and 5-celled with parietal placentae in *Allanblackia*.

Pollination The disc in *Allanblackia*, *Garcinia* and *Symphonia*, thought to be staminodal in origin, and the numerous stamens suggest insect pollination, the shape of the flower providing 'open access' to short-tongued insects. The smell of rancid butter given off by *Pentadesma* is possibly attractive to bats.

Fruits *Calophyllum* and *Mammea* produce drupes, the seeds in the stones of the latter being embedded in colourless edible pulp, probably representing the sarcotesta. The remaining fruits are berries, or berry like, in the latter the flesh being derived from the pulpa, or pulpa and arilloid together. The seeds are generally large, the non-cotyledonous part making up the bulk.

Dispersal *Garcinia* spp. are reported to be dispersed by primates, and it would be interesting to know whether the arilloid of *Garcinia* has the high sugar and sweet-tasting amino acid content of the arilloids of the *Aglaia* spp. dispersed by primates, as opposed to the high lipid content of the arilloids of bird-dispersed species (see Chapter 18, Meliaceae). Dispersal by birds has not been reported in connection with members of the family in West Africa. *Garcinia kola*, *Mammea* and *Pentadesma* are dispersed by elephants, though *Mammea* and *Calophyllum* are also reported to be dispersed by bats.

Economic species The tall forest species yield resinous timber. With the exception of *Symphonia*, the fruits and/or seeds are eaten, or oil is extracted. The introduced mangosteen (from Malaysia) (*Garcinia mangostana*) produces its fruits parthenogenetically, a great advantage in an introduced species where pollinators may be lacking. It is, however, very difficult to establish and slow to bear. The white flesh round each seed is particularly delicious. The West Indian *Mammea americana* has been introduced for its fruits (eaten cooked); a liqueur can also be made from them. *Calophyllum inophyllum*, which grows as a coastal tree in East Africa and elsewhere, yields resin.

Field recognition *Allanblackia* (A. Black, nineteenth-century British botanist). *A. floribunda* (now *A. parviflora*) is the only West African species in this small tropical African genus. It has large (6 cm ∅), scented red-to-pink ♂ and ♀ flowers in terminate clusters at the ends of drooping branches. The flowers are 5-part, ♂'s with thick stamen bundles alternating with the 5 lobes of the fleshy disc, and lacking a pistillode; ♀ flowers have a 1-celled ovary with a ± sessile stigma and 5 parietal placentae and numerous seeds, surrounded by 5 disc glands alternating with 5 staminodal bundles. The berry is brown, pendant and large (up to 50 cm long and 14 cm wide) and contains large seeds (2–4 cm long), each with a netted arilloid visible on its surface. The surrounding flesh is pink and jelly like.

Garcinia (L. Garcin, eighteenth-century French botanist) is mainly an Old World genus with only 16 species (out of c. 400) in West Africa. These are small-to-medium sized trees and shrubs, in savanna as well as forest. *G. kola* is a forest species also cultivated for its seeds ('bitter kola'), and is exceptional in having hairy umbel-like cymes followed by orange-like fruits with orange pulpa. This species and the ones mentioned below all have ♀, ♂ and ♀ flowers, sometimes ♂ and ♀ on separate plants, with 4-stamen bundles alternating with a fleshy disc. The ovary is 2–5-celled with an apical ovule in each cell. Of the savanna species, three occur in habitats near water. *G. livingstonei* is a northern savanna tree up to 17 m high, and extends from Mali and Guinée throughout tropical Africa (in similar habitats), the only one of our species to do so. It is also our only species to have free stamens and leaves in whorls of 3 (also visible in ♀ trees), which later bear small red or yellow fruits, the 2 seeds in jelly-like pulp. *G. ovalifolia* and *G. afzelii*, 2 species of fringing forest and forest outliers, are shrubs or trees up to 10 m and 17 m high, respectively, both with paired leaves, those of *G. ovalifolia* with marginal nerves, those of *G. afzelii* prominently marked with long wavy resin canals.

K5 C5 A(5 × 5) G(5)

G. *ovalifolia* has axillary clusters of small yellow flowers on 1 cm pedicels, followed by yellow fruits 1–1.5 cm ∅. G. *afzelii* has branched inflorescences of greenish flowers c. 2 cm ∅, with papery petals, followed by brownish fruits c. 2 cm ∅. This is 1 of a group of species with sessile anthers, which in addition, are transversely partitioned, as they are in G. *punctata* (Nigeria–Angola and Zimbabwe).

Mammea (West Indian name). The only species is M. *africana*, which has gland-dotted leaves and axillary clusters of a few white ♂ flowers, or solitary ⚥ flowers, c. 3 cm ∅, on robust stalks. The flower buds are, unusually for the family, enclosed by the 2(–3)-part calyx. C4(–6), round numerous stamens joined only basally, the ovary 2-celled with 1–2 basal ovules in each cell and a ± sessile 2–4-lobed stigma. The fruit is thick-stalked drupe, woody, 7–8 cm long, with a thick reddish-yellow rind and 2–4 large stones in yellow flesh within. Inside the stone, the seed is surrounded by an edible arilloid.

Pentadesma (Gr. 5 bundles) is a very small tropical African genus with only 1 species, P. *butyracea* (butter tree) in West Africa. The leaves are frangipani-like, but dotted and streaked with resin glands and canals. The flowers are waxy white, drooping on thick pedicels, 5-part and smelling of rancid butter. The style persists as a point on the reddish-green pointed berry, which is c. 15 cm long × 10 cm wide. It contains yellow flesh, with several seeds embedded in it. When broken open, the embryos are dark red, and oil is extracted from them. This is a riverain and swamp forest species which may develop stilt roots. The branches are horizontal and whorled, and the slash yields thick yellow juice which dries to a reddish gum.

Symphonia (Gr. joined stamens). S. *globulifera*, a swamp forest tree also found in riverain forest and on the banks of creeks and lagoons, is the only species. It may develop knee and stilt roots with a yellow slash. The tree has a small, flat semi-deciduous crown of whorled branches, on which small, red, flower buds and flowers appear almost continuously on small side branches. The flowers are ⚥ 5-part, the staminal tube fitting closely round the pistil, the 5 stigmata alternating with the 5 groups of

Fig. 8.1. *Garcinia* spp. A–H. G. *livingstonei*. A. Flowering shoot × 1. B. ♂ flower × 2. C. ⚥ flower × 2. D. ⚥ flower without sepals or petals × 2. E–F. Stamen × 4. G. Flower bud × 2. H. Fruit × 1. I. G. *punctata* ♂ flower, stamens and pistillode. J. Floral diagram of 5-part flower (G. *granulata* or G. *gnetoides*). K. G. *mangostana* half fruit × ⅔ (K4 but G(6–7), with the same number of truncated stigmas applied to the woody pericarp. The arilloid is white, like that of the lychee). A–H from Bamps, Robson & Verdcourt, 1978, Fig. 5; I–J from Pierre, 1879–1907; K from Engler & Prantl, 1885, Fig. 114A.) (British Crown copyright. Reproduced with permission of the Controller, Her (Britannic) Majesty's Stationery Office & the Trustees, Royal Botanic Gardens, Kew. © 1985.)

anthers. The berry is rough and reddish, $c.\,3\,cm$ $\emptyset$, crowned with the persistent stigmata. It contains 1–2 arilloid seeds and only a little flesh. There are $c.\,20$ species in Madagascar, only *S. globulifera* in tropical Africa and only 2 further species in North America (Columbia) so that the West African species is also by way of being a 'bridge species'.

Bibliography

Bamps, P., Robson, N. & Verdcourt, B. (1978). *Flora of tropical East Africa. Guttiferae.* London: Crown Agents.

Engler, A. & Prantl, K. (eds.) (1885). *Die natürliche Pflanzenfamilien: Guttiferae,* 1st edn. Leipzig: Engelmann.

Pierre, L. (1879–1907). *Flore forestière de Cochinchine.* Paris: Octave Doin.

9

Sterculiaceae – cocoa family

Ecologically important species in West Africa are mainly forest trees, especially in the drier parts of lowland forest. Few species appear north of the northern boundary of the forest-savanna mosaic. Nearly all the genera are Old World only in distribution.

Members of the family may be recognised by their alternate, digitately compound/lobed/nerved leaves with deciduous stipules and stellate hairs, and small flowers arranged either in clusters, sometimes on old wood, or else in lax, panicle-like cymes. Flowers are complex and five-part, about two-thirds of the genera having ♂ flowers with petals, the remaining one-third having at least some ♂ and ♀ flowers lacking petals. Fruits are also distinctive (see below).

Kleinhovia hospita, a shade tree introduced from India, is occasionally seen.

Byttneria is a tropical genus of lianes and scrambling shrubs, the only genus of this habit in the family in West Africa. There are three species in forest, the most common and widespread being *B. catalpifolia* subsp. *africana* in secondary forest. It has suborbicular leaf blades up to 18 cm ∅ on petioles the same length. There are small, scented, white ♂ flowers in leaf-opposed cymes, the petals tailed, and the fruits are distinctively prickly capsules. The young shoots of the other species are also prickly.

Flowers ⊥ ⊕ 5-part, *either* ♂ and ♀ and apetalous with a petaloid calyx, K(5) (*Octolobus* K(8)), *or* ♂ with petals, the calyx inconspicuous (spathe-like in *Mansonia*); polygamy can occur in both groups. Corolla usually contorted, the petals sometimes hooded and/or tailed (*Theobroma, Byttneria, Scaphopetalum*), persistent in 2 species of *Dombeya*. Two petaliferous genera (*Mansonia, Triplochiton*) and most apetalous ones have an androgynophore bearing 1 or 2 whorls of stamens, or stamens in bundles of at least 2 (*Triplochiton*) or the anthers in an irregular mass (*Hildegardia, Sterculia*); in ♀ flowers, the stamens are present as staminodes in the same positions and, in *Mansonia* and *Triplochiton*, there is an additional whorl of intrastaminal petaloid staminodes; in the remaining genera, the stamens are joined, forming a tube with the alternating staminodes; stamens – often in bundles – opposite the petals (*Byttneria, Leptonychia, Scaphopetalum, Theobroma*); in *Nesogordonia* the bundles are free from each other and alternate with the petals; of the 4 genera of small shrubs and herbs, only *Melhania* has staminodes; anthers 2-celled. G(1, *Waltheria*)–(5)– (12)–(∞ spirally arranged, *Octolobus*), each carpel with 2 rows of axile ovules.

Fig. 9.1. A–C. *Cola millenii*. A. ♀ flower × 4, and B. ♂ flower × 4, the near part of the calyx tube cut away. C. Floral diagram of combined ♂ and ♀ flowers. D–E. *Cola acuminata*. D. Androgynophore and stamens of ♂ flower × 9. E. Floral diagram of combined ♂ and ♀ flowers. F. *Pterygota bequaertii* fruit × ½. G. *Hildegardia barteri* fruit × ½. H. *Mansonia altissima* fruit × ⅝. I. *Sterculia rhinopetala* fruit × ⅝. J. *Triplochiton scleroxylon* fruit × ½. K. *Tarrietia utilis* fruit × ⅓. (E from Hallé, 1961, Fig. 11.1; by permission of the Muséum national d'histoire naturelle.)

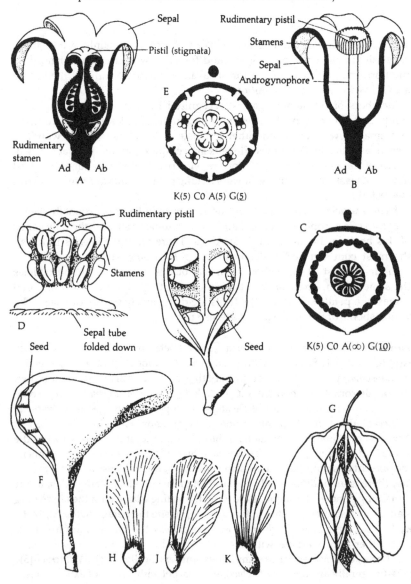

Pollination *Theobroma cacao* has been well investigated. The amelonado cocoa in West Africa is mostly self-compatible, but pollen is collected by biting midges (*Ceratopogonidae*) and by ants and other small insects (to a lesser extent) and deposited on the styles and stigmata in characteristic packets and smears. The midges feed (and breed) on the pulp of rotting cocoa 'pods' in Congo. There is no nectar in sterculiaceous flowers, but the colours and odours of *Cola* and *Sterculia* spp. suggest fly pollination, while *Hildegardia* is pollinated by bees collecting pollen. This could also occur in petaliferous species.

Fruit A schizocarp or capsule. In the schizocarps, the mericarps are either samaroid, as in *Mansonia* etc., (inflated in *Hildegardia*) or 'follicular', as in *Dombeya*, *Octolobus* and *Pterygota* etc. Only 1 mericarp in the whorl (or spiral) may develop. The seeds often have an arilloid, either complete (sarcotesta) or partial (usually basal). The wing of the seed of *Pterygota* etc. is said to be arilloid in nature. *Nesogordonia*, *Leptonychia* etc., and the smaller shrub and herb genera of the family, have 5-celled loculicidal capsules. In those of *Nesogordonia* the seeds are winged, while those of *Leptonychia* have a conspicuous sarcotesta.

Dispersal Winged fruits and seeds are clearly dispersed by wind, but for the follicles and capsules, such as *Byttneria*, birds are concerned. The latter are reported to disperse *Sterculia*, but ants have also been suggested; placental tissue forms an elaiosome. Primates and rodents will also gnaw arilloids, for example the sarcotesta of cocoa.

Economic species Kola nuts are derived from two species of *Cola*, *C. nitida* (Guinée–Ghana and introduced in Nigeria) and, to a lesser extent, *C. acuminata* (Togo–Angola). West African 'amelonado' cocoa (*Theobroma cacao*) was introduced from Fernando Po, where Amazon forastero seeds ('beans') had earlier been introduced. The important timber species include *Mansonia altissima* (black walnut), *Nesogordonia papavifera* (danta), *Sterculia* spp., *Tarrietia utilis* (nyankom, niangon) and *Triplochiton scleroxylon* (obeche). Obeche is used mainly in the production of pulp for paper manufacture, but the other species are used for structural purposes.

> **Field recognition** *Cola* (West African name). These are large-leaved, evergreen forest trees with apetalous flowers followed by large follicles with arilloid seeds. Of the group of species with digitately compound leaves, *C. chlamydantha* has the widest distribution, from Sierra Leone eastwards. It is up to 20 m high, with leaflets and petioles up to 60 cm long. Clusters of flowers with red calyces 2–4 cm long are formed on the trunk, ♂ ones with a single whorl of 20 anthers, ♀ ones with 9–12 carpels. These form large, thick-walled, green-to-red follicles, each with many red seeds. In the group of species with digitately lobed leaves, *C. millenii* (Côte d'Ivoire– S. Nigeria) is common in forest regrowth. It is 4–18 m high, with, as a rule, 5-lobed leaves, and clusters of red and green flowers in the axils

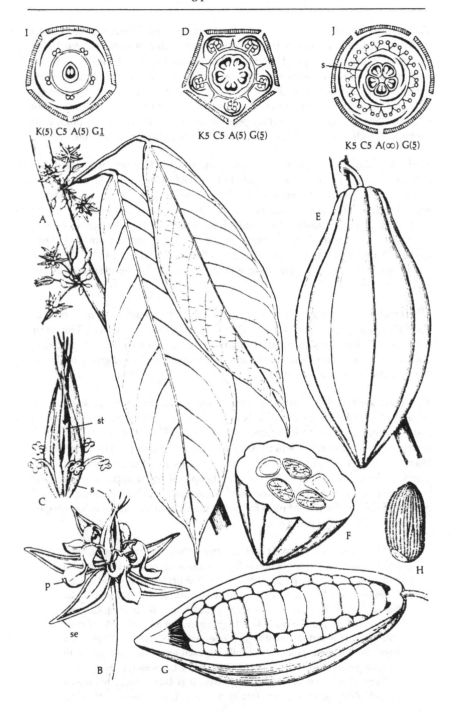

K(5) C5 A(5) G1

K5 C5 A(5) G(5)

K5 C5 A(∞) G(5)

of the leaves in the dry season. Groups of 6–11 large, deep pink-to-red follicles appear later. The lining of each follicle is a yellowish-red colour and there are several seeds. *C. laurifolia* is a riverside species up to 30 m high, with entire, elliptic leaves, young shoots covered with light brown stellate hairs, and panicles of tiny yellowish-white flowers also with brown hairs. ♂ flowers have a single whorl of anthers. Flowering is prolonged (November–June) and fruiting similarly so (January–July). In this period groups of 4–6 rounded, brown-to-red follicles, 5 cm long, appear. *C. marsupium* (S. Nigeria–Cameroun) is distinguished by its paired ant-inhabited pouches at the base of each leaf blade. *C. acuminata* and *C. nitida*, also kola-yielding species, also have entire leaves, but 2 whorls of anthers or staminodes. *C. verticillata*, the third kola species, has entire leaves in 4s. Only *C. cordifolia* is recorded for Gambia. *Cola* is a large African genus, 50 (two-fifths) of its species occurring in West Africa.

Dombeya (J. Dombey, eighteenth-century French botanist) is an Old World genus, with 50 species (out of 350) in tropical Africa. The 3 West African species are deciduous savanna shrubs and small trees, with slightly digitately lobed or ovate-orbicular leaves, digitately nerved and coarsely toothed or with small glandular teeth, thickly covered with stellate hairs beneath, at least when young. Cymes of pink or white ♂♀ flowers are produced in the dry season, followed by capsules. *D. buettneri* and *D. quinquesita* var. *senegalensis* are the most widespread species, the former with white flowers with a red centre (at the base of the petals and staminal tube) on leafy shoots, in heads. The ovary is 5-celled. In *D. quinquesita* var. *senegalensis* (also in Gambia) the flowers are at first pink, then white, produced before the leaves appear. The petals persist and become papery round the 3-celled capsule.

Hildegardia (St Hildegard, twelfth-century physician and mystic). *H. barteri* is the only West African species, a pioneer tree – up to 30 m high – in regrowth, and especially on rocks and slopes. Its seeds need light in order to germinate. It flowers spectacularly in the dry season, as, and after, the leaves fall, producing bright red flowers in large, branched panicle-like cymes. Each flower has a tubular calyx, out of which projects a fused mass of anthers above small carpels on the androgynophore in ♂ flowers, and 5 carpels on the androgynophore in ♀ flowers. The groups of 5 inflated, red samaras begin to appear about Christmas time, and the heart-shaped, entire, but digitately nerved leaves appear about the

---

Fig. 9.2. A–H. *Theobroma cacao*. A. Flowering shoot × ½. Flowers are normally pendulous. B. Flower × 3. C. Staminodes, stamens and pistil × 7; D. Floral diagram. E. Fruit × ⅓. F–G. Opened fruits × ⅓. H. Seed × ¾. I. *Waltheria indica*, floral diagram. J. *Triplochiton scleroxylon*, floral diagram. s, staminode; se, sepal; st, stigma; p, petal. (A–C, E–H from Larsen, 1973; D, I–J from Hallé, 1961, Figs. 11.9, 11.10, 11.14, by permission of the Muséum national d'histoire naturelle.)

beginning of March. Eight of nine species of this genus have restricted Old World distributions.

Sterculia (L. *stercus*, dung, the flower odour of a species). Deciduous trees with apetalous flowers, the anthers fused into an irregular mass. These are followed by (4)–5 woody follicles, each splitting along the top and then boat-shaped, with dark-coloured seeds, mostly with sarcotestas, on very thin funicles. A large tropical genus of *c.* 300 species, there are only 4 species in West Africa, and only 2 of these (*S. setigera* and *S. tragacantha*) are widely distributed in the rest of tropical Africa.

S. setigera (also in Gambia) is up to 12 m high, and is unusual in being a savanna species, even in open woodland savanna in the sudan zone. It has digitately 3-lobed leaves, persistently hairy on both surfaces. In the first half of the year, it produces red or red-streaked, saucer-shaped flowers, 2–2.5 cm ⌀, on last year's wood, as the new leaves appear. Large follicles, up to 10 cm long, appear later. They are hairy both outside and in, with thinnish walls (under 5 mm thick) and bristles along the margins where the usually arilloid seeds are attached.

The 3 forest species are up to 40 m high, with entire, pinnately nerved leaves, flowers on leafy shoots and thick-walled follicles. *S. oblonga* (timber – yellow sterculia) has apparently free yellowish-white sepals, 1 cm long, followed by very large follicles (at least 10 cm long) with walls 1 cm thick. Inside are several seeds, each covered by its yellow sarcotesta. *S. rhinopetala* (timber – brown sterculia) has tiny, cupular yellowish-white flowers in the late rains, followed by small follicles (up to 7 cm long) with walls *c.* 5 mm thick, with seeds with bright red sarcotestas. *S. tragacantha*, African tragacanth (which has pinkish gum, like the gum tragacanth of commerce) has tiny, cupular reddish flowers over a long period (October–June), then small, bright red-to-brown follicles with walls *c.* 5 mm thick, bristly inside, and containing naked seeds.

Tarrietia (Sudanese name) (African red cedar; timber – niangon, nyankom). *T. utilis* (now *Heritiera utilis*; Voorhoeve (1965)) is the only species in the genus and is confined to West Africa from Ghana westwards. It is an evergreen upper-storey forest tree, with plank-like stilt roots at maturity, and large digitately compound leaves on long petioles. Tiny, pale brown hairy flowers are produced in lax inflorescences in the rains, followed by samaras.

Triplochiton (Gr. three tunics, the three whorls of sepals, petals and staminodes) (obeche – the timber). A deciduous emergent with digitately lobed leaves, producing lax panicles of flat, white, petaloid flowers in the dry season while leafless. The 30(+) stamens are joined basally in pairs round the top of the androgynophore, surrounding 5 white, petaloid staminodes and 5 carpels. The samaras that follow are often galled. This is a small tropical African genus of only 2(–3) species.

Waltheria (Walther, eighteenth-century German botanist). A largely tropical American genus, from where the 2 West African species were probably introduced. The more widely spread species, *W. indica* (also in Gambia), is a common weed throughout West Africa and the tropics in general. It is a woody herb up to 2 m high, with simple crenate leaves and close leafy cymes of tiny flowers with yellow petals. Bracts and calyces have long, soft colourless hairs. The unilocular ovary becomes a minute, 2-valved hairy capsule with 1 seed, the large brush-like stigma withering away, but the hook in the style persisting on top of the 1-celled capsule.

Bibliography

Cudjoe, F. (1969). A key to the family Sterculiaceae in Ghana. *Technical Note of the Forest Products Research Institute, Ghana*, 7.

Enti, A. A. (1968). Distribution and ecology of *Hildegardia barteri* (Mast.) Kosterm. *Bulletin de l'Institut fondamentale d'Afrique noire*, sér. A, 30, pp. 881–95.

Hall, John B. & Bada, S. O. (1979). The distribution and ecology of obeche (*Triplochiton scleroxylon*). *Journal of Ecology*, 67, pp. 543–64.

Hallé, N. (1961). *Flore du Gabon: Sterculiacées*. Paris: Muséum national d'histoire naturelle.

Jenik, J. (1969). The life form of *Scaphopetalum amoenum*. *Preslia, Praha*, 41, pp. 109–12.

Larsen, K. (1973). *Kormofyternes taxonomi*. Copenhagen: Akademisk forlag.

Voorhoeve, A. G. (1965). *Liberian high forest trees*. Wageningen: Centrum voor Landbouwpublikaties en Landbouwkumentatie.

10

Bombacaceae – silk cotton family

A small pantropical family, represented in West Africa by three genera of ecologically prominent, large deciduous trees with 'whorled' branches.

Members of the family may be recognised by their alternate digitately compound leaves with deciduous stipules, large, robust, red or white, five-part flowers (produced while leafless), each with one-celled curled anthers on filaments variously united. The fruits are large capsules releasing seeds in wool, or indehiscent.

There are no genera of unusual habit in West Africa. In the sterile condition confusion with *Ricinodendron* (*Euphorbiaceae*) might be possible, but the latter has persistent stipules.

Balsa (*Ochroma pyramidale*) and *Pachira* spp. have been introduced, the former for its timber, balsawood (from which the *Kontiki* raft was made) and *Pachira* for its seed oil. Both seem to be planted only for decoration, and make fine specimen trees.

Flowers $\perp$ ($\div$ in *Ceiba*) $\oplus$ $\male$, 5-part. Calyx lobed (*Adansonia*) or $\pm$ cupular. Corolla contorted. Stamens united to form a staminal tube, which is joined basally to the corolla, with filaments in bundles arising from its upper edge (free in *Adansonia* and *Ochroma*); anthers 1-celled, folded (linear in '*Bombax*' sp.). G($\underline{2-5}$), with as many cells (opposite the petals) and stigma lobes as carpels; ovules 2 to many on axile placentae.

Pollination By bats in the white-flowered genera *Adansonia* and *Ceiba*, where the flowers have a characteristic stale smell. The red-flowered *Bombax* spp. are pollinated by birds, but the white-flowered '*B.*' *brevicuspe* needs investigating. The deciduousness of the calyx in red-flowered species may give access to nectar. In *Ceiba*, nectar is produced by the basal inside surface of the calyx in great quantity, and is lapped up by fruit bats as they crawl over the inflorescence. In *Adansonia*, the great ball of anthers and strong staminal tube are grasped and nectar licked from round the base of the latter.

Fruit In *Adansonia*, the fruit is indehiscent with seeds in dry acid flesh (pulpa), while in *Bombax* and *Ceiba*, the inner lining of the pericarp forms hairs (floss or wool) in which the seeds are eventually dispersed when the capsules open.

Dispersal Seeds of *Bombax* and *Ceiba* are dispersed by wind, those of *Adansonia* by baboons, which remove fruits to a distance and crunch them open. The seeds germinate from baboon dung. In East Africa, elephants are reported to bring about dispersal.

Economic species All species supply a range of locally useful products, though not timber, the wood being too soft. It does, however, supply fibre. Kapok and silk cotton (seed floss) have been important commercially, but seed oil is more valuable today, though not commonly extracted.

> **Field recognition** *Adansonia* (M. Adanson, eighteenth-century French naturalist) (baobab). There are 10 Old World species, but only *A. digitata* occurs in West Africa (though throughout tropical Africa also) and has been introduced into tropical America. The massive water-storing trunks are recognisable from a distance. The leaves are long, petiolate and 5-foliolate. The flowers are large and pendulous on 30-cm long peduncles, the staminal column with its numerous anthers protruding. The fruit is a green, velvety hesperidium up to 35 cm long, spherical or elongated and swollen in shape, remaining on the tree until it next becomes leafless. This is a species of drier savannas, and often planted.

> *Bombax* (L. silk, the floss round the seeds) (red) silk cotton. Two forest and 1 savanna species are confined to West Africa, though other species are found in both Old and New World tropics. These are trees with at most small buttresses, the trunks often thorny, and with 4–8 foliolate leaves. Solitary, erect, nearly sessile flowers are produced while the tree is leafless in the dry season. The petals and stamens (in 2 series of bundles) fall together, and the fruit is a 5-valved capsule, with numerous seeds embedded on floss. In forest *B. buonopozense* is red-flowered, with a deciduous calyx; the floss is white-to-grey, while *B. brevicuspe* is white-flowered with a persistent calyx, long narrow petals, largely free filaments, and reddish-brown floss. This species is now known as *Rhodognaphalon brevicuspe*. It also differs from *Bombax* spp. in having 1 whorl of stamen bundles. In savanna, *B. costatum* is a rather smaller tree (3–15 m), prominent in all open savanna woodland and specially obvious on rocky hills. It is red-flowered, with a red, deciduous calyx and white floss.

> *Ceiba* (? Spanish(–American) name) (kapok). This is an emergent species with large buttresses and massively bracketed branches, sometimes thorny (in savanna), generally preserved in farmland, but also colonising forest regrowth, and regenerating readily from upright branches when these are planted. It is particularly obvious in forest-savanna mosaic and forest outliers. Leaves have 8–15 leaflets with particularly numerous lateral nerves. The flowers are small and white, produced in the dry season, in downward facing clusters, with at most 15 anthers each, in 5 bundles. These are followed by cucumber-shaped capsules (slow to open) containing seeds in white-to-grey floss.

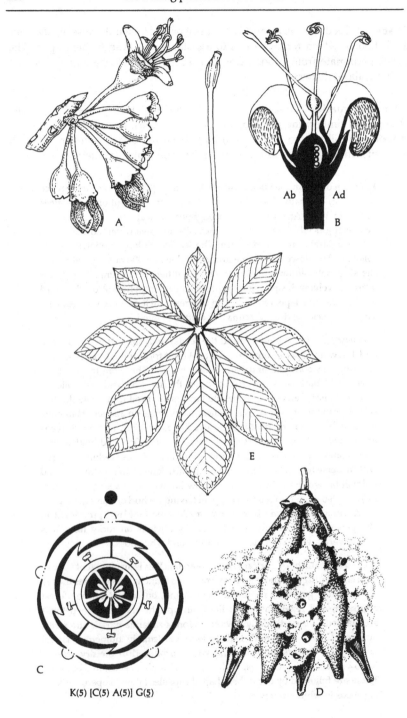

A

Ab   Ad

B

E

C

K(5) [C(5) A(5)] G(5)

D

Bibliography

Baker, H. G. & Harris, B. J. (1959). Bat pollination of the silk cotton tree, *Ceiba pentandra* (L.) Gaertn. (*sensu lato*), in Ghana. *Journal of the West African Science Association*, **5**, pp. 1–9.

Wickens, G. E. (1983). The baobab – Africa's upside-down tree. *Kew Bulletin*, **37**, pp. 173–209.

Fig. 10.1. A–D. *Ceiba pentandra*. A. Inflorescence × $\frac{1}{3}$. B. Half flower × $\frac{1}{2}$. C. Floral diagram. In B and C, what are apparently single stamens are bundles, each of 3 stamens. D. Fruit × $\frac{1}{2}$. E. *Bombax buonopozense* leaf × $\frac{1}{3}$. (Drawn by Mrs C. E. Darter.)

11

Malvaceae – cotton family

A tropical and temperate family, represented in West Africa by large woody or fibrous perennial herbs up to 3 m high, less often by shrubs and rarely by trees. Most species are deep rooting, withstand drought, and are prominent in drier savannas.

The decorative introductions include several *Hibiscus* spp.: *H. mutabilis* (blushing *Hibiscus*) and *H. rosa-sinensis* (rose of China, shoe flower) from China, *H. schizopetalus* (frilled *Hibiscus*) from East Africa.

Members of the family may be recognised by their alternate, stipulate simple leaves with digitate venation (pinnate in *Malvastrum* and *Sida*) with stellate hairs on young growth (scales in *Thespesia*). Ptyxis is mostly conduplicate, conduplicate-plicate in *Abutilon* and *Pavonia*. On the back of the leaf, glands occur in patches, depressions or pits (concealed in small swellings) (absent in *Pavonia*). The flowers are axillary, five-part and often large, lasting a day only, their colour often changing during that time. The petals are veined or basally blotched, and the column of anthers in the centre of the flower is prominent.

Flowers ⊥ ⊕ ♂ (rarely dioecious) 5-part. The flower is usually surrounded by 3(–13) united bracteoles, adnate to the calyx and forming an epicalyx in *Abelmoschus, Pavonia* and *Urena* etc., free from the calyx and forming an involucel in most genera and often associated with extrafloral nectaries (*Cienfuegosia*); bracteoles free in *Gossypium* and *Thespesia*; in *Malachra*, there is an involucre round each head of flowers, while bracteoles are lacking in *Abutilon* etc. K(5) as a rule, 5-toothed or lobed but cupular in *Gossypium* etc.; spathaceous and falling at anthesis in *Abelmoschus*; odd sepal adaxial where there are nil or many bracteoles, abaxial where there are few bracteoles. C5, mostly contorted, joined basally to the staminal tube. Stamens (∞) with kidney-shaped 1-celled anthers carried on short filaments on the outer face of the staminal tube, to the top only in *Abutilon, Azanza, Sida* and *Wissadula*, otherwise there are 5 staminodes opposite the sepals at the top of the tube. G(3–20) with as many cells and styles, *or* twice as many cells and styles (*Malachra, Pavonia, Urena*); ovules 1 to many on axile placentae.

Pollination Some *Hibiscus* spp. (*H. cannabinus, H. sabdariffa*) are largely or entirely self-pollinated, but as a rule, some cross-pollination (promoted by protandry) is followed by a final period of self-pollination. Pollination in

H. trionum, investigated in Australia, showed that the five styles bend downwards towards the anthers during the day unless cross-pollination of at least one stigma takes place (in this case with a fairly large number of pollen grains), when the styles straighten up again. Even pollen of some other malvaceous species was effective. The styles reacted simultaneously, even though only one stigma was pollinated, also in emasculated flowers. Cross-pollination is thus strongly favoured, while the possibility of late self-pollination is reserved (Buttrose *et al.*, 1977). Hovering birds are reported to bring about pollination of *H. rosa-sinensis* in Zanzibar, but butterflies and perching sunbirds were ineffective in Côte d'Ivoire (Prendergast, 1982). Honey bees are active among many species, but do not collect *Gossypium* pollen.

Fruit A loculicidally dehiscent capsule (*Gossypium* etc.) or a schizocarp with indehiscent D-shaped mericarps (cocci) (*Malachra, Malvastrum, Pavonia, Urena* etc.) or dehiscent few-seeded 'follicles' (*Abutilon, Wissadula*). Seeds scaly or hairy.

Dispersal Seeds of *Gossypium* and *Cienfuegosia* have lint and are adapted for dispersal by wind. The bristly or hooked mericarps form burr fruits and are dispersed on the coats of animals. The seeds of other capsules and follicles may be spread by a censer mechanism, but the seeds of some *Abutilon* spp. and *Hibiscus* spp. in Africa are reported to be distributed by birds, those of *Thespesia* by water after the fruit disintegrates.

Economic species Cotton is the seed coat lint of four species of *Gossypium*, the fifth West African species *G. anomalum*, possessing only the short undercoat of 'fuzz' hairs and forming 'cotton wool'. *G. hirsutum* and *G. barbadense* provide upland and sea island cotton, respectively, both introduced from America. They are tetraploids with higher yields than the Old World diploids *G. arboreum* and *G. herbaceum*, which they have supplanted. Cotton demands considerable use of pesticides, because its nectaries attract so many insects, and attempts to breed nectary-less races, also lacking glands on the seeds, have succeeded. At the same time, the polyphenolic allelochemical gossypol, ordinarily present in the seed and poisonous to non-ruminant animals, has also been eliminated, so that the seed is now available as a source of protein for human consumption. Fibre is obtained from the annual stems of the kenaf (*Hibiscus cannabinus*) and Congo jute (*Urena lobata*) and from diverse species of *Abutilon, Hibiscus, Pavonia* and *Sida* on a more local scale. Roselle or sour-sour (*Hibiscus sabdariffa*) produces fleshy red calyces which are added to soups and stews. Almost every part of the okra (*Abelmoschus esculentus*, formerly *Hibiscus esculentus*) is used, not merely the green capsules, while *A. moschatus* (formerly *H. abelmoschus*) is grown locally for its musk-scented seeds, used in perfumery. *Sida* spp. and *Malvastrum coromandelianum* are recognised as weeds.

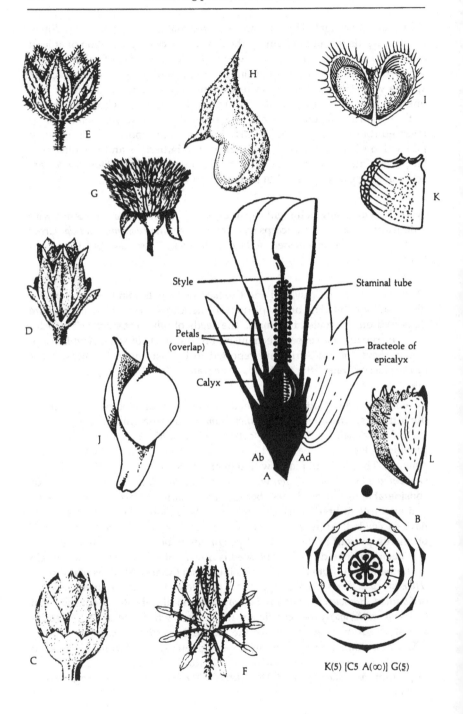

Style ———

Petals (overlap) ———

Calyx ———

——— Staminal tube

——— Bracteole of epicalyx

Ab   Ad

A

K(5) [C5 A(∞)] G(5)

Field recognition *Abutilon* (Arabic name). Hairy perennials of drier savannas (*A. angulatum* in Gambia), with cordate leaves. Flowers, without bracteoles, often yellow, anthers ± on the top of the staminal column and ovary of 10–20+ carpels with 2–3 seeds each. *A. mauritianum* (Sénégal and Ghana–Cameroun) appears also in forest, and has 20–30 2–3-seeded carpels, very pointed, the 'follicles' turning black and spreading at maturity. All *c.* 10 West African species occur commonly in similar habitats in the rest of Africa at least.

Azanza. The name may be Mexican, but the 2–3 species are found in tropical Africa and Indomalaysia. *A. lampas* has been introduced from the latter area, and is found in gardens. *A. garckeana* is a very hairy sudan zone shrub or tree with digitately lobed leaves and, at the end of the rains, large yellow flowers with a cadacous calyx round a cupular calyx. The ovary has 3 stigmata and cells, forming a green, hairy capsule, which is chewed for the sticky substance it contains.

Cienfuegosia (B. Cienfuegos, sixteenth-century Spanish botanist) is a small tropical and subtropical genus (except Asia) of undershrubs with angular branches bearing gland-dotted leaves, occurring in drier grasslands. One of the two West African species is the monoecious sudan–sahel *C. digitata*, which has digitately compound leaves and yellow flowers with a red centre carried on erect young shoots. Each flower (on a glandular pedicel) has a short epicalyx, 5-lobed calyx and single style and a (3–)5-lobed capsule with hairy seeds. The other species is *C. heteroclada*, a dioecious pyrophyte, occurring rather further south, producing pink or purple flowers on leafless stems after bush fires. The shoots, bearing digitately lobed leaves, appear later.

Hibiscus (classical name). There are *c.* 300 tropical and subtropical species altogether, about 10% of them appearing in West Africa. Two species lack an epicalyx (*H. sidiformis* and *H. lobatus*). Of the 2 tree (shrub) species (both in Gambia), *H. tiliaceus* is characteristic of coastal areas, particularly swampy ones. It has purple branches and leafy deciduous stipules leaving almost circular scars. The branches droop and root repeatedly. The flowers are solitary and yellow, with a dark red or purple centre. The remaining species have a persistent epicalyx. A number of species have bristles or prickles; *H. surattensis* is a weak-stemmed plant with recurved prickles, and *H. rostellatus* (also in Gambia) has similar prickles but on erect stems

Fig. 11.1. A. *Gossypium arboreum* half flower × ⅔. B–F. *Hibiscus* spp. B. Floral diagram of *H. rosa-sinensis*. C–F. Epicalyx and calyx, all × 1. C. *H. tiliaceus*. D. *H. panduriformis*. E. *H. trionum*. F. *H. surattensis*. G–H. *Abutilon mauritianum*. G. Fruit × 1. H. Mericarp × 3. I. *Kosteletzkya buettneri*, capsule in vertical section × 4. J. *Wissadula amplissima* var. *rostrata* follicle × 4. K. *Sida stipulata* mericarp × 8. L. *Urena lobata* mericarp × 4. (From Exell & Meeuse, 1961: C–F, Figs. 89.3, 89.4, 89.8 and 89.10; H, Fig. 93.10; I, Fig. 91.3, with permission from the Editorial Board of *Flora Zambesiaca*.)

Fig. 11.2. A–G. *Azanza garckeana*. A. Flowering shoot × ⅔. B. Undersurface of leaf × 4. C. Vertical section of flower × 1. D. Petal × ⅔. E. Stigmas separated × 2. F. Fruit × 1. G. Seed × 1. H. *Kostelezkya buettneri* staminal tube and style branches × 4. (From Exell & Meeuse, 1961, Figs. 88 and 91.5, with permission from the Editorial Board of *Flora Zambesiaca*.)

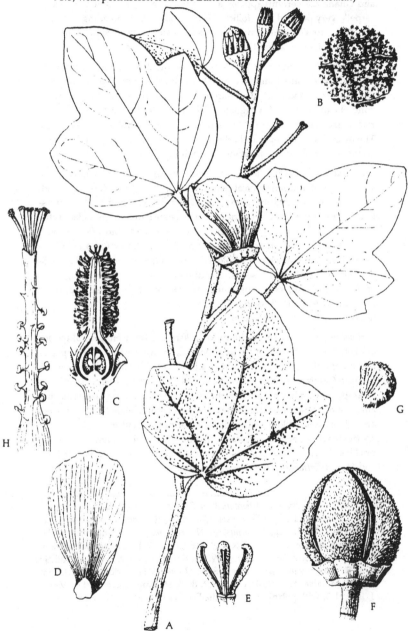

up to 3 m high. *H. panduriformis* is covered with bristles, *H. physaloides* with pricking hairs, and *H. micranthus* has spiny stipules; the epicalyces and calyces of these species are also characteristic, that of *H. trionum* being inflated. *H. vitifolius* has 5-winged capsules in which the carpels split from the receptacle at maturity.

Sida (classical name). Weeds with serrate, pinnately nerved leaves and small pale flowers lacking bracteoles. The fruit is a schizocarp with irregularly dehiscent mericarps up to 10 in number, each with an apical seed. *Sida stipulata* is a common shrubby weed of secondary growth. There are about 200 species in all the warmer parts of the world, 11 of them in West Africa (Ugborogho, 1983). *S. linifolia*, with linear leaves, is the Gambian species.

Thespesia (Gr. divine and wondroûs). *T. populnea* (Portia tree) is the only West African species, though the rest of this small genus is widespread in the tropics. It is a coastal tree or shrub, often planted in villages, scaly on all parts, with ovate long petiolate leaves and solitary yellow-to-purple flowers. Each has 3–5 deciduous bracteoles round a cupular calyx, which persists round the woody indehiscent fruit. The seeds are large, hairy and float on water.

Wissadula (Senegalese name) is a mainly American genus, *W. amplissima* var. *rostrata* (now *W. rostrata*) is the only West African representative. It is a savanna undershrub with cordate leaves which are dark green above and densely hairy below. The flowers are small, yellow and dark centred, carried in terminal panicles in the late rains to early dry season. Bracteoles lacking, the calyx 5-lobed. The fruit consists of 5 thorny-tipped 2-seeded 'follicles', the upper seed slightly hairy, the lower very much so.

Bibliography

Buttrose, M. S., Grant, W. J. R. & Lott, J. N. A. (1977). Reversible curvature of style branches of *Hibiscus trionum* L., a pollination mechanism. *Australian Journal of Botany*, **25**, pp. 567–70.

Exell, A. W. & Meeuse, A. (1961). *Flora zambesiaca: Malvaceae*, vol. 1, pt. 2. London: Crown Agents.

Newton, L. E. & Kisseado, S. V. A. (1978). Observations on the weed *Malvastrum corchorifolium* (Malvaceae) in Ghana. *Nigerian Field*, **43**, pp. 154–60.

Prendergast, H. D. V. (1982). Pollination of *Hibiscus rosa-sinensis*. *Biotropica*, **14**, p. 288.

Rao, P. R. M. (1978). Seed and fruit studies in Malvaceae 1. *Thespesia populnea*. *Phytomorphology*, **28**, pp. 239–44.

Ugborogho, R. E. (1983). Evolution of *Sida* L. (Malvaceae) in West Africa. *Bulletin du Muséum national d'histoire naturelle*, Paris sér. 4, **5**, sect. B, *Adansonia*, pp. 93–102.

12

Euphorbiaceae – cassava family

A large and diverse temperate and tropical family, well represented in West Africa by woody species in both forest and savanna.

Members of the family may be recognised by their alternate, simple, mostly stipulate, pinnately nerved leaves, and by the frequent presence of latex, glands, stellate and peltate scales. Flowers are small, ♂ or ♀, usually five-part and in inflorescences which are often axillary and condensed. The ♀ flowers usually have a three-celled ovary forming a three-lobed fruit with one or two seeds in each cell. *Casearia* (now *Keayodendron*) *bridelioides* has been transferred from the Samydaceae, which, in turn, is now joined with the Flacourtiaceae.

Species introduced for decorative purposes from America include poinsettia (*Euphorbia pulcherrima*), sandbox tree (*Hura crepitans*) and three species of *Jatropha*. Snow bush (*Breynia nivosa*, now *B. disticha* var. *disticha* f. *nivosa* (Radcliffe-Smith, 1980) and croton (*Codiaeum variegatum*) have been introduced from Polynesia.

Some *Euphorbias* are recognisable at a distance, being shrubs or trees with fleshy, often angular, branches without leaves, which are minute and soon fall, but sometimes with spines. These *c.* 10 species occur in drier savannas, on inselbergs and sometimes as hedges (Marnier-Lapostolle, 1966; Rauh, Loffler & Uhlarz, 1969). Both these and the herbaceous species, e.g. *E. glaucophylla* on the shore, have latex and possess distinctive inflorescences (pseudanthia, cyathia). Similar inflorescences are seen in *Elaeophorbia*, of which there are two species of forest trees with fleshy angular branches, fleshy leaves and paired spines. *E. drupifera* (now included in *E. grandifolia*; Hall & Swaine, 1981) is common on the Accra plains. The inflorescence is cupular (the 'receptacle') each ♂ flower consisting of a single stamen with a jointed 'filament', the single ♀ flower consisting of a stalked pistil.

There are few climbers in the family. *Tragia* spp. are herbaceous twiners armed with stinging hairs, the calyces of the ♀ flowers usually persistent and enlarged round the fruit. *Dalechampia ipomoeifolia* is also a slender twiner, though in forest, with similar ♀ calyces, though the ♂ inflorescence is embraced by two large bracts (Webster & Webster, 1972), while *Manniophyton fulvum* is a shrub or woody climber in forest, with yellow flowers with petals. For *Phyllanthus* and *Tetracarpidium*, see below.

Suffrutices occur in the genera *Phyllanthus* and *Sapium*, and a few genera include herbaceous species (*Acalypha, Croton, Euphorbia, Micrococca* etc.) *Caperonia senegalensis* (also in Gambia) is a widespread weed of moist ground, with bristly capsules, and *Cyathogyne viridis* is a small rhizomatous forest herb (Nigeria–Cameroun only).

Genera with only opposite leaves include *Mallotus*, two species of dioecious forest shrubs and small trees, with stellate hairs. *M. oppositifolius* is common in forest regrowth (also in Gambia) and has leaves which are glandular on the edge and near the base. *Oldfieldia* is the other genus, *O. africana*, with five- to eight-foliolate leaves, being a large forest tree in Sierra Leone. This species also belongs to the small group with digitately compound leaves including, in addition, para rubber (*Hevea*) with 3-foliolate leaves, a large gland at the base of each leaflet, and *Ricinodendron heudelotii*, a widespread deciduous tree of secondary forest with three to five leaflets with glandular margins, and persistent stipules. Neither *Oldfieldia* nor *Ricinodendron* has latex. Digitately lobed leaves are fairly common, peltate ones, ± lobes, rare (*Jatropha podagrica, Ricinus communis, Manihot glaziovii*).

Stipules, spines and leaf glands may also contribute distinctive features. Stipules are hood-like in *Ricinus* (cf. *Rhizophora* and the *Moraceae*), gland-tipped or represented by stalked glands in *Jatropha* spp., or by paired spines in *Phyllanthus* sp. (see below) and *Erythrococca anomala*, a widespread shrub of secondary forest. Root spines, spine roots and stilt roots are fairly common, e.g. *Bridelia*.

Two genera have been removed from the family since the publication of the second edition of the *Flora of tropical West Africa*. These are *Hymenocardia* (now Hymenocardiaceae) and *Microdesmis* (now Pandaceae). For present purposes, these transfers will be ignored.

Flowers ⚥ (÷ in *Bridelia*) ⊕ ♂ and ♀, dioecious or monoecious; if the latter, ♂ and ♀ flowers in the same (*Hevea, Manihot, Ricinus*) or different inflorescences (*Alchornea, Hymenocardia*). K4–5, rarely more, sometimes united, often petaloid in apetalous species, but seldom persistent and enlarging round the fruit. C (present in about one-third of the genera) 4–5, rarely more, sometimes united. A 1 to many, free or variously united, occasionally branched (*Ricinus*) or folded inwards (*Croton*); anthers 2-celled (4-celled in *Macaranga*), separate and erect (*Erythrococca*) or pendulous (*Mallotus* etc.) or twisted ('worm-like', *Acalypha*). The disc is usually highly developed, mostly as extrastaminal glands, less commonly as interstaminal or intrastaminal ones, or intercarpellary ones (♂ *Securinega*); the pistil is usually surrounded by a disc. G(3), 3-celled sometimes 1-celled above with 3 styles, 1 or 2 pendulous ovules in each cell, on apparently axile placentae (but see Singh & Singh, 1975). The ovary may become 2- or 1-celled by abortion, but regularly non-3-celled ovaries are rare, though 1-celled as well as 3-celled ovaries are found in some species of *Acalypha* (Radcliffe-Smith, 1974) and *Drypetes* spp.; a pistillode may be present in ♂ flowers.

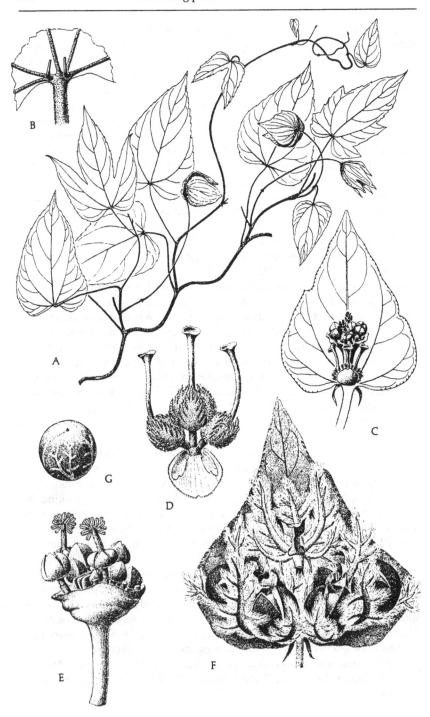

Pollination Little is known about non-economic genera. It has been suggested that species with long styles are likely to be wind pollinated, but the possession of so many features associated with wind pollination in temperate floras does not make this method likely in the tropics, especially not in forest. Observations on para rubber, cassava and castor oil suggest, on the contrary, that midges and thrips may be involved. All three are monoecious (the last two protogynous), and all have floral or extrafloral nectaries. Especially, midge activity has been observed in para rubber. In *Mallotus oppositifolius*, both ♂ and ♀ flowers are scented, and visited by syrphid flies and bees, and wind pollination, aided by the disturbance created by the bees, is also a possibility (Lock & Hall, 1981). Hagerup (1943) suggested that sahel species of *Euphorbia* and *Chrozophora* may be ant pollinated.

Fruit A schizocarp, splitting into 3 cocci (single carpels) often explosively (regmae), the cocci then opening (ventrally). The seeds commonly possess an oily outgrowth near the micropyle, a caruncle, or a strophiole (over the raphe, the attachment of the funicle to the seed). Dehiscent fruits may also possess seeds with brightly coloured sarcotestas (*Erythrococca, Mesobotrya*). Less common dry fruits include the 2-winged samaras of *Hymenocardia*, the 4-winged capsule of *Tetracarpidium* and the loculicidal capsule of *Spondianthus*. Fleshy fruits are uncommon but include drupes (*Drypetes*, the fruits of which look like oranges, but with 1–7 separated broadly triangular stigmas on top; *Elaeophorbia*, fruit yellow and ellipsoidal) and few-seeded berries (*Securinega*, fruit tiny, white, up to 6-seeded).

Dispersal Self-dispersal by explosion occurs; *Hura crepitans* seeds are recorded as being thrown 4–5 m. The fleshy fruits of *Antidesma, Bridelia, Macaranga, Securinega* and *Uapaca* all being reported to be dispersed by birds in Africa, those of *Uapaca* spp. also by monkeys. *Phyllanthus discoideus* capsules open to reveal a papery endocarp round each coccus. It cracks readily and the seeds, clad in a metallic blue-green arilloid, are attractive to birds. Seeds of *Bridelia ferruginea, Drypetes* spp. and particularly *Securinega virosa* germinate from baboon dung, those of *Ricinodendron heudelotii* from elephant dung. Genera with elaiosomes such as *Acalypha, Alchornea, Croton* and *Euphorbia* may be dispersed by ants.

Fig. 12.1. *Dalechampia ipomoeifolia*. A. Flowering shoot × ½, stipules not shown. The two involucral bracts may be green or white/yellow. B. Base of underside of leaf, with two extrafloral nectaries × 5. C. Inflorescence, one bract and its two stipules removed × 2. The ♀ cyme stands in front of the ♂ cyme. D. ♀ cyme of three flowers × 5 (involucel turned down, revealing pinnatifid sepals). E. ♂ cyme in cupular involucel × 5. F. Infructescence × 3. Two three-valved capsules have been formed, the sepals of all three flowers having enlarged and bearing stinging hairs. G. Seed × 5. (From Léonard, 1962, Pl. XIII.)

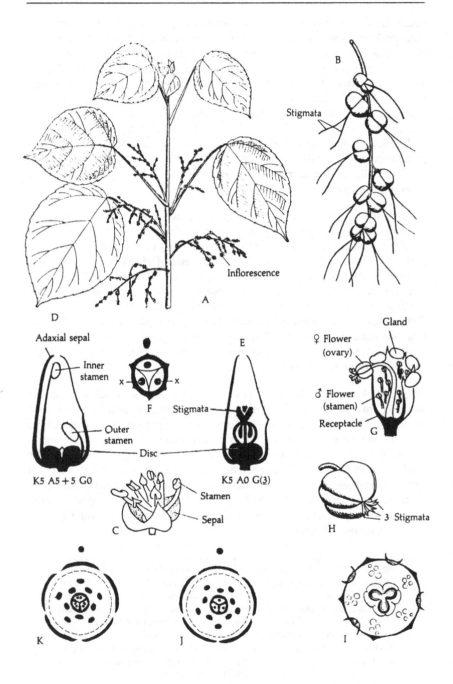

B

Stigmata

Inflorescence

A

D

Adaxial sepal

Inner stamen

x —— x

F

Outer stamen

Disc

K5 A5 + 5 G0

E

Stigmata

K5 A0 G(3)

Stamen

Sepal

C

Gland

♀ Flower (ovary)

♂ Flower (stamen)

Receptacle

G

3 Stigmata

H

K

J

I

Economic species Cassava (*Manihot esculenta*), castor oil (*Ricinus communis*) and para rubber (*Hevea brasiliensis*) have been mentioned. Castor oil is an Old World species, the others are introductions from South America, cassava in sixteenth century. Newer introductions are awusa nut (*Tetracarpidium conophorum*), from which oil can be extracted, and Indian gooseberry (*Phyllanthus acida*), with edible fruits.

Field recognition *Alchornea* (S. Alchorne, eighteenth-century Master of the London Mint). An African genus of dioecious shrubs and small trees, of which *A. laxiflora* is common in forest regrowth. It has large leaves, each blade with a pair of stipel-like outgrowths at its base and patches of glands in the axils of the basal nerves underneath. As the new leaves are expanding in the dry season, spikes of ♂ flowers appear on old wood, single ♀ flowers on new wood. ♂ flowers are yellow, apetalous and 4-part (stamens 7–8) with pendulous anthers, ♀ ones have a pistil with 3 long red stigmata, surrounded by minute sepals. *A. cordifolia* (also in Gambia) is met with in all kinds of forest, even in sudd in rivers. It lacks stipels though has similar leaf glands. The flowers are greenish white, in axillary drooping panicles, the pistil in ♀ flowers 2-celled, with 2 long, persistent styles. The genus lacks floral nectaries.

Bridelia (S. E. von Bridel-Brideri, nineteenth-century Swiss botanist). Monoecious shrubs and trees of both forest and savanna, West African species also being found in other parts of tropical Africa. Trunks and branches are spiny, the spines being adventitious growths (root spines), which later develop into (aerial) spine roots, some of these later developing into stilt roots. There are axillary clusters of 5-part flowers, with minute petals, ♂'s with 5 stamens joined in a central column surrounded by the disc, ♀s with a 2-celled ovary, only the 2 forked styles showing above the perigynous receptacle. The fruits are persistent black drupes, 1-celled in all species except *B. scleroneura*, and all except this species and *B. atroviridis*, the leaves have a marginal nerve. Only *B. grandis* is an upper-storey forest tree, *B. micrantha* (also in Gambia) and *B. ferruginea* being savanna species, the latter with stiffly hooked leaf tips.

Fig. 12.2. A–C. *Alchornea* spp. A. *A. cordifolia* flowering shoot × ⅓. B–C. *A. laxiflora*. B. Part of ♀ spike × 2½. C. ♂ flower × 9. D–F. *Manihot utilissima*. D. Longitudinal section of ♂ flower. E. Longitudinal section of ♀ flower. F. Diagram of cross-section of ovary. Floral diagram as in J. G–I. *Euphorbia* spp. G. Cyathium, part cut away. H. Fruit × 8. I. Inflorescence diagram of a cyathium. Circles are ♂ flowers. J. Floral diagram typical of the tribe Crotoneae. K. Floral diagram typical of the Phyllantheae. In *Phyllanthus discoideus*, with four-part flowers, sepals are orthogonal. (I from Eichler, 1875–8.)

Macaranga (Malagasy name). A large Old World genus of dioecious under-storey forest trees and shrubs with spiny branches bearing leaves with small and/or caducous stipules, and gland-bearing leaves. *M. heudelotii* (also in Gambia) is a small swamp forest tree with leaves which have a pair of glands at the base and apetalous flowers, ♂ ones in axillary clusters or panicles, 5-part with 4-celled anthers, ♀ flowers solitary with a single style to each 1-celled pistil. The fruit is stalked, 3 mm ∅ and covered with yellow scales. *M. barteri* and *M. hurifolia* both occur in forest regrowth, the former having pinnately nerved entire leaves, with 2 glands at the base; the latter has toothed leaves 3-nerved at the base. Inflorescences and flowers are similar, ♂ flowers having conspicuous bracts (more toothed in *M. hurifolia* than *M. barteri*). The fruits have a waxy glandular covering, the discoid stigma of *M. barteri* persisting on the side of the fruit.

Microdesmis (Gr. small bands) is a small Old World genus with 2 species in West Africa, of which *M. keayana* (formerly *M. puberula pro parte*) is widespread. It is a dioecious under-storey tree, also in secondary growth oblong leaves covered with transparent spots. The flowers appear in the dry season in small axillary clusters. The flowers are greenish yellow, 5-part, with petals, ♂ flowers with 5 free stamens with elliptical filaments round the pistillode, ♀ ones with a 3-locular ovary, which develops into a rough red drupe, and these appear in clusters. *M. puberula* extends from the east into S. Nigeria and can be distinguished by the broad filaments adhering to the 5-sided pistillode in the ♂ flower and by the 3-locular ovary in the ♀ one.

Phyllanthus (Gr. leaf flower, the flowers on (compound pinnate) leaf-like branches in some (non-West African) species). This is a very large genus in all the warmer parts of the world. The West African species are mostly monoecious shrubs. The leaves are often small, spirally or distichously arranged, or branching is 'phyllanthoid' (Bancilhon, 1971), there being 2 kinds of branches: lateral flowering ones, 'bipinnatiform' with distichously arranged leaves (or leafless), the whole shoot being shed, and permanent shoots with spirally arranged scales, e.g. *P. reticulatus*, a scrambling riverside shrub in forest and savanna, with a ♂ and 2–3 ♀ flowers in each cluster. The flowers are 5-part, the ♂'s with small extrastaminal glands and 2–3 of the stamens joined by their filaments. There is no pistillode. ♀ flowers have a cupular disc and an ovary with more than 3 cells. The fruit is a black berry with up to 16 seeds. The ovary in *Phyllanthus* is frequently subdivided, so that there are more than 3 cells but with only 1 ovule in each. *P. muellerianus* (now *P. floribundus*) also has phyllanthoid branching, but is a climber armed with recurved stipular spines and the flowers appear on axillary clusters of short racemes. In ♂ flowers the filaments are free, and the berries are red and few(6)-seeded. Two of the common woody weed species also have phyllanthoid branching (*P. amarus* and *P. urinaria*). *P. discoideus* (now *Margaritaria discoidea*; Webster

1979), however, does not. It is an upper-storey deciduous tree of drier forest and southern savanna areas, with flowers clustered on main shoots and at the bases of lateral shoots in the dry season. The flowers are yellowish green, 4-part with an annular disc, the ♂'s with free filaments, the ♀ ones in pairs, each with a (3–)4-lobed ovary, the fruit being dry and dehiscent, quite different from that of *Phyllanthus*, also in being dioecious.

Protomegabaria (Gr. first *Megabaria*) is an endemic West African genus of 2 species, of which P. *stapfiana* is locally common in swamp forest and the wetter parts of lowland forest. It is a dioecious tree, up to 25 m high, often producing aerial roots, which become stilt roots, propping up the leaning trunk. It has thin pinnate leaves in tufts at the end of the branches, and produces short axillary inflorescences of yellow apetalous 5-part flowers in the dry season. ♂ flowers are subtended by 3 series of bracts, the middle one cupular and enclosing the flower bud. Each ♂ flower has a 2-part pistillode. ♀ flowers have a pistil with 3 styles, surrounded by a disc which is adnate to the sepals. The capsule is hard, ± globular, 3–4 cm ∅, the 3 valves falling off.

Sapium (L. referring to the latex exuded by young shoots, flowers and fruits). Mostly locally occurring suffrutices, forest shrubs and trees, only S. *ellipticum*, an upper-storey tree by water in riverain and fringing forest, also in savanna, is widespread. The leaves are ± oblong, though very variable, but always with a markedly crenate wavy edge and a pair of glands at the base of the blade underneath. The species is deciduous and monoecious, producing narrow axillary spikes of yellowish-green apetalous flowers in the dry season. These bear distal clusters of ♂ flowers and proximal solitary ♀ flowers. The ♂ cluster has a pair of oval glands at its base and each flower has 2 stamens. The ♀ flower has a pair of prominent coiled styles and the 2-lobed fruits eventually turn yellow-red, and dehisce.

Uapaca (Malagasy name). A genus of tropical Africa and Madagascar, with 9 West African species. All are dioecious, with rather large pinnate leaves, gland-dotted below, in clusters towards the end of the branches. ♂ inflorescences look like single flowers at first sight, each having a white-to-yellow petaloid involucre round a pedunculate ball of 4-part apetalous flowers, each with a pistillode. ♀ flowers are solitary within the involucre, with 3-several branched stigmata, the pistil developing into a yellow or red drupe with 3 or 4 stones. U. *staudtii* and U. *paludosa* are swamp forest trees with large, leafy persistent stipules and stilt roots. U. *guineensis* and U. *heudelotii* are widespread forest trees, the latter always near water. Both have short-lived narrow stipules and stilt roots. U. *heudelotii* occurs continuously northwards into fringing forest, where it may be met with the savanna species U. *togoensis*, which also develops stilt roots in that habitat. The latter, however, lacks the tufts of red curly hairs in the young lead axils possessed by U. *heudelotii*.

Bibliography

Bancilhon, L. (1971). Contribution à l'étude taxonomique du genre *Phyllanthus* (Euphorbiacées). *Boissiera*, **18**, pp. 7–79 and 22 plates.

Bernhard, F. (1966). Contribution à l'étude des glandes foliaires des Crotonoidées (Euphorbiacées). *Mémoires de l'Institut fondamentale d'Afrique noire*, sér. A, **75**, pp. 67–156.

Eichler, A. W. (1875–8). *Blüthen-Diagramme*. Leipzig: Engelmann.

Hagerup, O. (1943). Myrebestøvning. *Botanisk tidsskrift*, **46**, pp. 116–23.

Hall, J. B. & Swaine, M. D. (1981). *Distribution and ecology of vascular plants in a tropical rain forest: forest vegetation in Ghana*. The Hague: Junk.

Léonard, J. (1962). *Flore du Congo etc: Euphorbiaceae*. Meise: Jardin botanique national de Belgique.

Lock, J. M. & Hall, J. B. (1981). Floral biology of *Mallotus oppositifolius* (Euphorbiaceae). *Biotropica*, **14**, pp. 153–5.

Marnier-Lapostolle, J. (1966). Les Euphorbes de l'ouest et du centre africain. *Cactus*, **87**, pp. 4–10.

Radcliffe-Smith, A. (1974). Allomorphic female flowers in the genus *Acalypha* (Euphorbiaceae). *Kew Bulletin*, **28**, pp. 525–9.

Radcliffe-Smith, A. (1980). A new status for the snow bush. *Kew Bulletin*, **35**, pp. 498, 980–1.

Radcliffe-Smith, A. (1987). Segregate families from the Euphorbiaceae. *Botanical Journal of the Linnean Society*, **94**, pp. 47–66.

Rauh, W., Loffler, E. & Uhlarz, H. (1969). Observations on some *Euphorbias* from tropical West Africa. *Cactus and Succulent Journal of the Cactus and Succulent Society of America*, **41**, pp. 210–20.

Singh, V. & Singh, A. (1975). Placentation in Euphorbiaceae. *Annals of Botany*, **39**, pp. 1137–40.

Swaine, M. D. & Beer, T. (1977). Explosive seed dispersal in *Hura crepitans* L. (Euphorbiaceae). *New Phytologist*, **78**, pp. 695–708.

Webster, G. L. (1979). A revision of *Margaritaria* (Euphorbiaceae). *Journal of the Arnold Arboretum*, **60**, pp. 403–44.

Webster, G. L. & Webster, B. D. (1972). The morphology and relationships of *Dalechampia scandens* (Euphorbiaceae). *American Journal of Botany*, **59**, pp. 573–86.

13

Caesalpiniaceae – pride of Barbados family

A large woody tropical family represented in West Africa by c. 60 genera, mainly trees. In this family, and the two that follow, the Mimosaceae and the Papilionaceae, the pod (legume) is regarded as the characteristic fruit. It is formed from the single superior carpel, and becomes dry, opening by splitting down both margins. Both pods and the seeds they contain are commonly referred to as 'beans'. In some classifications, the three families are united under the name Leguminosae (Fabaceae) (with subfamilies), thus forming the second largest family of flowering plants.

Members of the Caesalpiniaceae may be recognised by their alternate, stipulate, compound pinnate leaves without stipels (except *Amphimas* and *Pellegrinodendron*). The conduplicate leaflets may be alternate or paired, and there may (imparipinnate) or may not (paripinnate) be an odd terminal leaflet. A pulvinus occurs at the base of the petiole, a pulvinule at the base of the petiolule. Both have a wrinkled surface. Compound racemose inflorescences of zygomorphic five-part flowers each with a single superior carpel are usual. The fruit is generally a pod.

Unifoliate leaves are uncommon, most often being seen in three white-flowered savanna species: *Bauhinia* (now *Adenolobus*) *rufescens* (Schmitz, 1973) (with ♂ flowers) in riparian woodland, and, with ♂ and ♀ flowers on separate plants, *Piliostigma reticulatum* in sudan savanna and *P. thonningii* in forest-savanna mosaic, both also in northern guinea savanna. The leaves are two-lobed in all cases. Half a dozen other genera have species with entire unifoliate leaves.

A single pair of asymmetrical leaflets occurs in only few species, notably the swamp forest and riverside tree *Cynometra vogelii* (also in Gambia) with emarginate leaflets, the forest tree *C. ananta* (particularly in Ghana) with acuminate leaflets, and the swamp forest tree *Oxystigma mannii* (Nigeria–Cameroun), with leaflets rounded at both ends.

Bipinnate leaves are seen in a further eight genera, three of which are prickly or spiny (see below). None of these has the foliar glands typical of the Mimosaceae. Apart from flamboyante (see below), *Bussea occidentalis* is a large forest tree (Sierra Leone–Ghana) with conspicuous yellow flowers in the rains, followed by erect pods; *Burkea africana* is a deciduous drier savanna tree (from Ghana eastwards, in forest-savanna mosaic as well) with hanging panicles of small, scented, white flowers in the dry season, followed by brown, leaf-like samaras; and *Erythrophleum* is represented by three species of forest or savanna trees, with leathery pods which open down one side only, the seeds, on rather fleshy funicles, hanging from them.

Of the species with spines or prickles, *Parkinsonia aculeata* (Jerusalem thorn, from tropical America) commonly occurs near habitation in drier savannas.

Fig. 13.1. A–G. Leaf characters. A. Unifoliolate, bilobed. B. Paripinnate. C. Bipinnate, leaflets paired. D. Imparipinnate, leaflets alternate (but paired leaflets also occur). E. Bipinnate, leaflets alternate. F–G. Parts of leguminous leaves. H–K. Samaroid fruits. H. *Mezoneuron benthamianum* × ½. I. *Gossweilerodendron balsamiferum* × ½. J. *Burkea africana* × ⅔. K. *Distemonanthus benthamianus* × ⅓. (J from Brenan, 1967, Fig. 2.11.) (British Crown copyright. Reproduced with permission of the Controller, Her (Britannic) Majesty's Stationery Office & the Trustees, Royal Botanic Gardens, Kew. © 1967.)

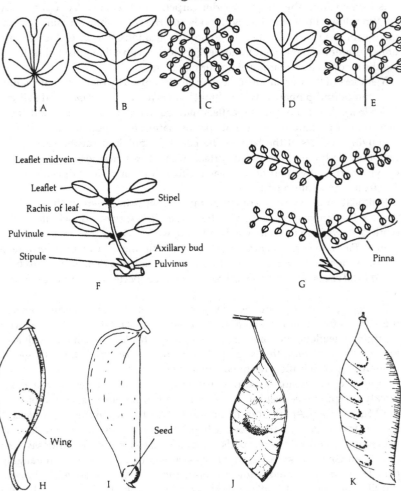

Each rachis forms a terminal spine, and, after the leaflets have been shed, the rachises remain green for a long time. The stipules also form spines. Superficial prickles occur on shoots and leaves in *Mezoneuron benthamianum* (also in Gambia) a climber in secondary forest and forest-savanna mosaic with red samaras winged along one side, also in *Caesalpinia bonduc*, a yellow-flowered scrambling shrub near habitation with short prickly pods containing one or two hard grey seeds. *Hylodendron gabunense*, an upper-storey forest tree, develops 'spines', possibly root spines, on its buttresses, but their exact morphological nature is not known.

Glands on the petiole or rachis are less common in the Caesalpiniaceae than in the Mimosaceae, being seen in less than half of the species of *Cassia* only. Glands on the blade are also rare, *Afzelia* being an example of the occurrence of a single gland near the base of the blade. Glandular dots or spots scattered over the blade are fairly common (*Daniellia* etc.). Only *Schotia* (now *Leonardoxa*) *africana* has domatia, in the swollen internodes.

Hylodendron and *Daniella* are also known for their large stipular hoods, which leave circular scars on the stems, cf. Chapter 17, Moraceae. Persistent intra-petiolar stipules, which may be fused basally in the axil, but are not hood like, are seen in *Crudia* and *Gilbertiodendron*, both genera of forest trees. *Crudia senegalensis* is a widespread under-storey tree with stipules up to 2.5 cm long, *C. klainei* (Côte d'Ivoire and Nigeria) is a small freshwater swamp tree, in which the stipules are 3 cm long, nearly as broad and purple in colour. The other two species are uncommon and have caducous stipules. *Gilbertiodendron* spp. may have large appendaged stipules, e.g. in *G. splendidum* (Sierra Leone–Ghana) also in swamp forest, has stipules up to 9 cm long by half as wide, each with a kidney-shaped appendage up to 10 cm ∅. Other species have smaller stipules and appendages, or only one or other persists, both fall early, or appendages are lacking.

Several genera have buds protected by scales, an unusual feature in a tropical flora. In particular *Hymenostegia afzelii* (which also has winged rachises), a widespread under-storey forest tree with persistent petaloid bracteoles, and the genus *Monopetalanthus* in Sierra Leone come to mind. *M. pteridophyllus* is a riverain tree with sessile paripinnate leaves with many pairs of small leaflets.

Introduced species include *Caesalpinia pulcherrima* (pride of Barbados) from America, and various *Bauhinia* spp. from Asia now placed mainly in *Pauletia* or *Perlebia*; Schmitz, 1973). *Peltophorum pterocarpum*, also from Asia, is decorative both in flower and fruit and makes a good specimen tree. Numerous *Cassia* spp. from elsewhere in the tropics have also been introduced, see below. Flamboyante (*Delonix regia*) comes from Madagascar.

Flowers ⚥, the receptacle often ± tubular, forming an hypanthium, .l. ♂ and 5-part. Bracteoles often distinctive, K then ± reduced. K5, free or joined, sometimes fewer, 0 or very small, rarely more than 5. C5, the adaxial petal internal, petals often of unequal size, or fewer or C0 (in at least a quarter of our genera), rarely more than 5. A5+5, (9)+1 (*Julbernardia* etc.) or fewer, then supplemented by staminodes, e.g. A3+5 staminodes (*Tamarindus*); A ∞ *Cordyla, Hymenostegia* etc., 11–16 in several genera; anthers 2-celled, opening by slits and dorsifixed (pores or slits in *Cassia* but basifixed, also basifixed in *Baphiopsis* etc.); a connective gland is present only in *Cordyla* cf. Chapter

Fig. 13.2. A–C. *Cassia siamea*. A. Floral diagram. The three adaxial stamens are sterile feeding anthers. The odd sepal is abaxial (as in those Malvaceae with few bracteoles) in the Leguminosae. B. Longitudinal section of flower × 2. C. *C. sieberiana* stamen × 2. D. Flower × 1½. E. *Berlinia grandiflora* half flower × ¾.

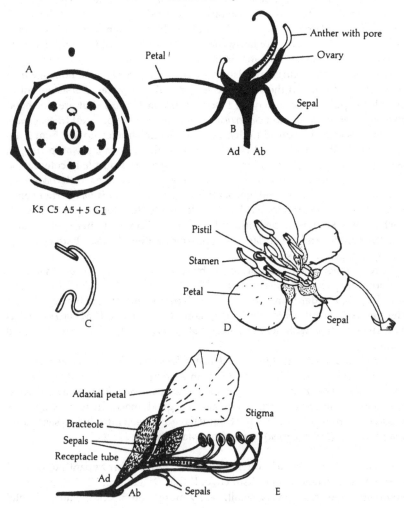

14, Mimosaceae. An intrastaminal disc is usually present. G1, 1-celled, stalked, the stalk sometimes joined to the hypanthium, with 2 marginal rows of ovules (ovules rarely in several series – *Piliostigma*), though sometimes few.

Pollination Insect pollination of some kind is most common. Nectar guides, consisting of either colour contrasts between anthers and petals, of ultraviolet-reflective differences between these two parts, have been described, and nectar is generally produced. Both inflorescences and individual flowers can assume the character of a 'brush', with, collectively, 10–∞ anthers, attractive to longer-tongued insects, or bats or birds in some cases. When the adaxial petal is especially large, a 'flag' type flower is formed, and pollination may be accomplished by bees, cf. Chapter 15, Papilionaceae, or butterflies – *Berlinia grandiflora* (not documented in Africa) and *Caesalpinia pulcherrima*. The pollination of *Cassia* is rather better known. These are pollen flowers (no disc), a proportion of the anthers being sterile and devoted to producing a food which is 'milked' while the bee vibrates its wings and throws up pollen, some of which lands on the insect's back and is carried away.

Fruit The dehiscent pod is common, but indehiscent pod-shaped fruits also occur. In these, the seeds may be embedded in pulpa (*Tamarindus* etc.). This is possibly also the case in the fruits of *Cordyla*, *Mildbraediodendron* and *Piliostigma* etc. Other indehiscent fruits include samaras (*Amphimas*) and nuts (*Cynometra* etc.), as well as drupes (*Detarium* etc.). Seeds sometimes have an arilloid (*Guibourtia* spp. *Copaifera* etc.) and endosperm is sometimes present, notably in *Cassia*.

Dispersal Dehiscent pods normally explode, and the seeds may be thrown many metres. Pulpa and other fleshy fruits and arilloid seeds are attractive to animals, e.g. elephants disperse *Cassia* spp. *Dialium aubrevillei*, *Detarium microcarpum* and *Swartzia fistuloides*, also, in East Africa, *Tamarindus indica*. Baboons eat pods of *Cassia mimosoides*. Samaras are dispersed by wind, but may also float for some time on water, as do *Caesalpinia bonduc* pods. *Afzelia* arilloids are reported to be attractive to birds.

Economic species Timber is very important, being obtained from many species; that from half a dozen species used to be exported. Three species of *Cassia*, including *C. senna*, are cultivated, the leaves and pods being used in the preparation of laxatives. Tamarind fruit pulp contains 10–15% (by dry weight) tartaric acid and is used like lemon juice as a condiment. *Griffonia simplicifolia*, a climbing shrub in coastal thicket, provides good dry fodder and regenerates readily if not overcut.

Field recognition *Afzelia* (A. Afzelius, eighteenth-century Swedish botanist) – is a small, Old World genus of forest and savanna trees, with bark which scales off in patches. *A. africana* is widespread in drier forest areas northwards into open savanna, forest outliers and fringing forest. It is evergreen, with paripinnate leaves with *c.* 5 pairs of leaflets on twisted petiolules, each leaflet with a gland on the blade near its base. The stipules are deciduous. Stiff flat panicles of scented flowers are produced in the dry season, each flower with a pair of deciduous bracteoles, K4 C1 (large, mainly white) A7+2 staminodes. The thick, black, straight woody pods open explosively to reveal large black seeds embedded in grooves in the spongy endocarp, each seed with a basal yellowish-red arilloid. The valves of this species curl after opening, but those of the other 4 species remain straight.

Berlinia (A. Berlin, eighteenth century Swedish botanist) is a small African genus, only 3 of the 10 species reaching as far west as Sierra Leone. Of these, *B. grandiflora*, an evergreen tree, is commonly seen near water, even occurring in riparian woodland. It has paripinnate leaves with 3–4 pairs of basally symmetrical leaflets on straight petiolules, and joined intra-petiolar stipules. In the dry season, panicles of white flowers with conspicuous valvate bracteoles appear, the bracteoles being persistent, densely lined with silky white hairs and enclosing the flower bud. The receptacle tube is well developed, with 5 subequal sepals, 1 large and 4 small petals (exceeding the sepals), and (9)+1 stamens on its rim. The pistil is joined to the adaxial side of the tube. The pod is reddish brown, slightly woolly and large (up to 30 cm long × 7 cm wide) with 2–5 seeds which are scattered explosively.

Cassia (classical name). A large tropical and warm temperate genus of woody perennials, about half the species found in West Africa having been introduced. The leaves are always paripinnate, but in only about half the species are the rachises or petioles glandular. The flowers are shallow, yellow or pink, with a brush of 2–3 kinds of stamens protruding prominently. The anthers are basifixed, or, if dorsifixed, also opening by pores or short slits. The pods are of diverse kinds, often with some kind of transverse marking. *C. sieberiana* (also in Gambia) is a widespread deciduous savanna tree, without foliar glands, producing pendulous racemes of yellow flowers in the dry season, as the new leaves appear. Three stamens have short filaments, 7 stamens are S-shaped. The pods are black and slim, up to 80 cm long, breaking up transversely. Each seed is in its own compartment, cross walls being formed by outgrowths of the placenta. *C. singueana*, a smaller also deciduous tree of drier savannas, also has 5–8 pairs of leaflets, but with a gland on the rachis between each pair. The inflorescences are erect and glandular, but the flowers resemble those of the previous species except for having 3 petaloid staminodes. The pods are yellow, sticky and semi-juicy, beaded outside and with compartments inside. Half a dozen species of undershrubs are widespread and common. *C. absus* (also in Gambia) is a sticky herb of drier savannas, with a gland on the rachis between each of the 2 pairs of leaflets. The flowers are some shade of yellow or red, with darker veins. *C. alata*, *C. kirkii*, *C. mimosoides* (also in Gambia), and *C. occidentalis* are found throughout West Africa, more or less as weeds. Only *C. alata* does not have glands on the petiole or rachis.

Daniellia (Dr W. F. Daniell, nineteenth-century collector in West Africa) is a small West African genus of deciduous forest and savanna trees, with paripinnate leaves of 4–12 pairs of large asymmetrical leaflets with glandular spots. The stipular hoods are also large (7–20 cm long) and deciduous. Each flower has 4 equal sepals and petals shorter than the sepals. *D. oliveri* is one of West Africa's most characteristic savanna trees, producing large flat panicles of scented white flowers in the dry season, each flower with a pair of caducous valvate bracteoles. Each flower has 4 equal sepals and a larger + 4 smaller petals, all shorter than the sepals. The 10 stamens are prominent on long filaments. The pods are dry, dehiscent and samara-like, the endocarp separating, so that there appear to be 4 'leaves', a purple-brown seed attached to one of them by a persistent funicle, which forms a small aril at the top. The other *Daniellia* spp. are blue-to-purple flowered forest trees with only 2 small petals and filaments united at the base. The widespread species is *D. ogea*, which has very (finely) hairy inflorescences.

Detarium (detar, Wolof name). Trees with imparipinnate leaves of basally asymmetrical alternate leaflets, each of these with numerous glandular dots and prominent marginal veins. Axillary clusters of white, 4-part apetalous flowers (without a disc) with deciduous bracteoles are followed by black drupes, *c.* 5 cm ∅, with green fibrous flesh. *D. microcarpum* occurs in drier savannas, and may, indeed, be regarded as characteristic of northern guinea savanna. It produces scented flowers in the rains, and the drupes are somewhat flattened. *D. senegalense* is a tall forest tree to 35 m, but also occurs in a smaller version in forest-savanna mosaic, including outliers. Short panicles of flowers are produced in the dry season, followed by ± spherical fruits. Both species are preserved when farmland is cleared. This is a small genus of northern tropical Africa, West Africa's third species being restricted to S. Nigeria–Gabon.

Dialium (Gr. classical name, velvet tamarind). Apart from 1 species in South America, this is an Old World genus, only 2 of West Africa's 5 species being widespread. These are forest trees with imparipinnate leaves and deciduous stipules, bearing terminal panicle-like inflorescences of white or greenish flowers in which the 1 petal (if present) together with bracts and bracteoles, is deciduous. There are 2 stamens with basified anthers, and some staminodes, a thick disc and a hairy ovary. The fruits are fleshy with a brittle shell. The slash of *D. guineense* (also in Gambia), also found in forest-savanna mosaic and even in fringing forest, produces red resin. It has leaves of *c.* 5 alternate leaflets and produces large, flat brown panicles of tiny flowers at the beginning and end of the rains. The fruit is flattened and black, 2–3 cm ∅, very hairy, with reddish flesh round 1–2 seeds, the flesh being eaten. *D. dinklagei*, confined to forest,

Fig. 13.3. *Daniellia oliveri*, all × ⅔. A. Leaf and stipule. B. Stipule round axillary bud. C. Stipule. D. Inflorescence. E. Sepal. F. Larger and smaller petals. G. Stamen. H. Pistil. I. Dehisced pods, one valve removed on right. (From Brenan, 1967, Fig. 24.) (British Crown copyright. Reproduced with permission of the Controller, Her (Britannic) Majesty's Stationery Office & the Trustees, Royal Botanic Gardens, Kew. © 1967.)

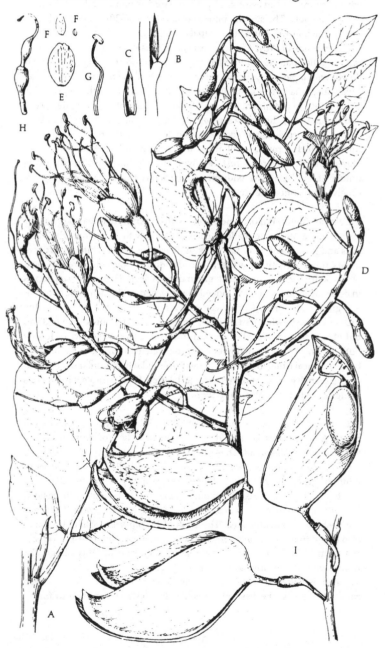

has at least 5 pairs of opposite leaflets, apetalous flowers in the rains, and its fruits are brown and ± spherical, with white flesh.

Isoberlinia (Gr. equal, and *Berlinia*) is another small tropical African genus. There are 2 species, both widespread and common in savanna, particularly the northern guinea zone. Both have paripinnate leaves with 2–4 pairs of basally asymmetrical leaflets and large, connate intrapetiolar stipules, which are later deciduous. Short racemes of scented white flowers, each with a pair of persistent, valvate dark-green bracteoles enclosing the flower bud, are produced in the dry season. Each flower has 5 ± equal sepals and 5 petals longer than the calyx. These are followed by large, flat, brown woody pods, which open explosively, the valves curling back and the disc-like seeds being ejected. Doka (*I. doka*) is a glabrous plant with bright red young foliage, which later becomes a glossy dark green. The sepals are white. *I. dalzielii* (now *I. tomentosa*) (white doka) is a whitely hairy plant, particularly associated with rocky hills and stony ground. The slash produces red resin.

Swartzia (O. Swartz, eighteenth-century Swedish botanist) is a genus of *c.* 100 species in both tropical Africa and tropical America, 2 species being confined to Africa. *S. madagascariensis*, which, in spite of its name, does not occur in Madagascar, is a common tree in the northern guinea and sudan zones, with imparipinnate leaves with 7–11 alternate-subopposite leaflets, which are rounded or notched at the apex. All young growth is densely covered with light-brown hairs, and the stipules are very small. Leafy axillary racemes of scented white flowers are produced at the end of the dry season as the new leaves unfold. Each flower has 1 petal and numerous basifixed anthers. The pod is hard, brown or black and indehiscent, up to 30 cm long and roughly finger thick. It is stalked, the irregularly lobed calyx persisting at the base of the stalk.

Bibliography

See also after Papilionaceae.
Brenan, J. P. M. (1967). *Flora of tropical East Africa: Leguminosae–Caesalpinioideae*. London: Crown Agents.
Cruden, R. W. & Hermann-Parker, S. M. (1979). Butterfly pollination of *Caesalpinia pulcherrima*. *Journal of Ecology*, **67**, pp. 155–68.
Hall, J. B. & Jenik, J. (1969). The dispersal of *Detarium microcarpum* by elephants. *Nigerian Field*, **34**, pp. 39–42.
Schmitz, A. (1973). Contribution palynologique à la taxonomie des Bauhinieae (Caesalpiniaceae). *Bulletin du Jardin botanique national de Belgique*, **43**, pp. 369–423.

14

Mimosaceae – *Acacia* family

A mainly tropical and woody family in which, in common with the previous family and the next one, the pod is the principal kind of fruit.

The family is relatively small in West Africa, its members being recognised by their bipinnate stipulate foliage with, in approximately half of the genera, one or more glands on the leaf rachis of petiole. The leaves often fold down if touched (Fleurat-Lessard, 1985) a response provoked by grazing herbivores, and likely to limit their predation. In so far as the existence of glands encourages the presence of an ant guard, these could also be construed as defensive. The leaflets are conduplicate in the bud.

The inflorescences are spikes or heads, brush like, and made up of small five-part actinomorphic flowers with prominent stamens. The pods, which follow, tend to hang in bunches.

Prickles and/or spines occur in a number of genera. Various *Acacia* spp. may possess stipular spines or prickles, and pseudo-galls may also be present. *Dichrostachys* has branch spines. The introduced weeds *Schrankia leptocarpa* (with four-angled stems, small balls of pink flowers and unjointed four-ridged pods) and *Mimosa pudica* (with terete stems, mauve flowers and flat-jointed pods) are prickly herbs. The indigenous *M. pigra* (also in Gambia) is a prickly riverside shrub forming tangles. Young shoots of the forest-emergent *Cylicodiscus gabunensis* and the swamp and riverain forest trees *Cathormion* spp. are also prickly. *Entada*, which includes some lianes with tendrils (formed from the distal pair of pinnae), also has one prickly species, *E. sclerata*, in forest regrowth (Caballé, 1980).

Very few species have only one pair of pinnae. Those mostly concerned are forest trees – *Cylicodiscus gabunensis*, *Newtonia duparquetiana* and *Xylia evansii*, but see also *Albizia* and *Pithecellobium*.

Neptunia oleracea is the only aquatic herb in the family. It has thick, spongy underwater stems bearing sensitive aerial leaves with two to three pairs of pinnae, and roots freely at the nodes. The five-part flowers in yellow heads and pods are superficially like those of a *Mimosa*.

Introduced species used decoratively include, from America, cassie flower (*Acacia farnesiana*), the flowers being deep yellow and highly scented (and used elsewhere in the perfume industry), and *Pithecellobium dulce*. This has twisted pods with flat, roughly triangular seeds with white arils. Both are planted as hedges.

Fig. 14.1. A–B. *Leucaena glauca*. A. Floral diagram. In most West African
Acacia spp., the odd sepal is adaxial. In four-part flowers, sepals are orthogonal (odd sepal adaxial, e.g. *Mimosa* and *Acacia nilotica* var. *tomentosa*)
(see Nongonierma, 1976, Pl. 23). B. Longitudinal section of flower × 10.
C–F. *Acacia* spp. C. *A. senegal* half flower × 3. D. *A. albida* fruit × ½.
E. *A. seyal* fruit × ½. F. *A. nilotica* var. *adansonii* fruit × ½. G. *Albizia adianthifolia* flower × 2. H. *Entada abyssinica* fruit × ⅕.

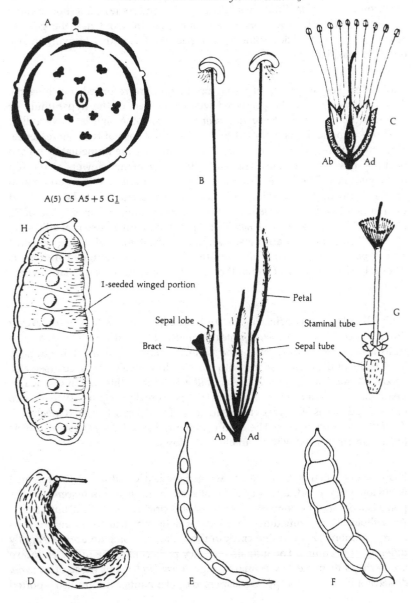

A(5) C5 A5 + 5 G1

1-seeded winged portion

Sepal lobe

Bract

Petal

Staminal tube

Sepal tube

Flowers ÷ ⊕ ♂ (rarely ♂ and ♀, or ♂ and neuter, the former in *Schrankia*, *Acacia* spp. and *Albizia* spp., the latter in *Dichrostachys* and *Parkia*), 5-part (4-part in West African *Mimosas* and some *Acacias*). K(5) valvate (imbricate only in *Parkia* and *Pentaclethra*). C(5) (C(4) in *Acacia* spp.), valvate. Stamens 10 (rarely 4 or 8 (*Mimosa*) or 5 with 5 staminodes (*Pentaclethra*), but in *Acacia* and *Albizia* very numerous), inserted on the hypanthium, apparently 'epipetalous'; joined in *Albizia*, *Cathormion*, *Pithecellobium* and some *Acacia* spp.; anthers versatile with a gland on top of the connective in about half the genera, at least in the bud; pollen sometimes in packets. G1, with 1 to many ovules on an adaxial placenta.

Pollination Few instances of nectar production are recorded, but it seems likely that nectar may be produced within the base of the hypanthium or staminal tube, although *Samanea saman* and a few other species have a disc. In *Parkia*, production is confined to flowers at the base of the inflorescence. Since the inflorescence is pendulous, the ♂ flowers are immediately on top of the ⚥ flowers, between them and a fringe of neuter flowers at the top of the peduncle. The flowers are protandrous, their scent, anthers and pollen as well as the nectar attracting bats. *Acacia tortilis* is reported to be pollinated by a sunbird (*Nectarina puletrella*). *Parkia* pollen grains are in packets of 32 grains, and the packets in *Calliandra* are provided with an external adhesive. Shallow brush flowers are most common in the family, the long staminal tube type of *Albizia*, requiring the attention of longer-tongued insects, is unusual. The function of the anther gland is uncertain, particularly where it is caducous.

Fruit Usually a pod (dehiscing down both edges), but a follicle (dehiscing down one edge, *Cylicodiscus*, *Piptadeniastrum*) and several kinds of indehiscent fruits also occur, as in *Amblygonocarpus* and *Tetrapleura*, which have initially juicy fruits of distinctive shape with cross-walls between the seeds (see also *Parkia*, *Prosopis* and *Samanea*). *Aubrevillea* produces thin oblong samaras. Lomenta occur in a number of genera, the pod breaking up transversely into one-seeded pieces which are either samaroid (*Entada*) or achenoid (*Cathormion*, *Acacia kirkii*) in character. The seed commonly has a long persistent funicle (± aril) and is marked with an areole (see below).

Dispersal of samaras, lomenta and winged seeds (*Cylicodiscus*, *Fillaeopsis* etc.) is obviously by wind, and scattering of seed by explosive dehiscence of the pod also occur. The lomenta of *Mimosa* are prickly, and form burr fruits. An arilloid occurs sometimes on seeds of dehiscent fruits, but an areole is more common. This is a fine crack in the testa, enclosing an area of slightly different appearance. The indehiscent juicy pods containing mealy endocarp are perhaps dispersed by primates, e.g. *Parkia bicolor*, but those of *Samanea dinklagei* and *Tetrapleura* spp. are dispersed by elephants, who are also reported to disperse *Acacia tortilis* subsp. *spirocarpa* in East Africa. Other *Acacia* spp. may be dispersed by ants, the funicle acting as an elaiosome.

Economic species A few of the tallest forest species are extracted for export timber, and the gum exuded by *Acacia senegal* (and some other *Acacia* spp.) is the basis of the gum arabic industry. *A. seyal* provides the shittim wood of the Bible. The family is potentially important as a source of fodder and food, fodder from such species as *Acacia albida* (which bears its foliage in the dry season), food from African locust bean (*Parkia* spp.), both fruit pulp and seed, the latter rich in protein; aidan tree (*Tetrapleura tetraptera*) fruits are similarly useful, the seeds being rich in oil. The introduced *Leucaena glauca* (also known by its Filipino name, ipilipil) (now *L. leucocephala*) is a fast-growing evergreen tree tolerant of low rain fall, producing fire wood and green manure. It is predicted that plantations of this species would help to ameliorate the climate of the sahel. It produces plentiful seeds which germinate readily.

> **Field recognition** *Acacia* (Gr. thorn). This is a large genus, found in all the warmer parts of the world, but mainly in the triangle Polynesia–Australia–Africa. It is best represented in West African eastern savannas, little so in forest or in the west, with only 3 (tree) species in Gambia (see below) and 3 scrambling species in Sierra Leone. *A. dudgeoni* and *A. gourmaensis* appear to be our only endemics, other species occurring elsewhere in Africa at least. *Acacia* spp. are woody and armed with prickles or spines, the leaves generally having several pairs of pinnae, each with several pairs of small leaflets (usually 1 pair only in *A. gourmaensis*, a species of swampy ground in forest-savanna mosaic) and glands on the upper side of the petiole and rachis. Inflorescences are usually axillary, either spikes (or spike-like) or spherical heads of small whitish or yellow flowers with numerous glandular anthers (except *A. albida*) being followed by dehiscent pods.
>
> It has been suggested that sufficient differences exist between the 3 already recognised subgenera to warrant raising them to the status of genera (Pedley, 1986). West African species with, for example, stipular spines and a floral disc would then remain in the genus *Acacia* (species 6, 9–10, 12–17 and 19 in *Flora of West tropical Africa*, 1958), while species 1–5, 7–8 and 18, with prickles and no floral disc etc. would become *Senegalia* spp. The third subgenus, mostly with phyllodes (*Racosperma*) does not occur indigenously in Africa. Nongonierma (1979) has made an extensive study of the West African species, without proposing these changes, but re-establishing the genus *Faidherbia* for *Acacia albida* (species 6) (Nongo-nierma, 1976), which differs in numerous ways from the rest of the genus in West Africa.
>
> The 2 most widespread species are the scramblers *A. pennata* (now *A. kamerunensis* mainly; Ross, 1979) and *A. ataxacantha*, both with scattered prickles. *A. pennata* has panicles of heads and grows in forest margins, *A. ataxacantha* forms spikes, and grows in disturbed ground and on rocky

hills. Rather few species occur as far south as forest-savanna mosaic. Apart
from *A. gourmaensis*, there is *A. dudgeoni*, *A. polyacantha* subsp. *campylacan-
tha* (also in Gambia), *A. sieberiana* var. *sieberiana* and *A. nilotica* var. *adansonii*
(now *A. nilotica* subsp. *adstringens*; also in Gambia). The first 2 species
have infranodal curved prickles, *A. dudgeoni* in 3s, 2 pointing upwards,
the middle 1 downwards; in *A. polyacantha* both prickles point downwards.
In the last 2 species, the (stipular) spines are paired, long and straight.
Both species have heads of flowers, those of *A. sieberiana* whitish, but
of *A. nilotica* bright yellow. The pods are also different, being woody and
tardily dehiscent in *A. sieberiana*, but leathery and indehiscent in *A. nilotica*.
In somewhat drier savanna, *A. albida* is a large tree of open savanna,
A. hockii a small one especially seen on inselbergs. Both have short,
straight-paired spines and yellowish flowers. *A. albida* produces spikes
of flowers (lacking anther glands) with the new foliage at the end of the
rains, having been leafless in the rains, while *A. hockii* produces heads
of flowers in the rains, the short, sticky green pods appearing about the
time *A. albida* flowers. The fruits of this latter species are indehiscent,
spirally coiled and reddish-yellow in colour. In the sudan and sahel zones,
A. senegal and *A. seyal* appear. The latter is a tree up to 12 m high, with
stipular spines and multicoloured (green, yellow, red) bark, which is shed
in patches. *A. senegal* is a smaller tree with a short, grey trunk with infranodal prickles in 3s, long spikes of yellowish-white scented flowers in
the rains, and papery dehiscent pods.

Albizia (F. del Albizzi, eighteenth-century Italian collector) is found in
all the warmer parts of the Old World, and is represented in West Africa
by *c.* 10 species. These are unarmed deciduous trees, in both forest and
savanna, all occurring elsewhere in tropical Africa at least. The leaves
have paired leaflets with a diagonal vein, the stipules are caducous, and
there are at least 2 glands, one on the petiole, the other between the
terminal pair of leaflets. The inflorescences are complex, made up of corymb-
like clusters of white and red flowers, each with numerous stamens united
below to form a tube, 1 or 2 flowers in each section of the inflorescence
larger but with a shorter staminal tube, and neuter. The 4 most widespread
species are found in forest and forest-savanna mosaic, particularly in
regrowth and on forest margins. *A. ferruginea* is a forest emergent, also
seen in forest-savanna mosaic, with leaves which have more pairs of leaflets (10–14) in the distal pinnae than the proximal ones, greenish flowers
with the calyx covered outside with rusty hairs, and the staminal tube
at most barely the corolla, though the filaments may be up to 5 cm long.
Pods are reddish-brown, glossy and veined, up to 24 cm long. *A. lebbeck*
(also in Gambia) is a smaller tree, originally introduced but now naturalised
and associated with farmland and villages. The leaves have only 2–4 pairs
of pinnae, with ± equal numbers of pairs of leaflets. The flowers are
also greenish, but the calyx is covered with grey-white hairs and the
filaments are only up to 3 cm long. The pod is yellowish or pale brown,
and often longer than that of *A. ferruginea*. In the 2 remaining

widespread species, the staminal tube is red and projects well beyond the corolla. *A. zygia* is found in regrowth in drier forest and forest-savanna mosaic but is restricted to fringing forest further north. It has 1–3 pairs of pinnae per leaf, the leaflets larger towards the end of each pinna. The stipules and bracts are minute, and fall very early. The pods are rather shiny and hardly veined. *A. adianthifolia* has *c.* 10 pairs of small oblong leaflets per pinna, the distal ones slightly smaller than the basal ones. The pods are very hairy and veined. This species occurs in the same areas as the preceding species, though not persisting as far north.

Parkia (M. Park, eighteenth-century British explorer). *P. biglobosa* is the African locust bean. This is a small pantropical genus of which there are 3 species, 1 in savanna and 2 in forest. These are unarmed deciduous trees with large leaves (up to 25 pairs of pinnae and 30 leaflets on each side) and substantial spherical heads of red flowers (lacking anther glands), pendulous on long peduncles. Flower orientation is unusual, the odd sepal being adaxial. More than one pod may hang from each somewhat enlarged infructescence axis (receptacle); each pod is leathery and indehiscent, containing, at least when young, mealy pulp in which the seeds are embedded. The savanna species is *P. biglobosa* (including *P. clappertoniana*); *P. bicolor*, the forest species, is found especially near water. The former has subopposite leaflets up to 3 cm long with a straight main nerve, and a single circular gland at the base of the petiole. Neuter flowers have 10 short anther-less filaments protruding. The pods are brown, cylindrical and filled with yellow pulp, which adheres to the seeds, the testa not being easily separable. *P. bicolor* has opposite leaflets up to 1 cm long with an S-shaped main nerve, and a single elliptical gland just above the base of the petiole. Neuter flowers have filaments protruding from the flower by 2 cm, and bearing 'anthers', so that the inflorescence is fringed at its base. The pods are reddish and shining, later black, constricted between the seeds, containing bitter pulp to which the testae adhere. *P. filicoidea*, a taller forest tree extending northwards into fringing forest, is less common than the above species (but see Brenan, 1970). It has larger leaflets and the fruit pulp is pink and sweet.

Pentaclethra (Gr. 5, and bolt, ?the stamens). *P. macrophylla* (oil bean) is the only West African species (of a genus of 3 species in tropical Africa and America). It is a large unarmed preserved (and planted) forest tree, with opposite leaflets, the young foliage red, turning dark green and forming a dense crown. The leaves lack glands. In the dry season, panicles of long spikes are formed on old wood, the flowers yellow-to-pink, each with imbricate calyx lobes and 5 stamens (without anther glands) alternating with 5 staminodes. The flat woody pod becomes 60 cm long × 10 cm wide and is very elastic, exploding first on the ground. The flat shining brown seeds are edible.

Piptadeniastrum (Gr. similar to *Piptadenia*). The only species is *P. africanum*, which is found across tropical North Africa down the west coast

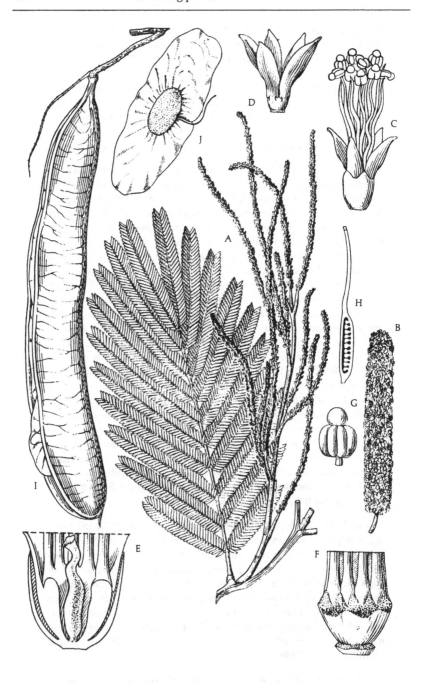

to Angola. It is an unarmed forest emergent with alternate leaflets, the leaves lacking glands. In the rains, panicles of narrow brush-like spikes of minute yellowish-white flowers appear, the anthers with glands, the peduncles persisting after flowering. The pods are pendulous and narrow, up to 36 cm long, opening down 1 side only to release winged seeds. This species is preserved in farmland, but can also invade fallowland, in addition to being able to germinate in shade.

Prosopis (classical name). *P. africana* (also in Gambia) is the only African species, there being over 30 species in America and 2 more in the Middle East. The West African species is a common, deciduous, unarmed savanna tree, with pale drooping foliage of opposite pointed leaflets, a gland being present between each of the 2–4 pairs of pinnae per leaf. Thick, solitary brush-like spikes are produced on new growth in the latter half of the dry season. The flowers are yellowish-white, the anther glands concealed under the folded connective. The pod is indehiscent, cylindrical, blackish and shining. While immature it is juicy, but later dries out so that the seeds rattle.

Bibliography

See also after Papilionaceae.

Brenan, J. P. M. (1959). *Flora of tropical East Africa: Leguminosae–Mimosoideae*. London: Crown Agents.

Brenan, J. P. M. (1970). *Flora zambesiaca*, vol. 3, pt 1, p. 13. London: Crown Agents.

Caballé, G. (1980). Caractères de croissance et déterminisme chorologique de la liane *Entada gigas* (L.) Fawcett & Rendle (Leguminosae–Mimosoideae) en forêt dense du Gabon. *Adansonia*, sér. 2, **20**, pp. 309–20.

Fleurat-Lessard, P. (1985). The motor cell of *Mimosa* pulvini: Structural and ultrastructural features. *Bulletin of the International Group for the Study of Mimosoideae*, **13**, pp. 87–114.

Hopkins, H. C. & White, F. (1984). The ecology and chorology of *Parkia* in Africa. *Bulletin du Jardin de botanique nationale de Belgique*, **54**, pp. 235–66.

Kenrick, J. & Knox, R. B. (1982). Function of the polyad in reproduction of Acacia. *Annals of Botany*, **50**, pp. 721–7.

Nongonierma, A. (1976). Contribution à l'étude biosystématique du genre *Acacia* Miller en Afrique occidentale, 2. *Bulletin de l'Institut fondamentale d'Afrique noire*, sér. A, **3**, pp. 487–642.

Fig. 14.2. *Piptadeniastrum africanum*. A. Shoot with young inflorescence × 1. B. Part of inflorescence × 1. C. Flower × 10. D. Corolla × 10. E. Base of half flower × 30. Ovary immature. F. Base of stamens, petals removed × 30. G. Anther × 30. H. Mature pistil, vertical section × 10. I. Fruit × $\frac{2}{3}$. J. Seed with persistent funicle × 1. (From Brenan, 1959, Fig. 5 pro parte.) (British Crown copyright. Reproduced with permission of the Controller, Her (Britannic) Majesty's Stationery Office & the Trustees, Royal Botanic Gardens, Kew. © 1959.)

Nongonierma, A. (1979). Contribution à l'étude biosystématique du genre *Acacia* Miller en Afrique occidentale, 10. *Bulletin de l'Institut fondamentale d'Afrique noire*, sér. A, **40**, pp. 723–60.

Pedley, L. (1986). Derivation and dispersal of *Acacia* (Leguminosae), with particular reference to Australia, and the recognition of *Senegalia* and *Racosperma*. *Botanical Journal of the Linnean Society*, **92**, pp. 219–54.

Ross, J. H. (1979). A conspectus of the African *Acacia* species. *Memoirs of the Botanical Survey of South Africa*, **44**.

15

Papilionaceae – cowpea family

A large family, present in both temperate and tropical parts of the world. Like the two families preceding, the pod is a common form of fruit, but it is less often woody than in the Caesalpiniaceae, and less often found in bunches than in the Mimosaceae. In West Africa, there are few tree species, but climbers, shrubs and especially undershrubs as well as herbs are common.

Mainly food and fodder species have been introduced, but the decorative species include several American and Asiatic species of *Erythrina*, *Clitoria* and *Pueraria*. Madre (*Gliricidia sepium*), from the West Indies, is popular as a hedge plant, as well as being useful for yam sticks.

Members of the family may be recognised by their alternate, stipulate compound leaves, which are often glandular and possess both pulvini and pulvinules. Leaflets are conduplicate and have stipels in about two-thirds of genera, but they often fall early and immature leaves must be examined. *Stylosanthes erecta*, a common seashore shrub, lacks stipels, as do several genera with paripinnate leaves (marked with an asterisk below). The flowers, described as 'papilionaceous', are distinctly zygomorphic, with the adaxial petal outside its neighbours. Nine of the 10 stamens are frequently joined in an abaxial trough, the tenth adaxial (vexillary) stamen, opposite the standard (vexilla) being free.

Compound imparipinnate or digitately trifoliolate leaves are common, but, although unifoliolate, compound paripinnate and digitate leaves are less common, they occur in so many genera as to be useless on their own for the purposes of identification. Other characters, of flower, fruit, ecology or geographical distribution will be needed.

Paripinnate leaves occur in the genera *Abrus*, *Aeschynomene**, *Arachis**, *Cyclocarpa**, *Kotschya**, *Sesbania* and, in Cameroun, *Smithia**. *Abrus precatorius* (crab's eye creeper) is one of three species of woody twiners in regrowth in which the rachis ends in a bristle. Racemes of pink flowers are followed by clusters of short pods which open to reveal red and black seeds (crab's eyes, see below). *Aeschynomene* are undershrubs with lomenta not covered by the calyx. *Arachis hypogaea*, the groundnut, is presumably well known. *Cyclocarpa stellaris* is a small, spreading undershrub of very dry savannas, with sensitive leaves, and *Kotschya* are small shrubs with reflexed flowers.

Unifoliolate leaves are seen in species of *Crotalaria* (see below), *Desmodium*, *Indigofera* and *Tephrosia*, all large genera contributing weed species. Among tree species, several with unifoliolate leaves are of ecological importance.

Among these is camwood (*Baphia nitida*), an under-storey forest tree with small axillary clusters of scented white flowers and flat, straight, several-seeded pods. *Haplormosia monophylla* is an upper-storey freshwater swamp and riverain forest tree, with stipels, and racemes of blue flowers followed by large leaf-like samaras.

In brackish swamps, two unifoliolate species occur regularly. On the seaward fringe there may be *Dalbergia ecastaphyllum*, while further inland *Ormocarpum verrucosum* occurs. Both are shrubs with white flowers, in the rains and the dry season, respectively. The fruits of the two are also quite different, those of *D. ecastaphyllum* being nearly coin-shaped nuts, those of *O. verrucosum* being lomenta of five portions, looking like a string of brown, striated angular oval beads, each with a few small lumps. Both kinds of fruit float in water. Stamen characters are also different (see below). *O. bibracteatum* is a small sudan-savanna tree with small, round, grey leaflets (leaf imparipinnate) and a prolific production of red and purple flowers in the dry season. The corollas are persistent round the coiled lomenta.

Spines are rare in the family, but another mangrove species, *Drepanocarpus lunatus* has stout, recurved stipular spines. It has mauve flowers in the dry season, followed by sickle-shaped nuts, which also float in water.

Prickles are also infrequent. Apart from *Erythrina*, described below, *Aeschynomene elaphroxylon*, a riverside and freshwater swamp shrub, has prickles associated with its stipules, and the prickles may possibly be spines. The flowers are deep yellow and are followed by coiled hairy lomenta.

Flowers ÷ ♂ .1. 5-part, papilionaceous; the long receptacle tube of *Arachis* is exceptional. K(5) cupular or splitting in various ways (K(4) *Mucuna*). C5 with little structural but some positional variation (e.g. standard abaxial in *Canavalia*, *Centrosema* and *Clitoria*). A10 in some woody species (*Baphia*, *Haplormosia*, *Sophora*); A(10) in *Arachis* and *Stylosanthes* at least basally, in *Crotalaria* at least apically; *Abrus* A(9), forming a tube; the commonest pattern is A(9)+1, the 9 forming a trough (not a tube), the free stamen being the adaxial one; in *Canavalia*, *Millettia* and *Leptoderris*, the basal part of the androecium is arranged like this, while above it becomes A(10) (tubular); the reverse arrangement is seen in *Indigofera*, *Sesbania* etc.; *Ormocarpum* has A5+5; alternate stamens may be of different shapes and alternate filaments of different lengths or attachment to the anthers (basifixed, or dorsifixed and then versatile); some genera display several androecium types. G1, resembling that in the 2 preceding families.

Pollination Frequently by bees, some species of which possess the body weight necessary to depress the keel, exposing pollen and stigma, the nectar-collectors among them also having the behaviour and proboscis length necessary to reach the nectar (secreted either by an intrastaminal disc or by the inside bases of the filaments). Species with a complete staminal tube lack nectaries and are visited for their pollen. Pollen is frequently released in the bud, but degree of self-fertility varies. *Arachis, Canavalia* and *Voandzeia* (now *Vigna*) *subterranea* (Verdcourt, 1980) are largely self-pollinated (though seed set is improved by insect visiting) while *Psophocarpus* shows low self-fertility and insect visits are more important. *Vigna unguiculata* behaves in this way in humid climates, but is largely self-pollinating in drier climates. *Crotalaria juncea* is totally dependent on insect pollination, since the stigma has to be rubbed before pollen will germinate. Some other *Crotalaria* spp. are self-pollinated. Only *Erythrina* spp. have been observed to be pollinated (by sunbirds) in West Africa. Bats have yet to be seen pollinating.

Fig. 15.1. A–D. *Gliricidia sepium*. A. Floral diagram. B. Flower in side view × 1½. C. Parts of the flower × 2; s, standard; w, wings; k, keel; ss, staminal sheath; fs, free stamen. The rim of the calyx has been cut away to show the bases of the petals, which have been removed. The base of the pistil is hidden in the staminal sheath. D. Standard petal in face view × 2. *Gliricidia* resembles *Millettia*, except in having true racemes and a glabrous ovary. E–F. Leaf forms. E. Digitately trifoliolate leaf. F. Pinnately trifoliolate leaf.

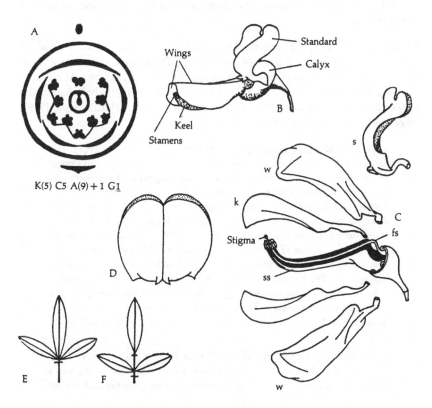

Fruit Samaras are particularly common among woody species, including *Hap-lormosia* (see above). Lomenta are characteristic of a whole group of species (*Desmodium, Stylosanthes*), but the small mango-like drupes of *Andira inermis* (also in Gambia), a pink-flowered tree of drier savannas, are unique in the family in West Africa. Pods may be inflated but are seldom explosive.

Dispersal The brightly coloured seeds revealed when pods open are probably dispersed by birds. Examples include the black seeds of *Abrus precatorius* with a red strophiole, and *Erythrina* seeds, which have a red sarcotesta. The metallic blue sheen of the seeds of some *Rhynchosia* spp. also probably attracts birds. Wind- and water-dispersed samaras and lomenta are mentioned above. *Teph-rosia* (and *Rhynchosia*) are reported to be dispersed by ants. Seeds of species of *Desmodium, Indigofera, Tephrosia, Teramnus* and *Uraria* germinate from baboon dung.

The ground nut habit, of seeds being buried in the soil in which they can germinate, is achieved in *Arachis* and *Kerstingiella* (now *Macrotyloma*, Maréchal & Baudet, 1977) by the growth of tissue immediately under the ovary, which, after fertilisation, produces a stalk pushing the developing fruit into the earth. In *Voandzeia*, the tip of the peduncle assumes this function, dragging the developing fruits downwards. The seeds of these species are thus kept near the parent plant.

Economic species include pulse, vegetable, green manure, cover and fodder crops. Only groundnut is a major export crop. There are a great many weed species. In Columbia, applications of leaves of *Canavalia ensiformis* as a mulch near colonies of leaf-cutting ants on three consecutive nights has cleared the area of these pests for between four months and five years (Mullenmax, 1979). Effects on other leaf predators remain to be investigated.

> **Field recognition** *Afrormosia* (now *Pericopsis*; Van Meeuwen, 1962). The timber of *A. elata* is still known as afrormosia. This comes from an upper-storey evergreen forest tree (Côte d'Ivoire–Ghana), while *A. laxiflora* (now *P. angolensis* subsp. *laxiflora*; Yakovlev, 1971) is a widespread savanna tree which extends across the north of tropical Africa both in open woodland savanna and in fringing forest south of the sudan zone. Both species have drooping branches bearing imparipinnate foliage with persistent sti-pels. Loose panicles of yellowish or greenish-white flowers appear in the late dry season or early rains, each flower with 10 free stamens with versa-tile anthers. The leaf-like, very persistent samaras have a wing up each side. *A. laxiflora* is covered in fine grey hairs and has minute stipels, and one wing of the samara is better developed than the other. It might be confused with that of *Xeroderris*, but the slash of the latter yields red sap.
>
> *Crotalaria* (L. a rattle, the ripe pod with loose seeds) – a very large genus, $\frac{4}{5}$ of its species in Africa. In West Africa there are *c.* 50 species of erect (or occasionally prostrate or climbing) annual and perennial herbs

and undershrubs, in both forest and savanna, often of weedy habit. The leaves are usually 3-foliolate, but some species have simple or unifoliolate, or digitately 5-foliolate leaves, which may also be exstipulate and/or exstipellate. Yellow flowers, in terminal or axillary racemes, with A(10), the sheath open only at the base, and alternating round and long anthers on, respectively, long and short filaments are characteristic. The style and keel bend sharply upwards towards the tip and the keel is open only at the tip, a so-called piston mechanism operating in pollination. The pods are inflated, with heart-shaped seeds which rattle when the pod is ripe. *C. alata* is an introduced Indian species, the terminal inflorescences with prominent orange bracts. This species, *C. retusa* (a weed of coastal and forest regrowth) and *C. calycina* (in savanna) are annuals with simple entire leaves, the keel twisted through at least 180° at the tip. In *C. alata* the stipules are prolonged down the internode below as wings, but in the other 2 species the stipules are minute. *C. calycina* is covered by long silky golden-brown hairs and has very long, lanceolate leaves.

C. glauca is an annual weed species with simple leaves and a straight keel, lacking stipules altogether, while *C. mucronata* (now *C. pallida* var. *pallida*) an undershrub occurring wild on riverbanks and also cultivated as a cover crop, has 3-foliolate leaves which are exstipulate or has small stipules falling early. *C. falcata* (now *C. pallida* var. *obovata*) is a coastal species. *C. ochroleuca* is a similar species but of open damp situations.

Dalbergia (after N. & C. G. Dalberg, eighteenth-century Swedish botanists). Apart from *D. ecastaphyllum* (see above), the genus consists in West Africa of woody climbers, shrubs and a savanna tree species, *D. melanoxylon* (African blackwood) with branch spines. The climbers tend also to have spines, which harden into hooks on contact with the support. The leaves are imparipinnate with alternate leaflets and lack stipels. The flowers are generally white, and various kinds of androecium are seen, though the anthers are always basifixed and open terminally (by a pore or slit). *D. ecastaphyllum* and 4 other species, mainly seen in Sierra Leone, produce nuts, but usually the fruit is a 1-seeded *Lonchocarpus*-style samara. *Dalbergia* is a pantropical and subtropical genus, but only a few West African species extend into southern Africa, and only *D. ecastaphyllum* to the West Indies and South America.

Dalbergiella (Gr. little *Dalbergia*) consists of 2 species only, confined to tropical Africa. One species, *D. welwitschii* (Angola–Guinée) is a liane in riverain forest with imparipinnate leaves with oblong, (sub)opposite leaflets rounded at both ends without stipels. Bottlebrush-style inflorescences of scented yellow-white flowers with a (9)+1 androecium of versatile anthers appear in the dry season. The inside base of the staminal trough produces nectar. The fruit is a flat, 1-seeded samara.

Erythrina (Gr. red, the flower colour). Excluding *E. excelsa*, a forest tree in Cameroun, these are deciduous savanna and forest trees, all with prickles on trunks, branches and leaf rachises, large trifoliolate leaves

Fig. 15.2. *Dalbergia melanoxylon*. A. Flowering shoot × ½. B. Half flower × 8½. C. Stamens × 7. The line down the middle represents the slit present in the unopened tube. D. Anther × 20. E. Fruit × 1. (A and B from Chalk & Burtt-Davy, 1939, Figs. 14.1, 14.3; C–E from Gillett, Polhill & Verdcourt, 1971, Figs. 21.6, 21.7, 21.9.) (British Crown copyright. Reproduced with permission of the Controller, Her (Britannic) Majesty's Stationery Office & the Trustees, Royal Botanic Gardens, Kew. © 1971.)

with at least 1 pair of glands at the top of the petiole and conspicuous racemes of red flowers (pink only in E. *mildbraedii*) in which the wings and keel are very small, the (9)+1 androecium being exposed. A nectar sac is formed at the base of the keel. The pod is constricted round each seed. After the pod opens, the red seeds are displayed on long funicles, and the endocarp separates as a papery layer. E. *senegalensis* (also in Gambia) and E. *sigmoidea* are savanna species, the former, with leaflets longer than wide, flowering in the dry season, the latter with leaflets wider than long, flowering in the rains. The genus is widespread in the tropics and several species have been introduced. West African species are rarely found elsewhere.

Leptoderris (Gr. slender, and *Derris*) is an African genus of woody climbers, only L. *brachyptera* being widespread. It occurs in open situations, e.g. regrowth and river banks, and has imparipinnate leaves with large opposite leaflets with undulate, thickened margins and persistent stipels. The small white flowers, the wing petals being adnate to the base of the keel, have an androecium with (9)+1 basally, (10) distally, with versatile anthers. The fruit is a long samara, winged down 1 side only.

Lonchocarpus (Gr. lance-shaped fruit) is a tropical genus (Australia to America, excluding Asia) represented in West Africa by three deciduous species, also found in other parts of Africa, and by Yoruba indigo (L. *cyanescens*) a straggling shrub, which is confined to West Africa. The leaves are imparipinnate with opposite leaflets and narrow stipels. Racemes or panicles of blue-purple flowers appear with the new leaves, in each flower the wing and keel petals apparently joined basally, the tubular androecium with versatile anthers. The fruit has 1(−5) kidney-shaped seeds but is indehiscent and without wings, accompanied by the persistent calyx. L. *cyanescens* is a coastal and forest species, often planted and occurring also in forest-savanna mosaic. It has erect panicles of pink flowers which turn blue by the end of the day, followed by papery veined fruits. L. *sericeus* (Sénégal lilac) is a tree of deciduous (often riverain) forest, northwards into fringing forest. All young growth appears to be covered with brown velvet. L. *laxiflous* is a tree of drier savanna, with drooping panicles and on woody branchlets and papery fruits.

Millettia (after C. Millett, a nineteenth-century officer of the East India Company). This is largely an Old World genus of which there are about 20 species, mostly trees and woody climbers. Leaves are imparipinnate, ± stipels, with opposite leaflets. The white, pink, blue or purple flowers, in small clusters along the inflorescence axis (false raceme; Tucker, 1987), each have a (9)+1 androecium basally, forming a tube above, bearing dorsifixed anthers. The ovary is surrounded by a disc. The fruits, the best means of distinguishing the genus from a number of others are pods, at length dehiscent, twisting back to reveal 3–6 squarish flat seeds with a well-developed arilloid. Of the widely distributed species in West Africa, M. *chrysophylla* is a white-flowered climber or tree of riverain forest, with leaflets densely covered with silky hairs beneath and a shortly toothed calyx. The remaining species have only moderately hairy leaves. *M. zechiana* is a shrub or small tree of forest regrowth with quite large (2.5 cm long) purple flowers in which the standard bears brown hairs on the back. *M. barteri* is a woody climber, also of secondary growth, with pithy stems and pink flowers turning purple, and *M. thonningii* is a deciduous tree of forest outliers and fringing forests. It may occur together with *Lonchocarpus griffonianus* (now *Millettia griffonianus*; Polhill, 1971) which has, however, persistent stipels.

Ostryoderris (Gr. *Ostrya* and *Derris*) are mainly forest shrubs, sometimes climbing, with imparipinnate leaves with persistent stipels and leaflets which are not acuminate. *O. stuhlmannii* (now *Xeroderris stuhlmannii*), differs from this pattern in being a savanna tree (of northern guinea savanna and the sudan zone) with thick branches bearing large leaf scars and with thick bark. It is deciduous, flowering while leafless in the dry season. The flowers are white with a (9)+1 androecium of versatile anthers, followed by long, 2-winged samaroid fruits. Unlike the remaining species (now transferred to *Aganope*; Polhill, 1971), the leaflets are acuminate and have minute caducous stipels. The former *O. brownii*, if correctly identified, belongs to *Andira inermis*.

Pterocarpus (Gr. winged fruit). All have circular or near-circular fruits, attached at the circumference, some of them samaroid ± bristles. This is a tropical genus, the half-dozen tree species being found more or less extensively in the rest of tropical Africa. Leaves are imparipinnate and exstipellate, with alternate-subopposite leaflets. The slash exudes red gum or juice. In the four widely distributed West African species, erect axillary panicles or racemes of yellow flowers with versatile anthers appear while the trees are leafless, or with the new leaves. In the savanna species *P. erinaceus* (also in Gambia), the flowers are followed by bristly fruits, while in the other savanna species, *P. lucens*, the fruits are without bristles, but only roughly circular, deformed on one side by a stalk-like protrusion. *P. santalinoides* (also in Gambia) is an evergreen riverain forest species with deep yellow scented flowers followed by rimmed nuts, which float in water. The other forest species is *P. mildbraedii*, which is similar except for its prominent bracts and the smooth circular fruits 10 cm ∅.

Bibliography

Allen, O. N. & Allen, E. K. (1981). *The Leguminosae: a sourcebook of characteristics, uses and nodulation.* London: Macmillan Publishers.

Anon. (1979). *Tropical legumes: resources for the future.* Washington, DC: National Academy of Sciences.

Chalk, L. & Burtt-Davy, J. (1939). *Forest trees and timbers of the British Empire*, vol IV. Oxford: Clarendon Press.

Duke, J. A. (1981). *Handbook of legumes of world economic importance.* New York: Plenum Press.

Gillett, J., Polhill, R. & Verdcourt, B. (1971). *Flora of tropical East Africa: Leguminosae–Papilionoideae*. London: Crown Agents.

Maréchal, R. & Baudet, J. C. (1977). Transfert du genre africain *Kerstingiella* Harms. à *Macrotyloma* (Wight & Arn.) Verdc. (Papilionaceae). *Bulletin du Jardin botanique national de Belgique*, **47**, pp. 49–52.

Mullenmax, C. H. (1979). The use of the jackbean (*Canavalia ensiformis*) as a biological control for leaf-cutting ants (*Atta* spp.). *Biotropica*, **11**, pp. 313–14.

Polhill, R. M. (1971). Some observations on generic limits in Dalbergieae–Lonchocarpineae Benth. (Leguminosae). *Kew Bulletin*, **25**, pp. 259–73.

Polhill, R. M. & Raven, P. H. (eds.) (1981). *Advances in legume systematics*. Kew: Royal Botanic Gardens.

Summerfield, R. J. & Roberts, E. H. (eds.) (1985). *Grain legume crops*. London: Collins.

Tucker, S. C. (1987). Pseudoracemes in papilionoid legumes: their nature, development and variation. *Botanical Journal of the Linnean Society*, **95**, pp. 181–206.

Van Meeuwen, M. S. K. (1962). Reduction of *Afrormosia* to *Pericopsis* (Papilionaceae). *Bulletin du Jardin botanique de l'Etat*, Bruxelles, **32**, pp. 213–19.

Verdcourt, B. (1980). The correct name for the Bambara groundnut. *Kew Bulletin*, **35**, p. 474.

Waddle, R. M. & Lersten, N. R. (1974). Morphology of discoid floral nectaries in Leguminosae, especially the tribe Phaseoleae (Papilionaceae). *Phytomorphology*, **23**, pp. 152–61.

Yakovlev, G. (1971). Notae de genere *Pericopsis* (incl. *Afrormosia*). *Novitates systematicae plantarum vascularium*, **8**, pp. 177–81.

16

Ulmaceae – afefe family

A small, mainly northern hemisphere family of trees, represented by only four genera in West Africa, all of them also present in a large part of the rest of tropical Africa. Only two species extend to Gambia. No introductions have been made.

Members of the family may be recognised by their simple, alternate stipulate leaves which are ± asymmetrical, pinnately nerved or three-nerved at the base. Flowers are very small, apetalous, ♂, ♀ or ♂̨, in small axillary inflorescences. The two styles persist on the fruit whether this is a samara or a drupe.

Chaetacme aristata, a small southern savanna tree (Côte d'Ivoire eastwards), is readily recognisable by the nodal prickles accompanying the leaf and axillary branch. Its dark-green leathery leaves end in a tiny spine, the produced midrib, and the large stipular hood leaves an encircling scar on the main axis, cf. Chapter 17, Moraceae.

Flowers ⊕ ♂, ♀ or ♂̨, nearly always monoecious, 4–8 part. K(4–5–8) (free in *Celtis*), imbricate in *Celtis* and *Holoptelea* and ♀ and ♂̨ flowers of *Trema* and *Chaetacme*, valvate with the margins folded in (induplicate-valvate) in ♂ flowers of *Trema* and *Chaetacme*. C0. A as many as, and opposite, the sepals, episepalous, introrse (except *Celtis*). G(2) 2-celled (1-celled by abortion), with an apical ovule per cell; the 2 styles are ± branched, divergent, in the adaxial–abaxial plane of the flower, ± a pistillode in ♂ flowers.

Pollination By wind, as far as is known.

Fruit A drupe or samara (*Holoptelea*) with persistent styles.

Dispersal Drupes are dispersed by birds, samaras by wind.

Economics species Most species have local uses.

Field recognition *Celtis* (classical name). The 7 West African species are ± evergreen. The only savanna species is *C. integrifolia* (also in Gambia), common in fringing forest and riparian woodland. *Celtis* spp. have leaves

K5 C0 A5 G(2)

3-nerved at the base, with all main nerves sometimes originating there; if serrate, then at most so towards the tip of the leaf, and with small caducous stipules. Flowers have an imbricate, generally 5-part calyx, and appear in small axillary inflorescences. The fruit is a small drupe. *C. integrifolia* is distinguished by its leaves, which are rough on both surfaces. Inflorescences are produced in the dry season on new shoots, each one a spike of many ♂ flowers round a larger ♀ flower. The drupes are yellow, then brown, and the styles are well developed and forked at the ends. *C. zenkeri*, a widespread forest species, also found in Sierra Leone, is an under-storey forest tree, the leaves having prominent parallel cross-veins between the main veins. The drupes are red and wrinkled. The West African species are widespread in tropical Africa.

Holoptelea (Gr. whole or real, elm). *H. grandis* is the only West African species, the other being in Indomalaysia. This is a deciduous upper-storey or emergent tree of drier forest, also found in forest outliers. The leaves are pinnately nerved and there are small caducous stipules. While the tree is leafless in the dry season, clusters of small ♂ and ♀ flowers are produced on the previous year's wood. The flowers are 4–6-part, with an imbricate calyx and 2 or 3 times as many anthers. The pistil is relatively large, flat and with 2 horn-like styles. The fruit is a papery samara, the 2 styles persisting in the notch at the tip.

Trema (Gr. a hole, the pitted seed). *T. guineensis* (now *T. orientalis*) (also in Gambia) (afefe) is the most common species of the family, and is often regarded as a 'weed' tree, so regularly does it occur in regrowth throughout forest areas and in the damper parts of forest-savanna mosaic as well as forest outliers and fringing forest. It is an evergreen shrub or small tree with toothed ovate, rather hairy and rough leaves, 3-nerved at the base but otherwise pinnately veined, arranged distichously. Growth continues virtually all the year round, clusters of tiny white flowers appearing successively in leaf axils. ♂ flowers have 'infolded' sepals, like those of *Chaetacme*, the ♀ ones persistent imbricate sepals, the style arms very short. The fruit is a small black drupe. *Trema* is a small tropical and subtropical genus, the West African species being the only one in Africa.

Fig. 16.1. A. *Celtis* sp. floral diagram. B–E. *Holoptelea grandis*. B. Leaf × 1. C. Flowering shoot, only ♀ flowers remaining × 1. D. ♂ Flower × 8. E. Fruiting branch × 1. F. *Chaetacme aristata* ♂ flower × 10. G. *Trema guineensis* ♀ flower × 20. (A from Eichler, 1875–8; B–E from Polhill, 1966, Figs. 1.1, 1.2, 1.3, 1.6, 4.2, 5.2.) (British Crown copyright. Reproduced with permission of the Controller, Her (Britannic) Majesty's Stationery Office & the Trustees, Royal Botanic Gardens, Kew. © 1966.)

Bibliography

Eichler, A. W. (1875–8). *Blüthen-Diagramme*. Leipzig: Engelmann.
Polhill, R. M. (1966). *Flora of tropical East Africa: Ulmaceae*. London: Crown Agents.

17

Moraceae – mulberry family

A family of woody plants in all the warmer parts of the world, represented in West Africa by trees, some shrubs and shrubby epiphytes, and a few stranglers.

Members of the family may be recognised by their latex and alternate stipulate leaves. The leaves may be pinnately veined, relatively small, and conduplicate in the bud, when the stipules are also small (but see *Ficus carica* and *Artocarpus heterophyllus*) or the stipules may be intrapetiolar, large and hoodlike, leaving circular scars on the stem. The leaves are then rolled (supervolute) in the bud (*Ficus* etc.) or pleated (plicate) in the bud and later large and digitately divided or compound (*Artocarpus communis* (now *A. altilis*), *Musanga*). The leaves are usually leathery or rough. The flowers are small, ♂ or ♀, apetalous and often four-part, closely packed in cymose inflorescences, the inflorescence axis (the receptacle, cf. *Euphorbia*) often enlarged. It may be shaped like a spike, a disc or a bag. The fruit is false, sometimes a syncarp or a syconium.

The boundary between the Moraceae and the Urticaceae rests on rather few characters, and *Musanga* and *Myrianthus* are now generally considered to be part of the latter family as a consequence of their possession of clear (not milky) sap, which turns black, basal ovules, small achenes and mineral bodies (cystoliths) in the epidermal cells. In conformity with the second edition of the *Flora of West tropical Africa*, however, the two genera are here retained in the Moraceae.

The best-known introduced species are probably the breadfruit and the jackfruit, which are handsome trees grown for their fruit (see below). *Cecropia peltata* has been introduced from America and comparison with *Musanga cecropioides*, sometimes referred to as a *Cecropia* without ants, will be interesting.

Most of the 20 or so species of *Dorstenia* (now including *Craterogyne*) are perennial herbs in forest in S. Nigeria and Cameroun. Although some are somewhat woody, others have fleshy stems below ground, and are thus quite unusual in this family. The receptacles are discoid, of various shapes (boat-shaped, triangular, 3–4-lobed) with teeth or linear 'arms' (bracts) on the edge, naked ♂ and ♀ flowers (the latter with one or two styles) being embedded in the disc. *D. smythei* (now *D. turbinata*, and including *D. ledermannii* and *D. buesgenii*) is a shrub or small tree (Sierra Leone–S. Nigeria), while *D. walleri* (now *D. cuspidata*; Berg & Hijman, 1977) and *D. preussii* (Guinée–Sierra Leone) are fleshy herbs among rocks, the former in savanna and sometime epiphytic.

Flowers ⊥ ⊕, monoecious with ♂ and ♀ flowers in the same (*Bosquiea*) or different (*Artocarpus*) receptacles, or dioecious (*Morus*). K0–4 to several, often ± fused. C0. A1 to several, often 4, opposite the sepals; anthers extrorse then versatile, erect or folded inwards in the bud. G(2), 2-celled with 2 stigmata, but the abaxial cell usually aborts though the stigma remains; ovule single, apical.

Pollination Protogyny generally occurs, and both wind and insect pollination are known. The best-documented case is that of *Ficus*, where each species (or small group of species) attracts its own species of agaonid wasp, which uses the syconium as a brood place. *Dorstenia* may also include insect-pollinated species. Breadfruit, *Antiaris*, *Chlorophora* and *Morus* are wind pollinated.

Fruit An achene, mostly embedded in the flesh of a persistent calyx and/or fleshy receptacle (false fruit), in a syncarp when the calyces are fused (in breadfruit and jackfruits, where the receptacle also becomes fleshy). The fig is a syconium (bag-like, the flowers or achenes form the lining).

Dispersal Often by bats, which consume the flesh, the large achenes being rejected. Birds etc. disperse achenes of *Chlorophora*, *Musanga* and *Ficus* spp. *Dorstenia* achenes are squirted out of the receptacle as it dries and shrinks. The false fruits of *Antiaris africana* are said to be dispersed by birds, antelopes and monkeys. Elephants disperse achenes of *Ficus macrosperma*, *Myrianthus arboreus* and *Treculia africana*. The latter is also dispersed by buffalo (bush cow).

Economic species Breadfruit (*Artocarpus communis*) and its fertile variety, with achenes, the breadnut, are well known. The trees have deeply pinnately lobed leaves and encircling stipules. Jackfruit, *A. heterophyllus*, has entire leaves, but in both species the ♀ flowers are joined by their calyces, forming a rind over the syncarp. Iroko, the timber of *Chlorophora excelsa*, was exported.

Field recognition *Antiaris* (Malay name) is a small Old World genus, both West African species being usually monoecious forest trees, which also occur across northern tropical Africa. *A. africana* (now *A. toxicaria* subsp. *welwitschii* var. *africana* (also in Gambia)) is a deciduous tree of drier forests and forest outliers with rough leaves, caducous stipules and smooth, grey bark. The watery latex later turns brown. In the dry season axillary inflorescences appear in the axils of the scars of the leaves just shed, 1 to few in each. The ♂'s are discoid and stalked, with 4-part flowers with erect stamens with introrse anthers. The ♀ flowers are solitary, each encased in its receptacle, which bears small scales. The fruit is an ellipsoid, velvety, false drupe containing yellow pulp round the achene. The other form, now var. *welwitschii*, is evergreen and confined to wetter forest areas. It has leaves which are finely hairy at most. The fruit is yellowish-red, swollen at 1 end and thinly fleshy.

166   Flowering plants in West Africa

Fig. 17.1. A–C. *Artocarpus* spp. A. *A. communis* young flowering shoot × ¾. x–x are the stipule scars of leaf 1. s–s stipules (of leaf 2). B. Syncarp (with rind of fused calyces) of breadnut, part of flesh and rind removed, × ⅙. C. *A. heterophyllus* part of cross-section of young jackfruit × 1. The stigmata protrude to the outside through the rind when flowering is in progress. The sepals are free within the rind. D. Floral diagram. E–F. *Ficus* spp. E. Two figs (syconia), the right-hand one cut in half vertically, × 1¼. F. Flower × 14. G–I. *Musanga cecropioides*. G. ♂ inflorescence × 1. H. ♂ flower × 15. I. ♀ inflorescence × 1. J. Part of ♀ inflorescence × 10.

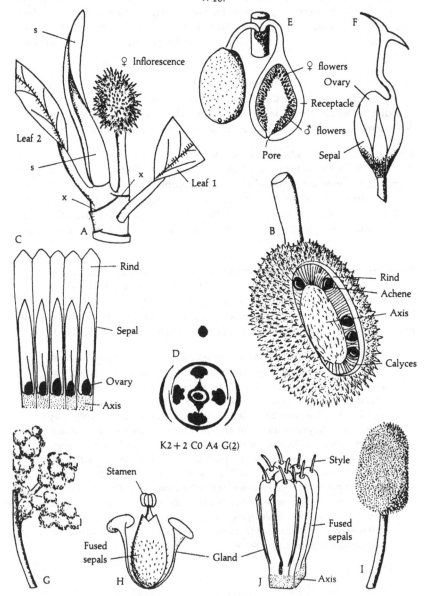

Bosqueia (?L. A. G. Bosc, eighteenth-century French naturalist, or Span. *bosque*, tree) is a monospecific genus. *B. angolense* (now *Trilepisium madagascariense*) being confined to tropical Africa and Madagascar, and occurring in West Africa from Guinée to Angola. It is an upper-storey evergreen forest tree with shining foliage, long, pointed buds and encircling stipule scars. The slash exudes a great deal of white latex, which eventually turns red. The inflorescence is stalked, axillary and bisexual, ± discoid, resembling the swollen end of the peduncle. The inflorescence buds are fat and covered by a bract, which splits to form a spathe-like structure. Each inflorescence contains many ♂ flowers of 1 erect stamen each, round a single ♀ flower, with a tubular 5-part calyx, only the purple anthers and paired styles projecting. The whole thing looks very much like a single flower. The false drupe is dark brown, the remains of the stamens and styles persisting on top.

Chlorophora (Gr. yellow-green and bearing, the inflorescences), also known as *Maclura*, now *Milicia*. This is a small 'bridge' genus (Madagascar, tropical Africa and America) with 2 species in West Africa. Of these, *C. excelsa* (Côte d'Ivoire–Sudan–Mozambique–Angola) is an irregularly deciduous, dioecious forest emergent, also found in forest outliers and preserved in farmland in forest-savanna mosaic. The stipules are large but not encircling and the leaves are prominently pinnately veined. The slash is hard and yellow, and yields white latex. The ♂ trees have dense dark crowns and produce slender, pale pendulous spikes of 4-part flowers with infolded anthers. The ♀ spikes are shorter and thicker (3–4 cm long by 14–18 mm wide) and both kinds seem scaly when young, the down-turned bract tips producing this effect. Later the anthers and long single styles protrude. The fruit is a slightly longer and fatter version of the ♀ inflorescence, green, fleshy and wrinkled, the achenes embedded in the flesh. *C. regia* occurs Gambia–Ghana, and in Ghana it can be distinguished by its leaves, which have 10 or fewer main veins in each half leaf, subsidiary veins being slender and faint. *C. excelsa* has 14 or more main veins in the half leaf, and thick subsidiary veins, making hairy pockets on the back of the leaf.

Ficus (L. fig) is a very large pantropical genus of which there are c. 70 species in West Africa, roughly one-third of them widely distributed, not only here but in the rest of tropical Africa as well. The genus is readily identified by the tufts of glandular hairs at the base of the midrib or in the angles of the basal lateral nerves, and by the syconia, which contain either both ♂ and ♀ flowers, together with gall flowers (sterile ♀s), or ♀ flowers and gall flowers on one plant, ♀s on another. The flowers are 1–7-part, often 4-part. The species are difficult to separate. All are trees or shrubs, many of the latter growing as facultative epiphytes, and some of these become stranglers, e.g. *F. anomani* (now *F. craterostoma*) in forest and *F. platyphylla* in savanna (also in Gambia). The pink figs of this latter species are much sought after by baboons. Most species readily produce aerial roots which hang down in festoons, and *F. congensis*

Fig. 17.2. A–E. *Chlorophora regia*. A. Shoot with ♀ inflorescences × ½.
B. ♂ inflorescences × ½. C. ♂ flower × 4. D. ♀ flower × 4. E. Leaf
venation × 10. F–G. *C. excelsa*. F. Leaf × ½. G. Leaf venation × 10. H.
Bosqueia angolensis inflorescence in vertical section × 2. (A–G from Voorhoeve, 1965, Fig. 54; H from Berg *et al.*, 1984, pl. 31.3, by permission
of the Muséum national d'histoire naturelle.)

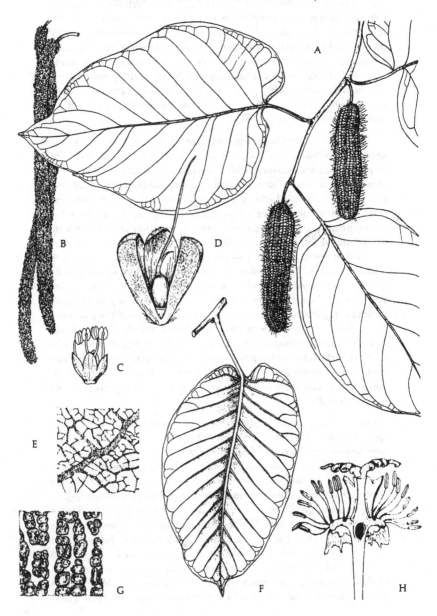

(now *F. trichpoda*), seen in swampy ground near rivers throughout forest regions and northwards into fringing forest, has stilt roots. *Ficus* spp. have encircling stipules, which are persistent in some species, and the foliage is evergreen in most species, which accounts for the use of some species as planted shade trees. The sap is by no means always latex like. *F. exasperata* (also in Gambia) is a deciduous species of drier forest areas in which both leaves and figs feel like very coarse sandpaper. There are *c.* 8 other species with this feature, and this is the group in which it is probably easiest to identify species to begin with. For example, in forest regrowth, there is *F. mucuso* with nearly circular leaves. Four of Gambia's 8 species (not *F. mucuso*) have this feature. *Ficus* is an inland genus, apart from areas where forest-savanna mosaic occurs near the coast.

Morus (L. mulberry). *M. mesozygia*, African mulberry, is the only West African species. The introduced mulberries (*M. alba* and *M. nigra*) are rarely seen. *M. mesozygia* is a dioecious, briefly deciduous tree of drier forest areas, with toothed papery leaves distinctively 3-nerved from the base, apparently unaccompanied by stipule scars. With the new leaves, stalked spike-like axillary inflorescences appear on last season's growth. Flowers are 4-part, ♂s in a pendulous spike with unfolded anthers and a pistillode, ♀s in an erect cluster, fused to each other by the sepals, which dry and persist round the achene, forming a dry syncarp. In the edible mulberries, the ♀ calyx becomes fleshy.

Musanga (Congolese name) (umbrella tree). The only species is *M. cecropioides*, a common and characteristic evergreen tree of forest regrowth in wetter areas, with stilt roots and digitately compound leaves up to 1 m across. The stipules are red and silky, up to 20 cm long and encircling the stem. Within, the leaves are pleated. There is no latex, and the pith is very attractive to ants. The ♂ trees are sympodial in growth, producing leaf-opposed (terminal) panicle-like inflorescences of small heads on new wood, each flower with joined sepals round a single stamen (erect in the bud) and accompanied by 2 glands. The ♀ inflorescences are short and club-like, on a long stalk, occurring in 1s and 2s in the leaf axils. Each ♀ flower has a fused calyx, a pistil with a basal ovule and a single style, and a pair of glands which secrete substances attractive to ants. The fruit is greenish and juicy, containing numerous achenes.

Myrianthus (Gr. numberless flowers) is a small tropical African genus, represented in West Africa by 4 species of dioecious evergreen shrubs or trees with very large, coarsely toothed leaves or leaflets. The stipules are hood like, but there is no latex. *M. arboreus* is found in forest regrowth and has very large digitately compound leaves. The ♂ inflorescences, produced in axillary pairs in the latter part of the dry season, are yellow, much-branched and panicle like. The flowers are 4-part, the stamens erect in the bud. The ♀ inflorescences are paired, stalked greenish clusters, each flower with a thick curled style projecting out of the fused calyx, and a basal ovule. The fruit is a syncarp of basally fused, yellow false drupes, up to 10 cm ⌀, the remains of the style projecting from each drupe. *M. serratus* is a riverain forest species with stilt roots and pinnately nerved leaves. There are few ♀ flowers per cluster, and usually only 2–3 are fertilised, so the syncarp is smaller, up to 3.5 cm ⌀.

Treculia (A. Trecul, nineteenth-century French botanist) is yet another small genus of tropical Africa and Madagascar, with 3 West African species. African breadfruit (*T. africana*) is an upper-storey, sometimes dioecious, evergreen forest tree, often planted and usually preserved in farming. The trunk is very fluted and, together with the older branches, bears the ♀ inflorescences. These are spherical or obovoid, stalked, the surface covered by the flat peltate tips of bracts between which the branched styles project. The ♀ flowers resemble those of *Artocarpus*, but are not joined and form no continuous rind. The single (–few) axillary ♂ inflorescences are considerably smaller, up to 7 cm ⌀, the anthers erect in the bud and protruding between the bracts. The fruit weighs several kilograms, and its achenes (okwe) are prized as food. The stipules are large and hood like, round distichous pinnately veined leaves. *T. obovoidea* (S. Nigeria–Cameroun) has smaller inflorescences in which the bracts are spine-like and black.

Bibliography

Berg, C. C. (1977). Revisions of African Moraceae (excluding *Dorstenia, Ficus, Musanga* and *Myrianthus*). *Bulletin du Jardin botanique national de Belgique*, **47**, pp. 267–407.

Berg, C. C. & Hijman, M. E. E. (1977). A precursor to the treatment of *Dorstenia* for the floras of Cameroun and Gabon. *Adansonia*, sér. 2, **16**, pp. 415–43.

Berg, C. C., Hijman, M. E. E. & Weerdenburg, J. C. A. (1984). *Flore du Gabon: Moraceés*. Paris: Muséum national d'histoire naturelle.

Breitwisch, R. (1983). Frugivores at a fruiting *Ficus* sp. in Southern Cameroun tropical wet forest. *Biotropica*, **15**, pp. 125–8.

Michaloud, G. & Michaloud-Pelletier, A. (1987). *Ficus* hemi-epiphytes (Moraceae) et arbres supports. *Biotropica*, **19**, pp. 125–36.

Newton, L. E. & Lomo, A. (1979). The pollination of *Ficus vogelii* in Ghana. *Botanical Journal of the Linnean Society*, **78**, pp. 21–30.

Ruiter, G. de (1976). Revision of the genera *Myrianthus* and *Musanga*. *Bulletin du Jardin botanique national de Belgique*, **46**, pp. 471–510.

Voorhoeve, A. G. (1965). *Liberian high forest trees*. Wageningen: Centrum voor Landbouwpublikaties en Landbouwdokumentalie.

18

Meliaceae – mahogany family

A family of tropical and subtropical trees and shrubs. Members may be recognised by their alternate compound pinnate exstipulate leaves with conduplicate ± opposite leaflets. Only *Turraea* in West Africa has unifoliate leaves, while *Melia* has bi- to tripinnate ones. The buds are either naked or protected by scales. Slash and wood are strongly scented and generally reddish. Lax, axillary, often large, panicle-like cymes of small, pale, scented flowers with a staminal tube are followed by capsules in most genera.

Decorative introduced species include Persian lilac (*Melia azedarach*) from India.

Turraea spp. are climbers in open situations in moist forest, are the only West African species of unusual habit.

Flowers ⊰ ⊕, ♂ and ♀ on the same or different plants, but ♂ in *Turraea*, polygamous in *Azadirachta* and *Melia*; mostly 5-part, 4-part in *Lovoa*, where the sepals are orthogonal and the petals diagonal. K(5) small. C5 (C(5) in *Turreanthus*). A(8) or (10), joined only basally in most *Trichilia* spp., *Pseudocedrela* and *Entandrophragma candollei*, A5 in *Cedrela* and *Toona*; in these 2 genera there is an adrogynophore bearing petals, stamens and pistil; anthers are borne on or within the rim of the staminal tube. G(5) 5-celled, rarely with more cells (*Turraea*) or fewer (G(3) 3-celled in *Trichilia*, 1-celled in *Heckledora*), with 2 to many ovules per cell on axile placentae (2 parietal ovules in *Heckeldora*).

Pollination The flowers are scented and an intrastaminal disc is present in all except *Turreanthus*. Insect pollination is indicated, the length and width of the staminal tube determining which insects can reach the nectar. In tropical America, sphingid moths pollinate *Cedrela odorata*. In West Africa, this may also prove to be the case with *Turraea* spp., where the staminal tube can be up to 3 cm long.

Fruit Commonly a capsule. This may be woody and explosive, with winged seeds attached to a persistent central column (columella) of placentae. Such capsules usually open septilicidally from the apex to the base. Loculicidal opening is seen in *Turraea* and *Turreanthus* etc., which have leathery capsules and fewer seeds, with a sarcotesta or other arilloid, e.g. in *Turreanthus* there are 2–5 yellow seeds in white sarcotesta pulp. It has been suggested that

Fig. 18.1. A. *Melia azedarach* half flower × 5. B–D. *Entandrophragma utile.*
B. part of staminal tube × 10. C. Seed × ¼. D. Fruit × ¼. E. *Guarea cedrata*
fruit × ¾. F–H. *Khaya* spp. F. *K. senegalensis* part of staminal tube × 10.
G. *K. grandifoliola* seed × ⅓. H. *K. senegalensis* fruit × ½, seeds and one
valve removed. I. *Lovoa trichilioides* fruits × ⅜. J. *Entandrophragma angolense*
floral diagram. K–M. *Trichilia* spp., segments of staminal tubes. K. *T.
heudelotii* × 7. L. *T. prieuriana* × 11. M. *T. roka* × 6. (J from de Candolle,
1894, Pl. 21 *pro parte.*)

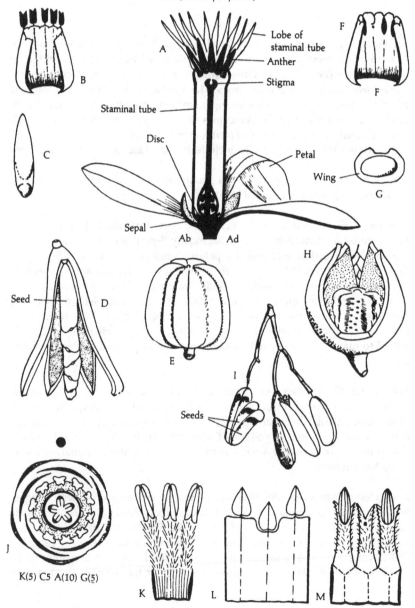

the wings of the species mentioned above may also be arilloid in origin. *Azadirachta*, *Melia* and *Ekebergia* have yellow (to reddish) drupes, those of *Melia* 5-stoned, and *Heckledora* has a pale-purple berry. The seeds of all these except *Melia* have a sarcotesta.

Dispersal by explosion, and subsequently by wind, of winged seeds has been mentioned (Pannell & White, 1985). Bats and baboons disperse the drupes of *Azadirachta*, and birds are reported to disperse those of *Ekebergia* and the seeds of *Trichilia*. Primate dispersal as in *Aglaia* spp. (Pannell & Kozioł, 1987) has yet to be reported.

Economic species produce timber, generally known as mahogany, a term which merely implies a hard red wood suitable for furniture, joinery and veneers. Introduced timber species include the American mahoganies (*Swietenia* spp.), the West Indian or cigar box cedar (*Cedrela odorata*, not a true cedar, *Cedrus* sp., though the wood smells similar), and toon (*Toona serrata*) and *Chukrasea tabularis* from Asia and Ceylon, respectively. *Azadirachta indica*, from India as the name implies, has been widely used in plantations, to provide firewood. It is now naturalised and very common in some areas, e.g. in the Accra Plains. The langsat, *Lansium parasiticum* produces the fruit most worth eating, and its sweeter-fleshed races would be worth introducing from Southeast Asia.

Field recognition *Carapa* (Guianese name) is a small, monoecious tropical genus of 7 species, *C. procera* (crabwood) being our only lowland species (Sénégal–Angola, some West Indian islands and Guiana). It occurs in swamp forest and the wetter parts of forest, also in similar areas of forest-savanna mosaic. It is a large, spreading, evergreen, short-trunked tree with imparipinnate leaves 1–2 m long, each with *c.* 12 pairs of oblong leaflets. Near its northern boundary, the habit becomes palm-like. The flowers have pink petals and a white *Khaya*-type staminal tube, and, in the early rains, large, leathery 5-valved capsules, each valve with a warted ridge running down its back. The 3–7 large, round reticulated seeds with a woody sarcotesta in each cell leave large scars on the central column. The capsule opens at apex and base simultaneously.

Entandrophragma (Gr. inner, male and partition, the small struts extending from the base of the staminal tube to the disc) is composed of 35 species in Africa south of the Sahara. The 4 West African species are monoecious, deciduodus emergents of drier forests yielding export timber. The buds have scales. The leaves are paripinnate and crowded at the end of the branches, the leaflets ± subopposite. As they appear, inflorescences are produced in the axils, each flower scentless, with a staminal tube with anthers above the rim and a very small disc. The fruit is a 5-valved woody capsule with seeds attached apically and with a basal wing. Species may be separated as follows:

Valves of fruit thick, over 5 mm, opening from apex to base	*E. utile* (utile)
Valves of fruit thin, under 5 mm Fruit opening from apex to base, seeds 3–10 per placenta	*E. candollei* (omu)
Fruit opening at both apex and base, valves ridged, falling away; seeds 3–4 per placenta	*E. cylindricum* (sapele)
Fruit opening from base to apex; valves not ridged; seeds 5–6 per placenta	*E. angolense* (gedunohor)

Guarea (Cuban name) – a largely American genus with *c.* 20 African species, all dioecious evergreen trees with naked buds in forest. Only 2 species are widespread, both being emergents with heavy crowns of imparipinnate leaves with 2 wings or ridges on the petiole. Short erect panicles of yellowish 4–5-part flowers with *Khaya*-type staminal tubes appear in the late rains (*G. thompsonii*, Liberia–S. Nigeria) or dry season (*G. cedrata*, Sierra Leone–S. Nigeria), followed by usually 4-valved capsules, the seeds with yellowish-red sarcotestas. The bark of both falls in scales, but the slash of *G. thompsonii* is yellow with white latex, the raw wood stinks and the fruits are reddish, while the slash of *G. cedrata* is reddish-yellow, though darkening, and the wood is highly scented, and the fruits are brownish.

Khaya (Senegalese name) (African mahoganies). This is a small genus of Africa south of the Sahara, with 4 West African species of large deciduous trees with scaly bark, a pinkish slash, buds with scales and paripinnate leaves. They are monoecious with narrow, erect panicles of whitish flowers in which the anthers lie within the staminal tube. The fruit is an erect, spherical woody capsule, containing seeds which are winged all round. After opening, the valves remain connected by fibres. Savanna mahogany (*K. senegalensis*) is found in open woodland savanna as well as forest outliers and fringing forest, also in riparian woodland. The flowers are 4-part, with a red disc round an ovary which tapers into a slender style. Large-leaved mahogany (*K. grandifoliola*) is an emergent of drier forest and forest outliers with 4 pairs of papery leaflets each up to 25 cm long × 10 cm wide. it has 5-part flowers in which the ovary narrows abruptly into the style. The capsules have 5 valves each over 5 mm thick. *K. ivorensis* (Lagos mahogany) is an emergent of wetter forest areas, with 4–7 pairs of leathery acuminate leaflets per leaf, 5-part flowers and 5-valved thin-walled capsules. In Côte d'Ivoire–Ghana and Cameroun, K. *anthotheca* (Sierra Leone–Ghana and Cameroun) overlaps with the species above, most likely with K. *grandifoliola*. It has, however, smaller, leathery leaflets and thinner-valved capsules.

Lovoa (River Lovoi, Congo). L. *trichilioides* (African walnut) is the only West African species (Sierra Leone–S. Nigeria) of this small African genus. It is a monoecious, evergreen forest emergent with imparipinnate leaves of 5–6 pairs of leaflets on a slightly winged rachis. Flowers can be found most of the year, in inflorescences up to c. 30 cm long, some of them terminal. Each flower is 4-part, with an *Entandrophragma*-style staminal tube. Bunches of black, pendulous,, 4-sided capsules then appear, the 4 thin valves splitting away from the base of the fruit and falling off. The seeds, 2 per loculus, each have a wing attaching them to the base of the fruit.

Pseudocedrela (Gr. false, *Cedrela*) is another small African genus of which 1 species, P. *kotschyi*, is present in West Africa, extending across northern tropical Africa. It is a medium-sized, monoecious, deciduous savanna tree, found specially on heavy soils in swampy areas as far north as in northern guinea savanna. The buds have scales. The young foliage is silvery-grey, the leaves being impari- or paripinnate, with up to 12 alternate asymmetrical leaflets with wavy edges. Narrow inflorescences of 5-part scented flowers appear in the dry season. Each flower has a staminal tube like that of *Trichilia roka*, but more rigid and glabrous. The capsules are erect and woody, like an upright *Entandrophragma* capsule, but the valves connected by fibres as in *Khaya*. The seeds are also like those of *Entandrophragma*, attached apically and with a basal wing; there is, however, a prominent nerve down 1 side of the wing.

Trichilia (Gr. 3-part, the 3-valved fruit etc.). This is largely an American genus, with approximately a dozen species in West Africa, some of them extending further into Africa. The West African species are evergreen trees, mostly in forest, with naked buds and a slash which exudes ± milky juice. The leaves are imparipinnate, and in their axils, on new wood, short lax inflorescences of greenish dioecious flowers with valvate corollas appear. The androecium is tubular at least at the base. The fruit is a small 3-valved capsule, generally pink or red changing to brown in colour, with black seeds not entirely covered by a red arilloid. Of the 3 widespread species, T. *heudelotii* (now T. *monadelpha*) is found in forest regrowth and damper forest areas. It has 3–6 pairs of abruptly acuminate leaflets. The filaments are stout and hairy and the fruit is obovoid, 3–4-valved and pale brown. T. *prieuriana* (also in Gambia) is an under-storey tree of drier forest and forest outliers, with a decidedly fluted bole and a dry slash. The leaves have 2–3 pairs of acuminate leaflets, the terminal leaflet longest, the length decreasing towards the leaf base. Flowers appear in the dry season, each with a completely tubular though lobed androecium with anthers on its rim. The capsules are pink, spherical and glabrous, and the red arilloid stands on top of the seed. T. *roka* (now T. *emetica*) is the savanna species in West Africa, occurring in open woodland savanna right into the sudan zone. All the younger parts of the plant, together

Fig. 18.2. *Lovoa trichilioides*. A. Flowering branch × ½. B. Flower × 4. C. Fruit × 1. D. Seed × 1. E. Seedling × 1. F. Columella × 1. (From Voorhoeve, 1965, Fig. 51.)

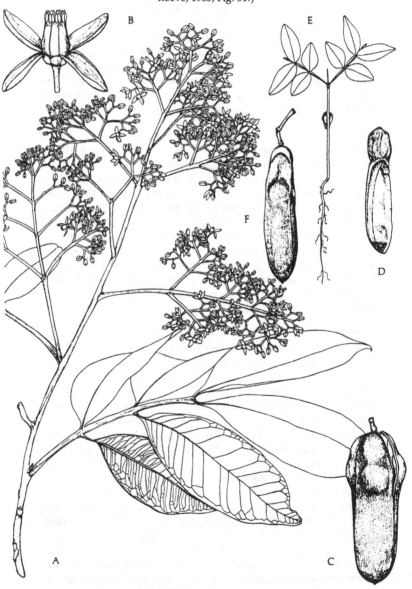

with mature petioles and leaf rachises, are covered with yellow woolly hairs. The leaves have *c.* 4 pairs of leaflets, which are rounded at both ends. The flowers are up to 2 cm long, with stout hairy filaments, each ending in 2 teeth. The fruit is spherical, densely hairy and slow to open.

Bibliography

Berhaut, J. (1975). A propos du *Trichilia emetica* Vahl. *Adansonia,* sér. 2, **15**, pp. 255–6.

de Candolle, C. (1894). Meliaceae novae. *Bulletin du Herbier Boissier,* **2**, 567–84.

Hall, J. B. (1976). A new species of *Turraea* (Meliaceae) from Ghana. *Adansonia,* sér. 2, **15**, pp. 505–8.

Pannell, C. M. & Kozioł, M. J. (1987). Ecological and phytochemical diversity of arillate seeds in *Aglaia* (Meliaceae): a study of vertebrate dispersal in tropical trees. *Philosophical Transactions of the Royal Society, London,* ser. B, **316**, 303–33.

Pannell, C. M. & White, F. (1985). Bird dispersal and some derived evolutionary trends in a tropical plant family, the Meliaceae. *Notes from the Forest Herbarium, University of Oxford,* **3**.

Pennington, T. D. & Styles, B. T. (1975). A generic monograph of the Meliaceae. *Blumea,* **22**, pp. 419–540.

Voorhoeve, A. G. (1965). *Liberian high forest trees.* Wageningen: Centrum voor Landbouw publikaties en Landbouwdokumentatie.

Wilde, J. J. F. E. de, (1968). A revision of the species of *Trichilia* P. Browne (Meliaceae) on the African continent. *Mededelingen Landbouwhogeschool Wageningen,* **68**(2), pp. 1–207.

19

Sapindaceae – akee apple family

A family of tropical and subtropical trees, shrubs and tendril lianes with, in West Africa, three species of these last representing large American genera – *Cardiospermum* spp. (balloon vines) and *Paullinia pinnata*.

Members of the family may be recognised by their alternate, compound pinnate leaves with conduplicate leaflets, and cymose spike or panicle-like inflorescences of minute, pale, apparently ♂ flowers. The petals bear scales or hairs on their inner faces, and there is an annular extrastaminal disc, or the disc or disc glands stand to one side of the stamens and pistil. The fruits are often three-valved capsules, with a single arilloid seed in each cell. Tissues are generally soapy when crushed in water.

The climbers have paired inflorescence tendrils and imparipinnate stipulate leaves, while the trees and shrubs have paripinnate exstipulate leaves, in which the rachis is often slightly prolonged. There is rarely only one pair of such leaflets (*Aphania*). The leaflets commonly decrease in size towards the base of the leaf, and the lowest pair may arise next to the axil (the leaf being ± sessile), and form pseudostipules (*Eriocoelum, Blighia*, compare Bignoniaceae).

There are 1–3-foliolate leaves in the genus *Allophylus* and 1-foliolate leaves only in *Dodonaea viscosa*, in both cases the leaves being exstipulate. *D. viscosa* is a glandular sea coast shrub with ⊕ apetalous four-part flowers.

The habit of a small erect palm is characteristic of some under-storey species of *Chytranthus, Deinbollia* and of *Radlkofera calodendron*. The leaflets of these species are in general very large, and the leaves particularly long (up to c. 1 m).

Introduced species yield fruit, and include the soapberry (*Sapindus saponaria*) the fruits of which do form a lather when crushed in water. The substances responsible are saponins, poisonous to cold-blooded animals, a property often made use of in the preparation of fish poisons. The Spanish lime (*Melicocca bijuga*) has black drupes, and the litchi (*Litchi sinensis*) has red, thin-shelled fruits with one seed in a white arilloid. Both fruits are edible.

Flowers ± .l. or ⊕, apparently ♂, ♀ or ⚥, monoecious or dioecious, 4–5-part. The flowers are very small and it is difficult to determine both their orientation and, in .l. flowers, the plane of symmetry. Only 3 genera are decidedly zygomorphic: in *Paullinia* (K5 C4), the third and fifth petals are fused (or one or other is occluded), and the stamens and pistil are rotated 72° to the left and displaced excentrically to the lower right, accommodating the disc glands. In *Cardiospermum* and *Allophylus* (and other 4-part flowers, including the ⊕

ones, e.g. *Dodonaea*), the sepals are orthogonal, the petals diagonal. But in *Cardiospermum*, the corolla, stamens and pistil are rotated 90° to the left, so that the plane of symmetry is at 90° to the adaxial–abaxial plane. In *Allophylus*, the whole flower appears to be twisted 90° to the left (the lateral sepal – 3+5 – becomes adaxial), the corolla, stamens and pistil being then further

Fig. 19.1. A–C. *Paullinia pinnata*. A. Flowering shoot × ½. B. Fruits ×½. C. Fruits after the valves have drupped × ⅝. D–G. *Cardiospermum* spp. D–E. *C. halicacabum*. D. Floral diagram. E. Seed × 6. F. Capsule × 1. G. *C. grandiflorum* seed × 4½. H. ⊕ flower, floral diagram. I. *Dodonaea viscosa* fruits × ½. J–K. *Eriocoelum* sp. J. Vertical section of flower × 5. K. Petals × 6. (From Exell, 1966: D–G, Figs. 101.A3, 101.A4, 101.B3, 101.A1; J–K, Figs. 105.3, 105.4, with permission from the Editorial Board of *Flora Zambesiaca*.)

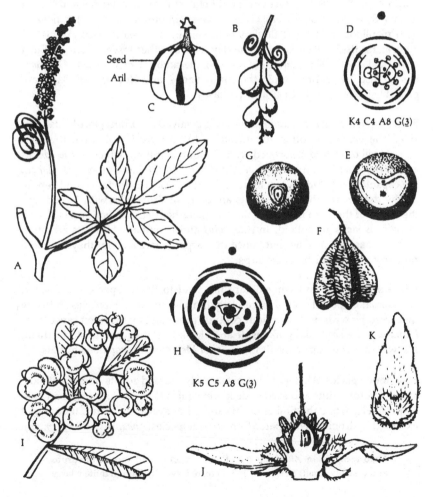

twisted through another 90°, so that the odd (abaxial) carpel ends up on the adaxial side of the flower. Another degree of zygomorphy exists in *Pancovia*, where the adaxial disc gland and excentric stamens and pistil (only) render the flower zygomorphic (in the adaxial–abaxial plane). Petals usually have a scale or hairs (absent in *Melicocca*), or C0 (*Majidea* spp., *Litchi*, *Placodiscus* etc.). Disc extrastaminal, annular in ⊕ flowers, to one side in .l. flowers and of separate glands. A8, rarely fewer (*Dodonaea, Zanha*), often more; apiculate only in *Laccodiscus*; the filaments folded in the bud in *Placodiscus* and *Zanha*. G(3) or (2), the median carpel abaxial or one or other carpel aborting; G(1–8) can occur, the higher numbers in *Chytranthus*; cells as many as carpels, each with one to two ovules on axile placentae; in a few genera the carpels are united only basally round the style (*Allophylus, Deinbollia*).

Pollination The disc produces nectar. Litchi has been investigated and shown to have three kinds of flowers: (1) ♂ flowers without pistillode, (2) ♂ flowers with pistillode, and (3) ♀ flowers with staminodes. All flowers are visited by bees for their nectar early in the day, and by other insects later on. Such visits make a considerable difference to the amount of fruit set. *Dodonaea* has no disc. Sex expression in the family varies, and although all plants are probably ⚥, there are both ♂ and ♀ phases.

Fruit Often a 3-valved (*Paullinia* etc.) or a 2-valved (*Aporrhiza*) loculicidal capsule. The valves are often somewhat fleshy, externally coloured, the seed in each cell black and associated with a coloured arilloid, which in *Laccodiscus* and *Lychnodiscus* completely covers the seed. In *Cardiospermum* and *Majidea* the capsule is inflated and membranous, in *Dodonaea* 2–3-winged and septilicidally dehiscent. The fleshy fruits are berry or drupe-like with 1 seed, but the origin of the flesh may be in the arilloid rather than the pericarp cf. *Litchi*. *Placodiscus* lacks an arilloid. In *Allophylus* and *Deinbollia*, the fruits are often described as apocarpous, but, since the carpels share a style, they are more properly considered as schizocarps.

Dispersal Balloon and winged capsules tend to float, especially on water, but animal dispersal seems more likely in most species. *Allophylus* fruits are dispersed by birds and the colour contrast presented by the open capsules of *Majidea* (brown outside, pink inside and blue seeds) is such as to attract them. Elsewhere in the tropics the fleshy fruits of *Melicocca* are dispersed by bats.

Economic species Akee apple is extensively planted for its edible arilloids, and the young fruits and seed endosperm could also be used for soap and oil, respectively, while the timber of this and other species is useful, though not exported. Fish poison is prepared from several species (*Aphania, Cardiospermum*).

Field recognition *Allophylus* (Gr. foreign race) – c. 10 species of shrubs and (usually small) trees with 1–3 foliolate leaves. Although quite a large tropical and subtropical genus, only 1 of the West African species, *A. welwitschii*, extends across Africa. *Allophylus* spp. produce pendulous inflorescences of the *Paullinia*-type, in the rains, made up of yellowish-white, .l., 4-part scented flowers with 4 disc glands. The petals have transverse hairy ridges on the inner face, and the fruit is red, of 1–3 fleshy mericarps. *A. africanus* (also in Gambia) is a deciduous shrub or small tree with 3-foliolate, flat, serrate leaflets on petioles up to 6 cm long. The flowers are often galled. This is a species of forest undergrowth and savanna areas protected from fire. *A. spicatus*, a shrub with simple spike-like inflorescences and smaller leaflets, occurs in open situations, e.g. on rocks, by streams and on termite mounds, especially in the Accra plains.

Aphania (Gr. inconspicuous) is a small Indomalaysian genus with 1 tropical African species, *A. senegalensis* (now *Lepisanthes senegalensis*; Leenhouts, 1969) (also in Gambia), an under-storey forest tree as far north as in fringing forest. It has 1–2 pairs of leaflets per leaf, each leaflet on a short thick petiolule. The rather woody panicles are found on new growth in the rains, all such new growth being heavily covered with yellowish hairs. Each branch of the inflorescence is of the *Paullinia*-type, but the flowers are ⊕ and 5-part, with broad petal scales, 7–8 stamens and only 2 carpels in ♀ flowers. The fruit is red and fleshy, 1 carpel generally aborting and remaining visible at the base of the developed carpel together with the style.

Blighia (W. Bligh, eighteenth-century British mariner). Of this small African genus, only 3 species are to be found in West Africa, and *B. sapida*, akee apple, is restricted to West Africa. All are evergreen forest trees, extending northwards into fringing forest, though *B. welwitschii*, a forest emergent, is usually seen in wetter forest areas. The leaves of all species have 1–5 pairs of leaflets, the uppermost pair the largest, and, in akee apple, the lowermost pair forming pseudostipules. In the dry season, *Paullinia*-type inflorescences of ⊕, 5-part, scented, yellowish flowers appear, ♂ and ♀ (or ♂⃠) flowers on separate trees. There is a pocket-like scale at the base of each petal. The fruit is an erect, red leathery capsule containing 3 large black seeds with fleshy, cupular, yellow arilloids round the base. In the akee apple there is, within the arilloid, a pink streak, which denotes the presence of hypoglycin A, a poison causing a violent reduction in the blood sugar level. Only arilloids from perfectly ripe fruits, ripened on the tree, should be eaten.

Cardiospermum (Gr. heart seed) – balloon vines are herbaceous climbers of regrowth, with tendrils and (bi-) imparipinnate stipulate leaves. Lax, pyramidal panicle-like inflorescences of yellowish-white 4-part flowers are produced in the rains. Each flower is transversely .l., with a 2-lobed disc placed laterally. The fruit is a balloon-like capsule with a small black seed

Fig. 19.2. *Aphania senegalensis*. A. Flowering shoot × $\frac{2}{3}$. B. ♂ flower × 6. C. ♂ flower, sepals and petals removed × 6. D. ♀ flower, three sepals and all petals removed × 6. E. Petal × 6. F. Vertical section of pistil × 6. G. Cross-section of ovary × 6. H. Fruits × $\frac{2}{3}$. (From Exell, 1966, Fig. 108, with permission from the Editorial Board of *Flora Zambesiaca*.)

in each of the 3 cells. *C. grandiflorum* has larger flowers, up to 1 cm long, with a pair of large horn-like glands, and a triangular glabrous capsule. *C. halicacabum* has flowers only a third of the size, with 2 small glands and 3-winged, hairy fruits.

Deinbollia (Gr. dreadful (?poisonous) fruit; connotation not clear). There are about 2 dozen species of this African–Madagascar genus in West Africa, mostly of very local distribution. *D. pinnata*, a shrub or small tree of forest regrowth, is the only widespread species, and, even then, it is confined to West Africa. It has leaves with at least 5 pairs of leaflets, on a short petiole. In the late dry season, terminal spikes, covered in brown hairs, are produced. The flowers are ⊕, 4–5 part (each petal with 2 minute scales) with *c.* 15 stamens. There are generally 4 carpels, united only basally by the style. The fruit then consists of 1–4 reddish-yellow, berry-like mericarps (with edible flesh).

Eriocoelum (Gr. woolly cavity) – a small West African genus of trees, of which *E. kerstingii* usually seen near rivers as far north as in riparian woodland, is the most widespread. The leaves have a pair of pseudo-stipules and 2–3 other pairs of leaflets, the rachis projecting between the top pair. Large terminal panicles of white flowers, the whole covered with brown hairs, appear in the dry season. Each flower is ⊕, the petals with flap-like scales, round a thin cupular disc and 8 stamens. The fruit is characteristic, with woody, curved, yellow valves, with a tuft of hairs inside the base of each valve, together with a single seed with a small cupular arilloid.

Lecaniodiscus (Gr. cup-shaped disc). Two species are known (Hall, 1980*a*), but the more widely spread species in West Africa is *L. cupaniodes*. It is an evergreen shrub or small tree of forest and forest outliers, with a spreading crown of paripinnate foliage with 4–5 pairs of leaflets per leaf. In the late dry season, axillary spike-like inflorescences covered with white hairs are produced on new growth. The flowers are ⊕, apetalous, 4-part and scented, with 10+ stamens. The fruits have a hard, velvety, yellow-to-red rind and a single black seed embedded in jelly-like flesh.

Paullinia (S. Paulli, Danish seventeenth-century professor of medicine and botany). Only *P. pinnata* is known in Africa (also in Gambia). It is a climbing shrub (Cremers, 1974), with stipulate imparipinnate leaves with winged petioles and rachises, and anomalous secondary thickening in its stems, resulting in a cable-like effect in long-established plants, though the species is more often seen as relatively young specimens in regrowth. Spike-like inflorescences of white, .l. flowers appear at various times of the year, each flower with 5 sepals and 4 petals, yellow-tipped petal scales, a lobed disc gland on one side, and 8 stamens and the pistil on the other. The fruit is a hanging red capsule, leathery at first, with black seeds developing almost complete yellow arilloids.

Bibliography

Cremers, G. (1974). Architecture de quelques lianes d'Afrique tropicale, 2. *Candollea*, **29**, pp. 57–110.

Exell, A. W. (1966). *Flora zambesiaca: Sapindaceae*. London: Crown Agents.

Hall, J. B. (1980*a*). Five new species of flowering plants from West Africa. *Bulletin du Jardin botanique national de Belgique*, **50**, pp. 249–66.

Hall, J. B. (1980*b*). New and little known species of *Placodiscus* (Sapindaceae) in West Africa. *Adansonia*, sér. 2, **20**, pp. 287–95.

Leenhouts, P. (1969). Florae Malesianae praecursores L. A revision of *Lepisanthes*. *Blumea*, **17**, pp. 33–91.

20

Anacardiaceae – cashew nut family

A largely tropical family of trees and shrubs with resinous tissues, probably best known through the cashew nut tree (*Anacardium occidentale*) and the mango (*Mangifera indica*). Both these species have unifoliolate leaves, whereas compound imparipinnate leaves are typical of the indigenous members of the family, with the exception of *Heeria*, with two widespread savanna species, one a suffrutex, the other woody. *Heeria* (now *Ozoroa*) also has opposite or whorled leaves, while alternate (and exstipulate) ones are otherwise typical of the family. Terminal and/or axillary inflorescences, of small, pale three- to five- part flowers, and resinous drupes with a single stone are characteristic of the whole family.

Nothospondias is now placed in the Simaroubaceae. There are new illust-rations of *Antrocaryon micraster*, *Lannea schimperi* and *Pseudospondias microcarpa* (Kokwaro, 1986).

Introduced species include two tropical American species of pepper tree (*Schinus* spp.).

Flowers ∸ ⊕ ♂, ♀ or ♂ on the same or different plants, 3–5-part; both mango and cashew nut (also *Fegimanra*) have .l. flowers, having often only 1 fertile stamen and having a single carpel. Most genera are polygamous as described above, but a few (*Haemostaphis* etc.) show dioecy. K small, 4–5 or (4–5). C4–5, mostly imbricate; in 4-part perianths, the sepals are orthogonal, the petals diagonal. Only *Haematostaphis* has 3-part flowers. A8 or 10, only 1 fertile with 4 staminodes in mango, and 1 stamen with a long filament accompanied by 9 with short filaments (though most may be fertile) in cashew; in *Trichoscypha* A4, and in *Heeria* A5 or 10 is found, in *Sclerocarya* and *Sorindeia* A10–20. The receptacle forms a prominent intrastaminal disc in most species (extrastaminal in mango) or, together with the pedicel, becomes swollen in the fruit (*Anacardium*, *Fegimanra*). G1, 1-celled with 1 style in mango, cashew and *Fegimanra* etc.; otherwise G(3–5), 1- or 3–5-celled with the same number of styles and/or stigmas; ovules 1 per cell, apical, or pendulous on a basal funicle or laterally adnate to the wall; where the style is single it is also terminal (except in mango), as in *Haematostaphis* and *Sorindeia*, but when 3–5 styles are present, they tend to be lateral, i.e. round the top of the ovary; of the genera with a 1-celled G(3) ovary on a regular basis, *Haematostaphis* and *Sorin-deia* have a single style, *Heeria* and *Trichoscypha* 3 styles. The cotyledons are thick and food storing.

Fig. 20.1. A–F. *Lannea welwitschii*. A. Fruiting shoot × ½. B. Inflorescence × ½. C. Flower bud × 7. D. Flower × 7. E. Flower in vertical section × 7. F. Disc and stigmas, sepals, petals and stamens removed, × 7. G. Floral diagram, e.g. *Heeria*. H. *Anacardium occidentale*. ♂ flower × 12. A–F from Aubréville, 1959, pl. 201; G from A. W. Eichler, 1875–8; H from Nicholls & Holland, 1929.)

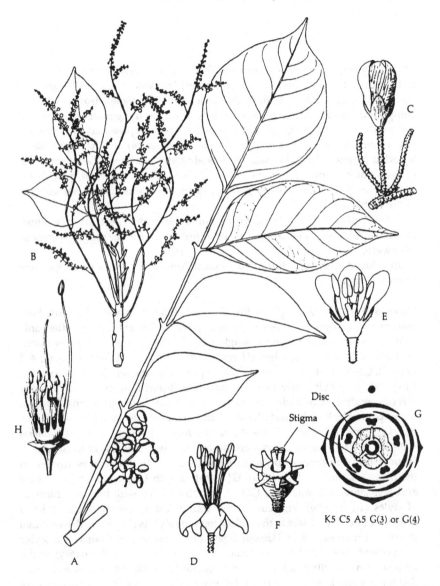

K5 C5 A5 G(3) or G(4)

Pollination has been best investigated in mango, in which the large erect panicles are made up mainly of ♂ flowers (in the upper part of the panicle) and of less than one-third as many ⚥ flowers. These are protogynous, and nectar is secreted by the disc. Numerous kinds of insects have been observed to visit mango flowers, and if these are excluded from inflorescences the ensuing fruit set is very poor. Flies of various species have been proposed as the main pollinators, but could quite well suck nectar without touching the stamens or stigmata. Bees on the other hand, have been observed usually to touch both. Cashew has also been studied but, unlike mango, bagged inflorescences set no fruit at all in spite of some evidence of self-compatibility. The services of a pollinator are required. Both other monoecious as well as the dioecious species in the family have open access flowers like mango and cashew, but nothing appears to be known about their pollination.

Fruit A drupe with a fleshy or thin and/or resinous, often edible, mesocarp, except in cashew, which has a nut. There is a single stone, which may be 1–5-celled, but abortion often takes place, and the number of cells (seeds) may be less than the number of styles or stigmata. Oil from kernels is more often edible than the kernel itself.

Dispersal of the mango is by bats, which are also attracted to the cashew apple (the swollen pedicel and receptacle). Both these fruits hang down clear of the foliage in the manner associated with bat dispersal. Cashew nuts can also be dispersed by water. Other fruits are yellow, black or red, those of *Lannea* and *Sorindeia* having been reported to be eaten by birds in Africa. Fruits of *Lannea acida* germinate from baboon dung. Fruits of *Antrocaryon micraster* are scented and dispersed by elephants, as are the fruits of *Sclerocarya birrea*.

Economic species yield fruit, in particular mango, hog plum (*Spondias mombin*) and cashew, the kernel of which is poisonous unless roasted thoroughly. No product is exported. Mangoes of the cultivar 'Alphonse' are particularly worth eating raw.

> **Field recognition** *Anacardium* (Gr. reversed heart, the shape of the cashew apple) is a tree up to 12 m high with obovate unifoliolate leaves frequently cultivated on lagoon banks and elsewhere, also in plantations in Nigeria. The ♂ and ♀ flowers occur on the same tree, in lax inflorescences twice a year. The flowers are 5-part, whitish, with (7–)10 stamens, the fertile one large. There is a single carpel, forming a nut, the receptacle and pedicel swelling up after fertilisation to form the cashew apple, to which the Greek name refers. This is edible when stewed in syrup.
>
> *Antrocaryon* (Gr. chambered nut). There are 2 species of this very small African genus in West Africa, both of them forest trees. *A. micraster* is widespread, and is a deciduous forest emergent with tufts of leaves at

the ends of the branches. The leaves consist of 8–10 opposite or suboppo-site pairs of leaflets with 20–24 nerves on each side of the midrib but no intramarginal nerve. Small panicles of ♂ and ♀ flowers appear in the late dry season, each 5-part with valvate petals and 10 stamens. The ovary is (3–)5-celled with 5 styles projecting sideways off the top of the ovary. The fruit has a very large 5-celled stone c. 4 cm ∅ and 3 cm high, inside scented, edible, yellow flesh, forming a 5-lobed drupe rather wider than tall.

Lannea (? Senegalese name) – a genus of savanna trees in Africa. The young growth is conspicuously hairy or sticky, especially so while the trees are otherwise leafless in the dry season, when axillary spike-like inflorescences of yellow flowers appear, crowded at the ends of the branches. The flowers are ♂ or ♀ (on the same or different trees), 4-part, the ovary with (3–)4 styles, developing into a red-to-black drupe. The leaves have 1–6 pairs of asymmetrical leaflets, usually under 5 cm broad. *L. welwitchii* is the only West African forest species.

   L. acida is often found on rocky hills in northern guinea and sudan savannas. It has curled stellate hairs resembling felt on the younger parts, which is reddish, but does not persist long. The fruits are ellipsoidal, yellowish-red. *L. microcarpa* is found in derived savanna and drier forest regrowth. It has short, close, simple hairs on young growth, and black ellipsoidal fruits. In Nigeria this species is replaced by *L. schimperi*, with reddish felted hairs persisting only beneath the leaf veins. The fruits are ± globular and red. *L. kerstingii* (now *L. barteri*; Kokwara & Gillett, 1980) may be confused with *L. schimperi*, but has spirally twisted, rather than flaking, bark. Stiff, straight hairs are carried on leaflet margins, and along the underside of the veins; fruits glabrous, nearly black, squared off at the end, the 4 persisting styles forming 'corners'.

Pseudospondias (Gr. false *Spondias*). Of the two species of this African genus, only *P. microcarpa* is found in West Africa (Gambia–East Africa and Angola). It is a tree, up to 20 m high, seen on forest edges and north-wards in patches of forest into fringing forest. It has a short, crooked bole, and dense foliage of leaves with c. 15 asymmetrical alternate leaflets. Large panicles of ♂ or ♀ flowers are produced on separate trees in the dry season. The flowers are (3–)4-part, the ♂ ones with (6–)8 stamens in which the anthers are wider than long, the ♀ ones with (3–)4 styles on the pistil, forming a persistent point on top of the small purple drupe.

Sclerocarya (Gr. hard nut). *S. birrea* is the only West African species in this small African genus. It is a deciduous tree, up to 12 m high, of drier savannas, especially the sudan zone, with pale foliage of leaves with 7–10 pairs of leaflets, clustered at the ends of the shoots. These are stout and prominently scarred. The spike-like inflorescences of ♂ or ♀ flowers (on separate trees) appear just before the leaves unfold. Each flower is 4-part, but ♂ ones have 12–26 stamens and ♀ ones (3–)4 styles. The fruit is a drupe, like a small mango.

Spondias (Gr. plum). *S. mombin* is the only indigenous species and is often planted as a live fence and for its fruits, which resemble small mangoes externally, though they contain (3–)5-celled stones. *S. mombin* is a deciduous tree up to 18 m high, with 5–8 pairs of asymmetrical leaflets per leaf, each leaflet with a prominent nerve parallel to its margin. Panicles of 4–5-part flowers are produced in the rains and again at the end of the dry season. ♂ and ♀ flowers have 8–10 stamens, ♀ and ♂ a 5-celled ovary with (4–)5 styles. Other species of *Spondias* worth cultivating are the Gambia plum (*S. purpurea*) and Polynesian otaheite apple (*S. cytherea*). *S. mombin* is common in regrowth and farmland throughout forest-savanna mosaic and in forest.

Bibliography

Aubréville, A. (1959). *Flore forestière de la Côte d'Ivoire*, 2nd edn. Nogent-sur-Marne (Seine): Centre technique de forestière tropical.

Eichler, A. W. (1875–8). *Blüthen-Diagramme*. Leipzig: Engelmann.

Kokwara, J. O. & Gillett, J. B. (1980). Notes on the Anacardiaceae of Eastern Africa. *Kew Bulletin*, **34**, pp. 745–6.

Kokwaro, J. (1986). *Flora of tropical East Africa: Anacardiaceae*. Rotterdam: A. A. Balkema.

Nicholls, H. A. & Holland, J. H. (1929). *A textbook of tropical agriculture*. London: Macmillan.

Sedgeley, M. (1982). Pollination biology of fruit and nut crops. In *Pollination and dispersal*, ed. N. B. M. Brantjes & H. F. Linskens, pp. 69–73. Nijmegen: University of Nijmegen.

Teichman, I. von, & Robbertse, P. J. (1986). Development and structure of the drupe in *Sclerocarya birrea* (Richard) Hochs subsp. *caffra* Kokwaro (Anacardiaceae), with special reference to the pericarp and the operculum. *Botanical Journal of the Linnean Society*, **92**, pp. 303–22.

21

Sapotaceae – sheabutternut family

Trees and shrubs with latex and fleshy fruits associated with a persistent calyx. In West Africa, most species are large, evergreen forest trees with simple, entire, alternate, leathery leaves with prominent pinnate venation. Flowers are small and white, in axillary clusters, each one three- to eight-part; flower structure in some species is complex. The flowers seldom open widely, never before senescence, and the corolla and staminodes absciss circularly round the base and fall as a piece. The flowers are followed by fleshy or woody berries, sometimes very large, with seeds having relatively large scars. The family is not yet reported growing indigenously in Gambia.

The introduced sapodilla (*Manilkara zapotilla*) from Central America has grey or brown fruits with sweet red flesh, and the star apple (*Chrysophyllum cainito*), from the same area, has berries with delicious white or purple flesh. *Mimusops elengi* (tropical Asia) is sometimes planted for shade.

Flowers ⊕ ♂. Simple 5-part flowers occur in *Aningeria* etc. (K5 [C(5) A5] G(5)), with 5 staminodes alternating with the stamens only in *Vincentella* and *Aningeria*. The more complex flowers of *Butyrospermum*, *Tieghemella* and *Manilkara* have 2 whorls each of sepals and petals, appendaged petals (each petal represented by 3 lobes) and petaloid staminodes. *Butyrospermum* and *Tieghemella* have 4-part flowers (K4+4 (the whorls dissimilar) [C(8) A8] G(8)), *Manilkara* has 3-part flowers (K3+3 [C(6) A6] G(6)), all have petaloid staminodes. In *Omphalocarpum* the flowers are of the simple 5-part type, but there is a group of 2–6 stamens opposite each petal lobe. G(5)–(many), with as many cells, each with 1 ovule on a basal or axile placenta.

Pollination Very little is known. Some species have a disc.

Fruit A fleshy or woody berry with 1 to many large shiny seeds, each with a proportion of its surface occupied by a dull scar (where it was attached to the placenta) of which the hilum is only a part). Most of the fleshy berries are 1-seeded and could be mistaken for drupes, e.g. *Butyrospermum*. The calyx persists round the fruit.

Dispersal may be by bats in some species, but *Bequaertiodendron*, *Butyrospermum*, *Manilkara* and *Mimusops* are reported to be eaten by birds, *Chrysophyllum taiense* by monkeys and *C. pruniforme*, *C. albidum* and *C. pentagonocarpum* by elephants, which also disperse species of *Aubregrinia, Kantou, Tieghemella* and *Omphalocarpum.*

Fig. 21.1. A–F. *Malacantha alnifolia.* A. Fruiting branch × $\frac{3}{8}$. B. Flowering branch × 1. C. Flower × $4\frac{1}{2}$. D. Corolla opened × $4\frac{1}{2}$. E. Pistil × $4\frac{1}{2}$. F. Seed × $\frac{3}{4}$, showing scar. G. *Chrysophyllum* floral diagram. H. *Manilkara* floral diagram. (A–F from Hemsley, 1968, Fig. 3 *pro parte*) (British Crown copyright. Reproduced with permission of the Controller, Her (Britannic) Majesty's Stationery Office & the Trustees, Royal Botanic Gardens, Kew. © 1968.) (G and H adapted from Eichler, 1875–8.)

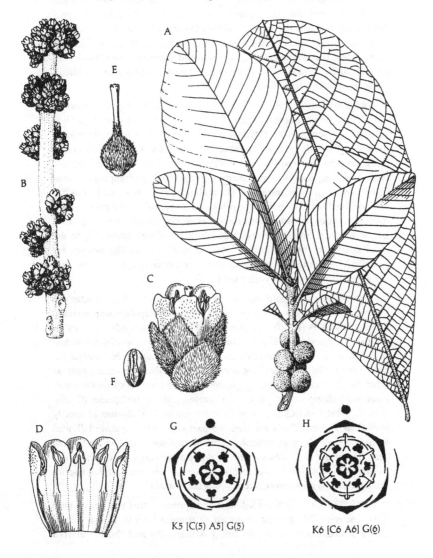

K5 [C(5) A5] G(5)   K6 [C6 A6] G(6)

Economic species produce mainly timber, e.g. *Tieghemella* timber, exported as makore; other species find local uses. *Butyrospermum* kernels yield an edible oil, shea butter; the extract is exported as shea oil. The fruit species were mentioned above. In addition *Synsepalum dulcificum* berries are used as a sweetener, the glycoprotein they contain causing food to taste sweet.

Field recognition *Aningeria* (Côte d'Ivoire name, also spelled *Aningueria*). There are 2 species, both West African but extending to the east, both forest emergents: *A. altissima* particularly in forest outliers, *A. robusta* within lowland forest, especially in riverain forest. The leaves have the prominent pinnate nervation common in the family, with the lateral nerves turning towards the tip only just within the margin. The flowers appear in the dry seaon, each on a pedicel 3–5 cm long, 5-part with needle-like staminodes. The berries are red, 1-seeded, and about 2 cm long. The leaves are exstipulated and gland-dotted, though this character is at times difficult to see, even with a hand lens.

Butyrospermum (Gr. butter seed). *B. paradoxum*, the only species, extends right across north tropical Africa. The West African subspecies *parkii* is a deciduous, somewhat twisted woodland savanna tree of forest savanna mosaic northwards as far as the sahel. The bark is black, thick and fissured into cubes, the slash red. The mode of growth of the branches is distinctive, short and long shoots being clearly differentiated, and stipulate leaves being carried in tight spirals on short shoots. Flowers occur in the dry season, the flower clusters being produced in the axils of the many nodes on a short shoot just before the new leaves appear. The flowers are pedicillate and 4-part, complex, the corolla lobes spreading widely only in the older flower, as the staminodes begin to curl inwards. The flowers are described as scented or strong-smelling by different authors. The fruit is a greenish-yellow berry with 1 seed.

Chrysophyllum (Gr. gold leaf, from the colour of the hairs under the leaves of *C. cainito*). A large genus, mainly of exstipulate upper-storey forest trees in West Africa, where most species have silvery, grey or reddish hairs under the leaves. About half of the 13–14 species are widespread in West Africa, though only *C. albidum* occurs in N. Nigeria. *C. welwitschii* is the only climbing shrub, and *C. subnudum* occurs only as a twisted under-storey tree. *C. giganteum* is the only species with loose, extended inflorescences (panicles or racemes), the other species (*C. albidum, C. delvoyi* and *C. pruniforme*) developing axillary clusters of shortly pedicillate flowers. These are always 5-part and simple, and are followed by 3–5-seeded, elongate reddish-yellow berries, each seed with a narrow scar. The fruits of *C. albidum* and *C. delevoyi* are eaten. In *C. pruniforme* the leaves have a multitude of fine lateral nerves, while in the other species the lateral nerves are fewer, subopposite and distant.

Malacantha (Gr. soft thorn). Like *Butyrospermum*, this is a genus with only 1 species, *M. alnifolia* (confined to tropical Africa). It is a tree or shrub of deciduous forest and forest outliers, the soft thorns referred to above being the T- or Y-shaped hairs borne on the younger parts – a hand-lens character. The flowers are produced on the previous season's growth just as, or immediately after, the leaves drop from the current growth in the dry season. They appear in sessile clusters on short lateral shoots, each flower simple and 5-part. The fruit is a hairy yellow-to-red berry, pointed, with 1 seed, each with only a narrow scar. The leaves are exstipulate and gland-dotted, the lateral nerves running out to the margin, turning towards the tip and forming the margin.

Manilkara (Malabar name) is a tropical genus of which only 1 or 2 species are found in West Africa, occurring from lagoon banks to riparian woodland and rocky hills in the sudan zone. *M. obovata* is the taller growing species of wetter areas, with stipulate obovate leaves; *M. multinervis*, the lower-growing tree of drier areas with oblong leaves, is often included under the former name. The tree often branches near the base, but becomes 15–30 m high. Clusters of flowers on pedicels *c.* 1 cm long appear in the axils of leaf scars in the dry season, each flower 3-part and complex, followed by a pale yellow berry, 2–3 cm ∅, with 1(–6) seeds. The leaves are characterised by being frequently bitten or galled, and the flowers open widely, the sepals turning right back (reflexed) and the corolla lobes spreading. The staminodes are short, forked and erect.

Omphalocarpum (Gr. navel fruit, referring to the depression on the top of the fruit). A genus of 3–5 species, confined to W. Africa, in which *Ituridendron bequaertii* may be included. These are upper-storey forest trees producing flowers on the main branches and trunk, followed by the hard-shelled berries, up to 25 cm ∅. Each berry has a thick, pale-brown woody rind crossed by numerous dark corky veins, round a ring of about 30 very large seeds with narrow scars, embedded in whitish pulp. The flowers which precede the fruits are 5-part and simple, except that a group of 5–6 stamens stands opposite each corolla lobe. *O. elatum* produces pedicillate flowers on the main trunks, the flowers having a group of 5–6 stamens per corolla lobe. The fruits are *c.* 8 cm ∅. The species is widespread in, but confined to, West Africa. *O. pachysteloides* (*Ituridendron bequaertii*) carries its flowers on leafless twigs, sometimes on older branches of the trunk. There are only 2–3 stamens per group in the flower and the fruit is *c.* 5 cm ∅. This species extends as far as Congo.

Pachystela (Gr. thick pillar, the style). Of the 4 species of this genus (Africa only), *P. brevipes* occurs in most of West Africa and much of the rest of tropical Africa. It is an under-storey forest tree, extending into forest outliers and fringing forest. It has very tough leaves, and is one of the few species in the family with persistent stipules. Five-part ± sessile scented flowers are produced in the axils of leaf scars on last season's growth, the corolla lobes spreading. The 1-seeded berries are reddish yellow, often beaked by the remains of the style. The scar on the seed covers about two-thirds of the surface area.

Fig. 21.2. *Tieghemella heckelii.* A. Flowering branch × ½. B. Flower × 3. The spreading lobes are the 16 petal appendages, the 8 corolla lobes being folded over the centre of the flower at anthesis. C. Fruit × ½. D. Seed × ½. E. Seedling × ½. (From Voorhoeve, 1965, Fig. 67.)

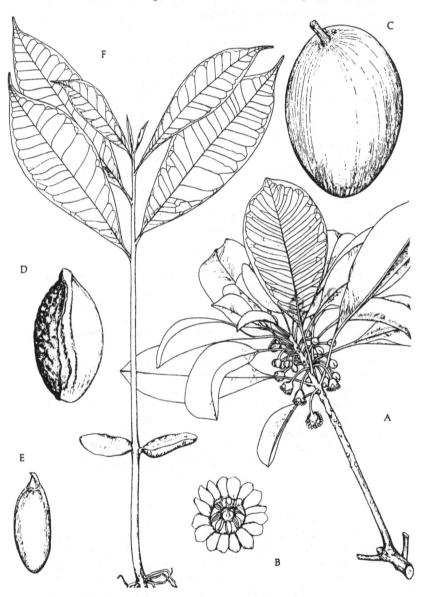

Tieghemella (P. van Tieghem, nineteenth-century French botanist). This is another monospecific genus, *T. heckelii*, being confined to West Africa. It is a forest emergent with horizontal upper branches and ± obovate leaves, with caducous stipules. The flower buds are globose, standing on 2 cm pedicels, in clusters of 1–3. Each flower is complex and 4-part, and is followed by large yellow berries, *c.* 10 cm ∅, each with 1–3 seeds *c.* 5 cm long, the scar covering about one-third of the surface; the flesh has an unpleasant smell and taste.

Vincentella (connotation uncertain) is a small genus in tropical and southern Africa, of which *V. passargei* is one of the few Sapotaceae in savanna, particularly northern guinea and sudan savanna near streams. The small stipulate leaves are clustered at the branch tips. Clusters of very shortly pedicillate 5-part flowers (with staminodes) appear in the axils of leaf scars, the ovary, and young growth in general, being heavily covered with rusty hairs. In older flowers the corolla opens wide, the tubular part being very short and the slender staminodes being scarcely distinguishable from the corolla lobes. The fruit is a yellow or reddish-yellow berry with edible flesh, containing a single seed with a small scar.

Bibliography

Eichler, A. W. (1875–8). *Blüthen-Diagramme.* Leipzig: Engelmann.
Hemsley, J. (1968). *Flora of tropical East Africa: Sapotaceae.* London: Crown Agents.
Voorhoeve, A. G. (1965). *Liberian high forest trees.* Wageningen: Centrum voor Landbouwpublikaties en Landbouwdokumentatie.

22

Apocynaceae – frangipani family

A large family well represented in West Africa, at least half of the genera containing species which are widespread throughout West Africa, many of these, in addition, in Gambia.

Introduced garden species give a useful introduction to the general appearance of members of this family. *Allamanda* spp., oleander (*Nerium oleander*), Madagascar periwinkle (*Catharanthus roseus*), frangipani (*Plumeria* spp.) and herald's trumpet (*Beaumontia grandiflora*) are all widely available.

Members of the family may be recognised by their opposite, simple, entire, generally exstipulate leaves (though often with interpetiolar ridges), possession of latex, and cymes of fragrant, gamopetalous 5-part flowers with contorted corolla lobes. The fruit is a berry of two carpels, two berries (of one carpel each), or a pair of follicles (or one by abortion) with plumed seeds. Most of the widespread species are climbers (some with branch tendrils) or under-storey trees in forest, a few only being open woodland savanna species.

Four genera diverge sufficiently from the 'family pattern' as to be recognisable on sight. The only species with spirally arranged leaves (and prominent leaf scars) is a fleshy stemmed deciduous shrub, *Adenium obesum*, found (and often planted) in the northern guinea and sudan zones of Nigeria.

Carissa edulis, in the same zone (Ghana–Nigeria) has branch spines, the only species in the family in West Africa to do so.

Two genera have whorled leaves. *Alstonia* spp. are trees to 30 m high, from swamp forest (*A. congensis*) through riverain forest (*A. boonei*) to forest outliers, with obovate leaves in whorls of four to eight. The branches are, expectedly, also whorled. *Rauvolfia vomitoria* is a shrub or small tree with leaves in threes, mostly found in forest regrowth. As with *Alstonia*, the branches are also whorled, and the nodes enlarged and lumpy. The minute flowers have, however, hardly any free corolla lobes, and the fruits are fleshy, while *Alstonia* fruits are pairs of follicles.

Flowers ⊥ ⊕ ♂ and 5-part. Sepals small, free or joined, sometimes with glandular scales (*Vahadenia, Cyclocotyla, Picralima, Hunteria*). Corolla lobes overlapping to the left or right when seen from outside (in the bud), a character revealed because the calyx is short. The open flower is bell or trumpet-shaped, or with the lobes spreading at right angles to the axis of the tube; the uppermost part of the tube (the throat) may be narrowed (*Landolphia*), rarely with

Fig. 22.1. *Alstonia boonei*. A. Shoot × $\frac{1}{2}$. B. Leaf tip × $\frac{1}{2}$. C. Inflorescence × $\frac{1}{2}$. D. Flower bud × 3. E. Flower × $1\frac{1}{2}$. F. Open corolla and pistil × $2\frac{1}{2}$. G. Anther × $9\frac{1}{2}$. H. Pistil × $4\frac{1}{2}$. I. Fruit × $\frac{1}{2}$. J. seed × $\frac{4}{5}$. (From De Jong, 1979, Fig. 1.)

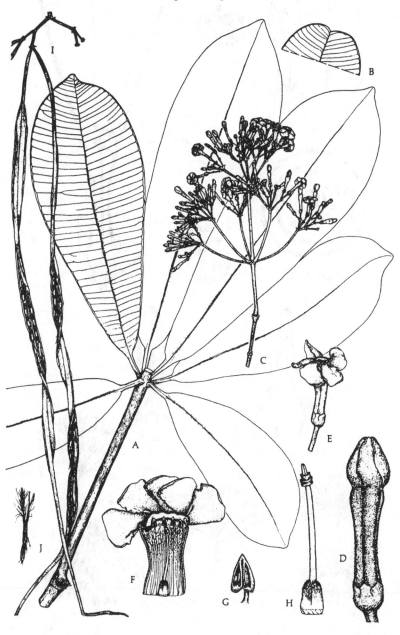

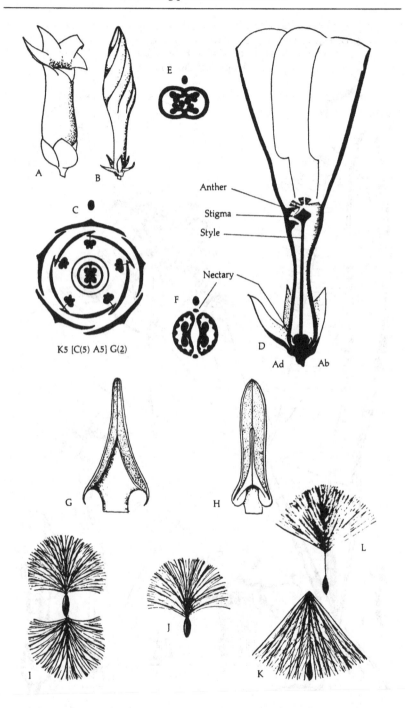

Anther

Stigma

Style

Nectary

K5 [C(5) A5] G(2)

Ad　　　Ab

an appendage between each pair of corolla lobes (*Strophanthus, Pleioceras, Oncinotis*). Stamens epipetalous, included within, or exerted from, the corolla tube; anthers fertile to the base but with sterile basal tails in *Callichilia* (including *Hedranthera*), *Tabernaemontana, Voacanga, Adenium* etc., in the last genus and some others, with sterile apical appendages also. Disc, between the corolla and pistil, rarely absent (*Alafia, Picralima, Hunteria, Pleiocarpa*) or of only 2 glands (*Catharanthus*). G(2), the carpels completely joined with axile or parietal placentae, or joined at most basally and by their styles; 1 carpel occasionally aborting; G(3–5) rare (*Pleiocarpa*); stigma drum-shaped (clavuncula) with receptive surfaces provided with sticky exudate on its sides.

Pollination Rowley (1980) gives a detailed account of the mechanism in *Adenium obesum*, which seems to be similar to that of *Nerium oleander*. Only long-tongued insects (Rowley suggests large bees for *Adenium*) have the proboscis length necessary to reach the nectar in the base of the flower, and access to it is provided by the slit between neighbouring stamens. In passing the stigma, pollen is deposited on its sticky surface, while on withdrawing, fresh pollen is picked up as the proboscis moves past the sticky mass of pollen lodging in the tip of the cone formed by the anthers. Cross-pollination is probably frequent, perhaps occurring regularly, as it is reported to do in *Tabernaemontana* and *Voacanga*. Motor tissue is reported in the stamen filaments in some genera, and Boiteau & Allorge (1977) make the point that *Alstonia, Catharanthus, Holarrhena, Plumeria* and *Rauvolfia* (all in the same subfamily) are self-pollinating and lack motor tissue in the filaments. Together with *Thevetia*, the stamens in these genera are free of the clavuncula, while in *Adenium, Baissea, Funtumia, Nerium* and *Strophanthus* (another subfamily) the stamens are joined to the clavuncula. *Thevetia peruviana* produces stigmatic exudate.

Fruit When the 2 carpels remain completely joined, the fruit is a syncarpous berry with 2 axile or parietal placentae and many seeds, *Thevetia* and *Allamanda*, respectively, or with 1 apparently parietal placenta (by abortion). When the fruit is dry, it consists of a pair of follicles (rarely 1) with plumed seeds. This fruit is theoretically schizocarpic, the mericarps follicular. The nature of the arilloid, present in some genera, is being investigated by Allorge (1985).

Dispersal of plumed seeds is by wind, especially of advantage in high-climbing lianes. Little is known about the dispersal of seeds of fleshy fruits, though *Carissa edulis* is eaten by baboons, *Rauvolfia* and *Tabernaemontana* by birds.

Fig. 22.2. A. *Funtumia elastica* flower bud × 6. B–D. *Thevetia peruviana*. B. Flower bud × 1. C. Floral diagram. D. Half flower × 1½. E. *Allamanda* sp. cross-section of ovary. F. *Catharanthus roseus* cross-section of ovary. G. *Funtumia elastica* tailed (sagittate) anther × 6. H. *Allamanda* sp. non-sagittate anther × 6. I–L. Plumed seeds.

In *Catharanthus*, placental flesh forms an elaiosome, and seeds are reported to be dispersed by ants.

Economic species Rubber is obtained from Lagos silk rubber (*Funtumia elastica*) and a number of subsidiary species. In general the latex of the family must be regarded as poisonous. This really means that the entire plant must be regarded as poisonous until proved otherwise. *Strophanthus* spp. yield cardiac genins, which, in the form of glycosides, are used to treat heart disease, while other species provide the starting point of the manufacture of cortisones. Alkaloids from Madagascar periwinkle are currently used in the treatment of leukaemia.

Field recognition *Alafia* (indigenous name?) is a genus of shrubs climbing to 20 m high in forest, in both tropical Africa and Madagascar. Approximately 10 of the 20 species extend into West Africa, and 2 of these *A. barteri* and *A. scandens* are widespread. The stems show interpetiolar ridges, and the leaves are dark green above, paler below, and, when dried, have the wrinkled or varicose appearance common to the leaves of so many members of the family. There are terminal scented cymes of white flowers, in which the corolla lobes overlap to the right. The anthers are included within the tube, which narrows at the mouth, and the base of each anther is sterile. There is no disc. The follicles are long, and slim, each seed with a terminal tuft of hairs. *A. barteri* has loose inflorescences of flowers with orbicular corolla lobes with hairy edges; in the bud the top part formed by the lobes is shorter and wider than the tube. *A. scandens* (also in Gambia), a species of lowland forest inselbergs, has short cymes of flowers with narrow white corolla lobes and a purple tube, the flower bud being entirely narrow.

Baissea Climbing shrubs of forest up to about 10 m high, occurring in both tropical Africa and Asia. The *c.* 10 West African species seem to be virtually confined to West Africa, and only *B. multiflora* occurs in Gambia. *B. leonensis* is, however, more widespread; it has terminal and axillary cymes of white flowers in which the tubes are short and wide, the corolla lobes long and slender and overlapping to the right. The anthers are included in the tube, each anther sterile at the base with a straight acute tail. A disc is present. The follicles are up to 50 cm long and contain seeds each with an apical tuft of hairs. The leaves are small, ovate and elliptical, and there are no interpetiolar ridges.

Clitandra (Gr. enclosed stamens). The only species is *C. cymulosa*, a high-climbing liane up to 90 m long, with branch tendrils and axillary clusters of very small flowers with pinkish tubes. The stems have clear interpetiolar ridges and the leaves have a marginal nerve well within the margin. The anthers are scarcely a third as long as the corolla tube and are inserted in its upper part. The fruit is a syncarpous berry.

Funtumia (vernacular name). Both species of this tropical African genus occur in West Africa, *F. elastica* (Lagos silk rubber) yielding excellent rubber while the product of *F. africana* is of poor quality. Both species are trees 10–30 m high, with ± elliptical leaves with slightly undulate and thickened margins, but no intramarginal nerve. Interpetiolar ridges are prominent. In *F. elastica* leaves there is a pit at the angle of each lateral nerve with the midrib, in *F. africana* at most a tuft of hairs. In addition the latex of *F. elastica* can be rolled into a ball, while that of *F. africana* cannot. Axillary clusters of white flowers, the corolla lobes overlapping to the right, are produced in the dry season and again in the early rains, each flowers with anthers tailed at the base and included in the corolla tube. The fruit is a pair of thick divergent follicles containing seeds with a plumed basal stalk pointing towards the base of the follicle.

Holarrhena (Gr. complete male). *H. floribunda*, the only species, is known as false rubber tree, and is found in drier forest areas northwards through forest outliers and closed savanna woodland into fringing forest and on inselbergs in northern guinea savanna, and in some situations in the sudan zone, particularly in savanna secondary growth throughout. It is a shrub or smaller tree to 15 m high, with small shining lanceolate leaves up to 15 cm long, produced in flushes in the dry season along with erect axillary cymes of white flowers. Each flower is scented, has corolla lobes overlapping to the right, the anthers, which are fertile to the base, included in the tube. The fruit is a pair of slim, long-lasting follicles up to 60 cm long, the seeds with an apical tuft of hairs. The species exists in 2 varieties, var. *tomentella* with leaves densely pubescent at least when young in Gambia, the extreme western forest-savanna mosaic and the sudan zone, var. *floribunda* with leaves ± glabrous even when young to the south.

Landolphia (Landolph, eighteenth-century French commander of the expedition to West Africa). This is a largely African genus of climbers with terminal branched tendrils (Cremers, 1973), which sometimes harden into hooks like those of *Strychnos*. There are dense cymes of small flowers of various colours, though often white, in which the corolla lobes overlap to the left and the tube is constricted at the throat, above the insertion of the anthers, by a thickening of the wall. The fruit is a syncarpous berry of the *Allamanda*-type. Interpetiolar ridges are present on the stems. Of the widespread species in West Africa, *L. calabarica* and *L. owariensis* (white rubber vine) have only terminal inflorescences. *L. calabarica* has rusty hairs on all young growth, the leaves, however, becoming glabrous except on the backs of the veins, these looping prominently to form an intramarginal nerve. The flowers are white with a tube 11–17 mm long and lobes 7–22 mm long, making the 'face' of the flower maximally 44 mm across. The tube is glabrous in the throat, but the lobes are hairy on the tips at the back. *L. owariensis* is a climber up to 90 m in forest or forest outliers, or a small tree or shrub in savanna. It has purplish-brown branches and scented flowers, which change colour from yellow through orange to brown with age. The corolla tube is only 4–8 mm long and the lobes are 2–9 mm long. The tube is hairy in the throat. The leaves become glabrous as they mature. The berry is brown with light spots, and about 5 cm $\emptyset$. This is the only species to extend across northern tropical Africa (as far as Tanzania). *L. hirsuta* has axillary inflorescences of yellowish flowers and its leaves are not gland-dotted beneath, while the remaining species with axillary inflorescences all have leaves gland-dotted beneath. *L. dulcis* with white, yellow or violet flowers, var. *dulcis* from Sénégal to Côte d'Ivoire and var. *barteri* from Guinée to Congo, is common in Sierra Leone. Var. *dulcis* has densely hairy young growth, var. *barteri* with slightly hairy growth.

Motandra (motile and male, the stamens) is another West African genus of 1 species, *M. guineensis*, a shrub, usually climbing, in forest. On drying, the leaves assume the crinkled varicose aspect frequent in the family. Terminal, slim inflorescences covered with rusty hairs appear in the dry season, the flower buds tapering and the flowers very small and white, the corolla lobes overlapping to the right, the anthers included in the tube and sterile at the base. The fruit is a pair of finger-thick follicles about 12 cm long, bearded with brown scaly hairs to an extraordinary degree. Under the pelt, the pericarp is ridged.

Oncinotis (Gr. the ear-like appendage between each pair of corolla lobes) A genus of Africa south of the Sahara and Madagascar, only 3 species extend into West Africa (as far as Guinea–Bissau in one case). They are climbing forest shrubs with interpetiolar ridges on the stems, and corolla lobes overlapping to the right. *O. gracilis* has young growth covered with rusty hairs, the shortly acuminate obovate leaves hairy on the veins beneath. The flower buds are very small, with slender tips contrasting with the plump tube beneath. The follicles are slim, scarcely little-finger $\emptyset$, and up to 20 cm long. *Oncinotis* differs from *Motandra* and *Baissea* in having 5 scales alternating with the corolla lobes, and from *Baissea* in having curved, blunt sterile anther tails instead of straight ones.

Saba (West African name) is a West African genus of 3 species, only *S. florida* extending widely through tropical Africa, Madagascar and even to the Comoro Is. They are tendril lianes with leaves which dry crinkled and varicose and have a thick margin, though there is no intramarginal nerve. All are robust climbers resembling *Landolphia* spp. though the corolla tube is not thickened at the throat. *S. florida* occurs in forest and fringing forest, producing terminal congested cymes of white flowers, yellow in the throat, in the dry season. The flowers are larger than those of a *Landolphia*, being up to 7 cm across the corolla lobes on a tube 1.5–3 cm long. The fruits are edible yellow berries.

Strophanthus (Gr. cord flower) – a genus of climbing shrubs of the Old World tropics, 4 of the 8 West African species being confined to parts of West Africa, the 4 widespread West African species also occuring at least as far east as Central Africa. All have corolla lobes overlapping to the right, paired appendages between each pair of corolla lobes, long-pointed sagittate anthers sterile at the base, and plumed seeds with an apical beak bearing hairs pointing towards the tip of the follicle (contrast

Funtumia). Three species have the corolla lobes distinctively produced into tails much longer than the corolla tube. These are *S. sarmentosus* (also in Gambia), a glabrous forest climber also found straggling over rocks on lowland forest inselbergs. The flowers are white, turning yellow, red inside, with petal tails 2–4 times longer than the corolla itself. The follicles are thick and woody, 20–30 cm long and lacking a beak. *S. preussi* is another glabrous climber with purple flowers turning yellow with a tube shorter than that of the preceding species, only up to 1.5–2 cm long, with tails more than 4 times longer than the corolla lobe itself. *S. hispidus* is a coarsely hairy climber with yellow flowers spotted with red or brown inside. The tube is only up to 1.5 cm long, the lobes and tails roughly as in *S. sarmentosus*. The follicles are 30–45 cm long. *S. gratus* is a species with a bell-shaped white-to-purplish corolla and no petal tails. This species can occur as a small, possibly evergreen tree, whereas the species are at most of shrub-like dimensions, when not climbing. The hairs on the seed plume (coma) are spreading, or point back towards the seed.

Tabernaemontana (J. T. Tabernaemontanus, sixteenth-century German botanist and physician). Forest shrubs and under-storey trees with inter-petiolar ridges and terminal inflorescences of large white flowers which are trumpet-shaped, fleshy and scented, with a conspicuous corolla tube, and corolla lobes overlapping to the left. The fruit is a pair of green berries joined basally (or a single berry) containing large arilloid seeds, and surrounded by the persistent calyx. This is a large tropical genus in which the subgenera present in West Africa are now recognised as genera (Boiteau & Allorge, 1981). *T.* (now *Camerunia*) *penduliflora* is a forest shrub in S. Nigeria eastwards, but only *T.* (now *Sarcopharyngia*) *crassa* and *T.* (now *Gabunia*) *glandulosa* are widely spread in West Africa. These have glandular scales inside the calyx, and also corolla lobes overlapping to the left, but unlike other West African species with such scales, the fruit is a pair of follicles. *T. crassa* is a smaller tree of wetter forest areas, and *T. glandulosa*, a glabrous forest climber, both have flowers with corolla tubes over 1.5 cm long and calyces at least 2(–10) mm long. The flowers of *T. crassa* opening the evening. Neither species is larger than a shrub in Nigeria. *T. pachysiphon* is locally common as a sapling in bush fallows in Nigeria, and has glabrous leaves and fleshy flowers with a short tube (1.5–3 cm long) but 7–8 cm Ø across the lobes.

Voacanga (Malagasy name). This is an Old World tropical genus, with 6 species extending into West Africa, only a single species entirely confined to this area. The widely distributed species are *V. thouarsii* and *V. africana* (both in Gambia). The former, flowering in the rains, resembles *Tabernaemontanus* a great deal except that the calyx falls early, leaving only a basal cup, and in both species, the corolla tube is inconscpicuous (*c.* 1 cm), hidden within the calyx at least to begin with. *V. thouarsii* is a savanna species, found in damper areas (beside streams and in swampy areas, where it can develop peumorhizae; Jenik, 1970). *V. africana*, a forest species, flowers in the dry season; both having green fruits mottled with white, containing numerous small arilloid seeds. The stems bear clear interpetiolar ridges.

Bibliography

Allorge, L. (1985). Contribution a l'étude des graines des Apocynaceae–Tabernaemontanoideae: origine de l'arille et ornamentale du tégument séminal. *Bulletin du Muséum national d'histoire naturelle, Paris*, sér. 4, 7, sect. B, *Adansonia*, pp. 433–51.

Beentje, H. J. (1982). A monograph on *Strophanthus* DC. (Apocynaceae). *Mededelingen Landbouwhogeschool Wagengingen*, **82**.

Boiteau, P. & Allorge, L. (1977). Morphologie et biologie florales des Apocynacées 1. *Adansonia*, sér. 2, **17**, pp. 305–26.

Boiteau, P. & Allorge, L. (1981). Révision des genres *Gabunia* et *Camerunia* (Apocynaceae). *Bulletin du Muséum national d'histoire naturelle, Paris*, sér. 4, 3, sect. B, *Adansonia*, pp. 23–40.

Boiteau, P., Allorge, L. & Sastre, C. (1978). Morphologie florale des Apocynacées 2. *Adansonia*, ser. 2, **18**, pp. 267–77.

Cremers, G. (1973). Architecture de quelques lianes d'Afrique tropicale. *Candollea*, **28**, pp. 249–80.

De Jong, B. H. J. (1979). A revision of the African species of *Alstonia* R. Br. (Apocynaceae). *Mededelingen Landbouwhogeschool Wageningen*, **79**(13).

Jenik, J. (1970). The pneumatophores of *Voacanga thouarsii* Roem. et Schult. (Apocyncaeae). *Bulletin de l'Institut fondamentale d'Afrique noire*, sér. A, **32**, pp. 986–94.

Rowley, G. D. (1980). The pollination mechanism of *Adenium* (Apocynaceae). *National Cactus and Succulent Journal, UK*, **35**, pp. 2–5.

23

Asclepiadaceae – blood flower or milkweed family

Shrubs and twiners, occasionally stem succulents resembling cacti superficially, e.g. *Caralluma dalzielii*, sometimes seen in gardens. Milky juice is always present, though not always copiously so (*Leptadenia*), while leaves are simple, exstipulate and whorled or opposite, with flat or curved ptyxis. Some genera are, however, leafless, and in a few others the leaves are very short lived. There are cymes of complex five-part flowers, in which each anther contains one or two pollen masses (pollinia) in each loculus, and the paired carpels are joined only distally by the styles. The fruit is a pair of follicles (or a single one by abortion), containing plumed seeds.

In West Africa, the family is well represented in savanna and sahel, but only a few species are common in forest. Many of our species occur also in similar habitats elsewhere in tropical Africa, but within West Africa most of the species have very local distribution.

Only sodom apple (*Calotropis procera*) and blood flower (*Asclepias curassavica*) seem to have been introduced successfully into West Africa, blood flower, generally seen as a small shrub, being common in gardens and very suitable for the investigation of the flowers.

Leptadenia and *Sarcostemma* are ecologically important genera on the sahel zone and driest savannas, possessing also distinctive habits. *Leptadenia* consists of two species of twining shrubs and an erect one. This last is *L. pyrotechnica*, which produces tiny leaves, a few at a time, on its new, wiry stems. These leaves soon fall, and the shrub is generally seen in the leafless condition, with rather thick, juicy branches. Of the two twining species, *L. hastata* (also in Gambia) occurs as far as the coast in suitable situations (not Sierra Leone). The leaves are persistent, petiolate and not in the least hastate (or even cordate) as the name might suggest. Short, axillary umbel-like cymes of small, bell-shaped, greenish-yellow flowers are produced, mainly in the dry season. They are followed by finger-shaped follicles, about 10 cm long. There is sticky juice in this genus, rather than milky sap.

In Ghana and Nigeria, *Sarcostemma viminale* is a scrambling or trailing shrub in similar situations to *Leptadenia* spp. but it is always leafless, with fleshy stems hairy at the nodes. The flowers are yellowish or greenish, fragrant, and followed by slender follicles.

Flowers $\doteq \oplus \, \male$ 5-part K(5), very short or absent. C(5), the lobes overlapping in various ways, or even joined at the tips (*Ceropegia*). Stamens and pistil are united distally to form a gynostegium, the stamens also being joined throughout their length, forming a short staminal tube below. In addition, the anthers are joined to the 5-sided stigmatic disc forming the tip of the 2 joined styles. These are free below, each arising from 1 of the 2 free carpels.

Fig. 23.1. A–E. *Calotropis procera*. A. Flowering shoot × $\frac{3}{8}$. B. Flower × $2\frac{1}{4}$, one petal removed. C. Gynostegium from above, with corona, × 3. D. Gynostegium in cross-section above the ovaries × 3. E. Gynostegium and ovaries in cross-section × 3. F–G. *Asclepias curassavica*. F. Floral diagram. G. Seed × 2. A–E from Burger, 1967, Fig. 56 *pro parte*. F–G from Rendle, 1938, Fig. 223C and G.)

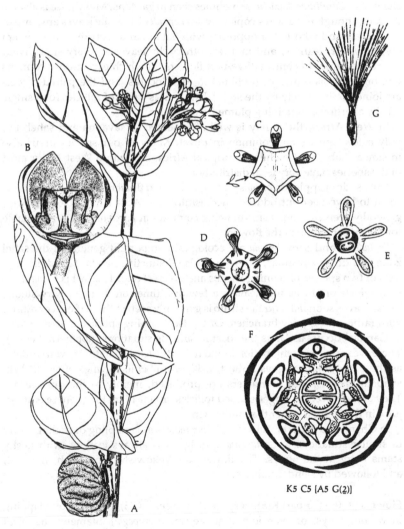

K5 C5 [A5 G(2)]

Within the gynostegium, each anther loculus forms a pocket, in which 1 or 2 pollen masses is situated. The pollinia (El-Gazzar *et al.*, 1974) from neighbouring loculi (of different stamens) are joined in pairs by a bridge-like connection consisting of a small projection from each pollinium (a caudicle), joined to a common pollen carrier developing in front of a stigma. Petaloid outgrowths from the backs of the stamens form a corona of lobes alternating with the petals. The form of the corona varies considerably and is often complex, and coronal characters are frequently used to separate genera. In *Asclepias curassavica*, the gynostegium and surrounding corona are clearly revealed, since the calyx and corolla are reflexed. The corona is yellow, of 5 'horns', each surrounded by a petal-like lobe. If such a coronal segment is pulled down, a needle can be inserted between the 2 stamens revealed, and an upward movement will draw out the 2 pollinia connected by their pollen carrier. An upward movement is necessary in this genus because the loculi are vertical, with pendulous pollinia. In nearly half the genera, the loculi are horizontal or even upward pointing, so the pollinia must be removed by an outward or downward movement as appropriate.

Pollination is extremely specialised, at most only a few species of insect being concerned with each plant species. Blood flower in its native area (tropical America) is pollinated by butterflies of the genus *Danaus*, but in general, asclepiads are probably fly pollinated, the insects being deceived into visiting them by odour, which promises decaying organic matter. *Ceropegia* spp. have trap flowers. In searching for food (nectar, generally secreted in the corona) the insect inserts a leg in the slit between two anthers, and, in withdrawing it, disturbs the pollen carrier, departing with a pair of pollinia attached. On reaching the next flower, pollination occurs.

Fruit One or 2 follicles, the styles having withered and freed them. The seeds bear a tuft of silky hairs round the micropyle.

Dispersal of plumed seeds is by wind.

Economic species are few. All extracts are poisonous, and, for example, fish poisons are frequently obtained. Sources of drugs may exist also. The seed plumes can be used as kapok, and sodom apple is cultivated in America for this purpose.

Field recognition *Calotropis* (Gr. beautiful keel). *C. procera*, from Asia, is the only species in West Africa (also in Gambia). It is a shrub or small tree with thick soft branches and cymes of white and purple flowers, in which the corona lobes are spurred at the base. The pollen masses are pendulous. The follicles are green and inflated, and generally appear in pairs. The leaves are large, thick and sessile, oblong-elliptic with a cordate base, and whitely felted underneath. On older specimens the pale, fissured softly corky bark is distinctive. The species is often cultivated or escaped from cultivation, and frequently appears in disturbed ground.

Gongronema (Gr. outgrowth and thread (at the base of the stamen)). Three species of climbing or erect shrubs, of which G. *latifolium*, a glabrous climber of forest and forest-savanna mosaic, is by far the most widespread. It has hollow stems bearing broadly ovate-elliptic leaves widely cordate at the base. From the late dry season onwards it produces minute greenish-yellow flowers in stalked panicles with raceme-like branches. The corolla is shortly tubular with almost erect lobes; the corona arises from the staminal column (see *Asclepias*) and has simple lobes without appendages. The follicles are slim, 7–8 cm long and papery, not woody.

Oxystelma (Gr. pointed corona). *O. bornouense* (also in Gambia), our only species is a wiry stemmed glabrous climber, with distinctively small (oblong), lanceolate pale leaves 2–8 cm long. The inflorescences are few-flowered axillary cymes appearing in the dry season and early rains; the flowers are relatively large, 3 cm ∅ with white with red centres, the corona of 5 narrow erect lobes, from the staminal column, and the pollen masses pendulous. Follicles ellipsoid, 5 cm long.

Pergularia (L. trellis). Of the 2 West African species, only *P. daemia* (also in Gambia), a stiffly hairy twiner or trailer, is widespread. The stems are actually scabrid to the touch, with deeply cordate, long-petiolate leaves, and axillary corymb-like inflorescences of greenish-white flowers, each up to 1 cm ∅. The follicles are thick at the base, tapering sharply to a somewhat prickly tip, and containing flat seeds toothed round the lower margin. The corona of this species is particularly complex, being double, the inner corona with both horns and spurs. This species probably occurs mainly to the south of the sudan zone, while the other species, *P. tomentosa*, a softly hairy scrambler, occurs throughout the sudan zone.

Secamone (Arabic name). *S. afzelii*, a widespread climbing and scrambling shrub of secondary forest, is the only common species (out of 4) of a part of the family with 2 pollinia per anther loculus, each with a scarcely developed caudicle, and with pale non-horny pollen carriers. *S. afzelii* has small, abruptly acuminate oblong leaves, and large, forking, pale-green glabrous inflorescences of tiny reddish-yellow flowers, followed by slender follicles up to 10 cm long.

Telosma (meaning uncertain). *T. africanum*, the only West African species, is a twining shrub in the wetter parts of deciduous forest and by rivers, with oblong-ovate leaves up to 10 cm long. Umbel-like cymes of small yellow flowers arise on stalks outside the axils of the leaves. Each flower has linear corolla lobes obviously contorted in the bud, later spreading widely open to reveal a densely hairy throat, this tubular portion inflated at the base. The corona lobes are expanded, with an outgrowth on the inner face of each.

Tylophora (Gr. outgrowth-bearing). There are 9 species, making this the second largest genus in West Africa, and *T. sylvataica* is the most widely distributed, being a herbaceous twiner in thickets and young secondary growth, with petiolate narrowly heart-shaped leaves up to 16 cm long. Flowers are minute, red to purple-brown in tiny lateral clusters. The corona may appear to be absent, but occurs as 5 tiny outgrowths on the staminal column. The slim curved follicles usually appear singly.

Xysmalobium (Gr. scale and small lobe). *X. heudelotianum*, the only West African species (also in Gambia), is an erect herb in savanna with narrow, shortly petiolate leaves rounded at the base. In the late dry season to early rains, there are umbel-like inflorescences of greenish-yellow flowers on extra-axillary peduncles, the corolla lobes spreading or reflexed and the flowers *c.* 1 cm ∅. The corona lobes are stalked, fleshy and pointed, and there are small lobes on top of the gynostegium. The follicles are slim, pointed and erect.

Bibliography

Bally, P. (1973). Notes on the flora of tropical East Africa. *Bulletin of the African succulent plant society,* **8,** pp. 157–60.

Bayer, M. B. (1978). Pollination in Asclepiads. *Veld & Flora,* **64,** pp. 21–3.

Burger, W. C. (1967). *Families of flowering plants in Ethiopia.* Stillwater, Oklahoma: Oklahoma State University Press.

El-Gazzar, A., Hamza, M. K. & Badawi, A. A. (1974). Pollen morphology and taxonomy of the Asclepiadaceae. *Pollen and Spores,* **16,** pp. 227–38.

Reese, G. (1973). The structure of the highly specialized carrion-flowers of the Stapeliads. *Cactus and Succulent Journal of the Cactus and Succulent Society of America,* **45,** pp. 18–29.

Rendle, W. C. (1938). *The classification of flowering plants,* vol. 2, 1st edn. Cambridge: Cambridge University Press.

Rowley, G. D. (1978). *Illustrated encyclopedia of succulents.* London: Salamander books.

Wyatt, R. (1980). The impact of nectar-robbing ants on the pollination systems of *Asclepias curassavica. Bulletin of the Torrey Botanical Club,* **107,** pp. 24–8.

24

Rubiaceae – abura family

One of the largest families, cosmopolitan, and the next largest West African family, in terms of species, after the grasses. Shrubs and under-storey trees in forest are particularly important.

Members of the family may be recognised by their opposite or whorled, simple, entire leaves accompanied by interpetiolar stipules or ridges. Ptyxis conduplicate as a rule, revolute in *Rothmannia* sp. The inflorescence is a cyme, sometimes head like. Flowers are ⊕, mostly four- to five-part, always gamopetalous, with an equal number of epipetalous stamens alternating with the corolla lobes, and with a two-celled inferior ovary.

Rather few species have been introduced. *Cinchona* spp., yielding quinine, and the garden ornamental *Gardenia jasminoides*, the most usually cultivated *Gardenia* sp. of temperate climates (originally from China and Japan), are sometimes seen. The red and white *Ixoras, I. coccinea* and *I. parviflora*, respectively, and, in Sierra Leone and Ghana, *Pentas lanceolata* (from East Africa), are more common. Many indigenous West African species deserve a place in the garden.

The tribe Naucleae (*Adina, Mitragyna, Nauclea* and *Uncaria*) showing similarities to the Combretaceae, is sometimes separated as a family, the Naucleaceae. Verdcourt (1958) proposed an entirely different intrafamilial classification to the one employed in the *Flora of West tropical Africa*.

Very few West African plants have obviously unique features. One is the genus *Uncaria*, climbers in forest with sensitive hooks formed from inflorescence branches. *Rutidea membranacea*, another forest climber, has recurved, but not hooked, axillary spines. *Attractogyne bracteata* (Hallé, 1967), also a forest climber (Côte d'Ivoire–?–S. Nigeria) has both sensitive hooks (formed from twining lateral branches) and stiff petiolar spines.

With few exceptions, the herbs in the family belong to only three tribes (Table 24.1), and in one of these, Spermacoceae, together with *Oldenlandia* (Hedyotideae), the interpetiolar stipules unite with the petiole on each side. In the rest of the Hedyotideae, the stipules are fringe like, the segments tipped with wax-producing glands (colleters).

Another form of stipular development is seen in some members of *Gardenia* and *Craterispermum*. The interpetiolar stipules unite round the stem, across the leaf axils, to form a supra-axillary sheath. This protects the terminal bud until growth is renewed. The sheath may persist (*Gardenia, Oxyanthus*), or fall early, leaving an encircling scar. *Mitragyna* has prominent appressed stipules, forming a stipular hood over the terminal bud.

Spines are consistently produced in rather few species. Apart from *Rutidea*, *Xeromphis nilotica* (now *Catunaregam nilotica*; Tirvengadum, 1979) a shrub of drier woodland savanna in N. Nigeria and eastwards, and *Dictyandra arborescens*, a climbing shrub in forest, have supra-axillary spines, while *Didymosalpinx abbeokutae* is an erect forest shrub with short blunt thorns on the oldest stems.

Bracts and bracteoles can also be distinctive. In *Polysphaeria* and *Tricalysia*, the bracteolar cup below the flower resembles a second calyx, while in *Sabicea* (now split into four genera; Hallé, 1963), the bracteoles form an involucre round the head-like inflorescences. In *Coffea*, except for two deciduous species, the bracteoles are leafy, or cupular with leafy lobes. In *Cuviera*, bracts, bracteoles, and calyx lobes resemble each other, being mostly 1–4 cm long, while

Table 24.1. *Some genera of herbs and undershrubs (Rubiaceae)*

Genus	Habitat	Savanna/grass	Sandy soils/seashore	Forest	Rocks	Moist places/near water	Weeds	Upland	Tribe
Borreria[a]		+	+	+	+	+	+	+	S
Diodia		+	+				+		S
Fadogia		+						+	V
Kohautia		+					+		H
Mitracarpus							+		S
Oldenlandia		+	+	+	+	+	+		H
Otomeria		+	+			+		+	H
Pentas								+	H
Pentodon						+			H
Richardia							+		S
Thecorchus						+			H
Virectaria			+	+	+	+		+	H
Cephaelis[b]			+						P
Geophila[c]			+						P
Lasianthus			+						P
Parapentas			+						H
Trichostachys			+						P

S, Spermacoceae; V, Vanguerieae; H, Hedyotideae; P, Psychotrieae.
[a] Sometimes included in *Spermacoce*.
[b] Also includes some shrub species.
[c] Four of the seven species may be placed in *Hymenocoleus*.

the bracteoles subtending the spike *Hymenodictyon floribundum* are red and leaf like, on long stalks.

A number of species exist in symbiotic relationship with other organisms. Species with domatia possibly housing ants are found in the genera *Canthium*, *Cuviera*, *Gardenia*, *Heinsia*, *Nauclea*, *Psychotria*, *Randia*, *Rothmannia*, *Uncaria* and *Vangueriopsis*. The domatia are in the stem in all cases except *Gardenia imperialis* where they are at the base of the leaves. The relationship has not been established in every case in the field in West Africa, and an association of *Pauridiantha* spp. with mites remains to be observed.

Bacterial nodules, seen as dark streaks or spots in the leaf, exist in many species of the genera *Pavetta*, *Psychotria* and *Tricalysia*, of which spp. 22–5 in *Flora of West tropical Africa*, have been transferred to *Sericanthe* (Robbrecht, 1978). Mineral concretions (cystoliths) are seen, by contrast, as light spots, e.g. in *Craterispermum*.

Aluminium accumulation, with no apparent harmful effects on the plant, and possibly an adaption to the aluminium-rich soils of West Africa, occurs in *Craterispermum* (with markedly light green-to-yellow leaves), *Geophila*, *Lasianthus*, *Morinda*, *Pentanisia* and *Psychotria*.

Flowers ⚥ ⊕ ♂, (2)–4–5–(12)-part. (K) sometimes truncate (forming a cup), but generally lobed, though the lobes sometimes unequally developed (1 lobe enlarged, petaloid, red or white in *Mussaenda*, except *M. elegans*); the lobes attached outside the calyx tube in *Pseudogardenia*, or leafy (*Heinsia* etc.), the calyx sometimes persistent on the fruit; united calyces/receptacles are found only in *Morinda* and *Nauclea*. Colleters are frequently present on the edges of the calyx lobes. (C) basally tubular, then distally either funnel shaped with spreading lobes, or the lobes ('limbs') at right angles to, or reflexed upon, the tube (e.g. *Ixora* and *Canthium*, respectively); lobes imbricate in *Adina*, *Nauclea* and *Uncaria* (1 lobe imbricate in *Heinsia crinita*), otherwise valvate or contorted; tailed lobes (*Corynanthe*, *Globulistylis*, *Hutchinsonia* and *Pausinystalia*) are rare. In 4-part flowers, the sepals are orthogonal, the petals diagonal. Stamens epipetalous, equal in number to, and alternating with, the corolla lobes; in *Dictyandra* the anthers are divided internally into chambers; the pollen is generally granular, but in tetrads in *Gardenia* and *Oxyanthus*, and a few other genera of the *Gardenia* tribe, in large masses only in *Massularia*. There is usually a large disc, round the base of the style, and it may be lobed, e.g. 2-lobed in *Virectaria*. G($\overline{2}$) (occasionally ($\overline{3}$)–($\overline{9}$)), but superior only in *Gaertnera*, which used to be placed in the Loganiaceae; predominantly 2-celled, with numerous ovules on axile placentae, *or* with 1 ovule per cell on a basal or apical placenta, but 3-celled in *Richardia*, and 1-celled with 2–9 parietal placentae in most members of the *Gardenia* tribe; 1-celled by abortion fairly often in some *Canthium* spp.; the stigma is capitate or 2-lobed, 3-lobed in *Richardia*, 5-lobed in *Sabicea*.

Fig. 24.1. A–B. *Morinda lucida* leaf-opposed inflorescence × ½. B. Fruits × ½. C–D. Diagrams of stipule arrangement. C. Intrapetiolar. D. Interpetiolar. E–G. *Macrosphyra longistyla*. E. Flower × ¾. F. Floral diagram. G. Fruit × ½. H. *Mussaenda erythrophylla* diagram of cross-section of ovary. I. *Kohautia grandiflora* fruit × 6, with persistent sepals. J. Diagrammatic vertical section through a two-celled ovary with a single ovule in each cell. If basal and apical placentation are included, also the occurrence of ovules in pairs, then this type of ovary is seen in more than half the tribes making up the family in West Africa.

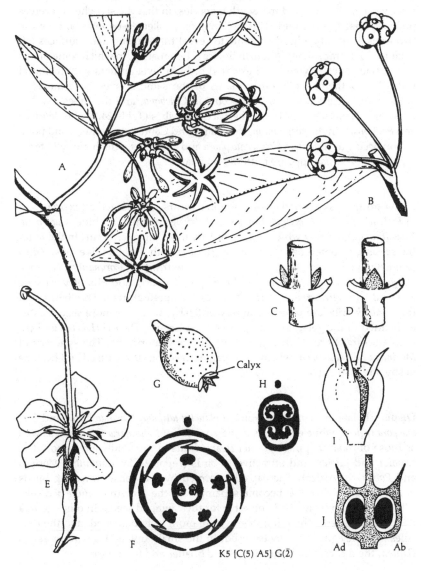

K5 [C(5) A5] G($\bar{2}$)

Pollination is by long-tongued insects in long-tubed flowers, but some flowers are so small, even though gamopetalous, that other insects must have access. Where the flower is white, robustly constructed, opens at night and is strong smelling (*Gardenia ternifolia*), bat pollination is likely. This also probably applies to the sickly scented corymb-like cymes of *Crossopteryx*. Cross-pollination is promoted in various ways, by protandry, by separation of style and stigma (anthers included, style exserted in all flowers, e.g. *Hymenodictyon* or anthers exserted, style included) and by self-incompatibility. A special form of protandry occurs in *Ixora* and many other species, in that the style head receives pollen released in the bud. This pollen is first collected by insects, then the two arms of the style head separate to reveal the two stigmatic surfaces (cf. Chapter 25, Compositae). A fourth device is heterostyly, in which some plants of a species have flowers with exserted or high level anthers with included styles below them, while other plants of the same species have included anthers and exserted styles. This occurs in *Cinchona*, and the tribes Psychotrieae and Naucleae, and in the 'herbaceous' tribes (Table 24.1) plus *Morinda* and *Pentanisia*. In *Anthospermum*, a small upland Cameroun genus, wind pollination occurs as it does in *Coffea canephora* and *C. liberica*, which are self-sterile diploids. *Mussaenda* spp. are butterfly pollinated.

Fruits are always crowned by the calyx or its scars, an enlarged calyx occurring in *Psilanthus* and *Thecorchus*. Berries and drupes, sometimes blue (*Gaertnera*, *Lasianthus*), but rarely white, and 2- and 4-valved capsules occur. In the latter the seeds are sometimes winged, i.e. *Uncaria* with 4 ribbon-like tails, *Adina* (seeds winged at each end), *Crossopteryx* (wing fringed), *Corynanthe*, *Hymenodictyon* and *Pausinystalia* (wing oval but split at 1 end), and *Mitrangyna* (wing at 1 end). In *Virectaria*, 1 of the 2 valves falls prematurely. The 'berries' of the *Gardenia* tribe are woody/leathery or fleshy, with 1 or more seed masses embedded in the gelatinous pulp from the placenta. Dicocci (*Diodia*) and tricocci (*Richardia*) are schizocarpic fruits of 2 or 3 mericarps. The syncarps of *Morinda* and *Nauclea*, in which the enlarged receptacles are fused and become fleshy, are distinctive.

Dispersal Baboons consume fruits of *Nauclea latifolia*, *Morinda lucida* and *Cremaspora triflora*, while elephants disperse those of *Massularia acuminata* as well as those of *Nauclea* spp. The other fleshy fruits in the family are a good deal smaller and juicier, and attractive to birds (reported for *Canthium*, *Morinda* and *Pavetta* in Africa). The capsules formed by the dense flowering heads of *Mitragyna* and *Uncaria* become separated by the growth of their pedicels. Wind dispersal is thus made possible for their winged seeds. In other genera with winged seeds, the inflorescences are more laxly branched, and the capsules are consequently fully exposed. In the small, upland Cameroun genus *Galium*, the dicocci are clothed in hooked bristles and form burrs.

Economic species Timber is the most important export, principally obtained from abura (*Mitragyna ciliata* and *M. stipulosa*) and opepe (*Nauclea* spp.). Part of the world's coffee production stems from West Africa, and this is mainly 'robusta' coffee from *Coffea canephora*, used for the manufacture of 'instant' coffee and in blends.

Field recognition *Gardenia* (A. Garden, eighteenth-century American naturalist and correspondent of Linnaeus). Mostly twisted savanna shrubs or small trees, all with persistent supra-axillary sheaths, and foliage which is sticky at least when young. Some spines may be carried. The flowers are large, fragrant and white, fading to yellow in 24 hours, 5–11-part, with a long tube widening into a funnel-shaped mouth, the contorted corolla lobes round its edge. The anthers are sessile, attached in the funnel by a short length of connective and not exserted. The style, by contrast, is exserted, with a massive stigma. The fruit is a kind of berry, only moderately fleshy and 1–(or partially 2)-celled, with 2–9 parietal placentae. The seeds cohere in a mass in the placental pulp. *G. imperialis* is a swamp and riverain forest tree with large, broadly ovate leaves up to 80 cm long × 20 cm wide. There are a pair of pouches at the base of each leaf blade, where ants keep and feed on scale insects. The flowers, in pairs, are 15 cm or more long, 5(–9)-part, with 2–3 placentae. The fruits are ellipsoidal, up to 5 cm long, with a 2-layered woody wall. The other widespread forest species is *G. nitida*, a shrub or small tree to no more than 4 m high, with solitary flowers. The corolla tube is very slender and only *c.* 10 cm long, but the corolla lobes are a third to half as long as the tube. The calyx is tubular with *c.* 6 leafy lobes, persisting and enlarging on the woody spindle-shaped fruit, which has 4 placentae. The 2 most widespread savanna species are *G. erubescens* (also in Gambia) and *G. ternifolia*, both with solitary terminal flowers, the corolla tube under 10 cm long. In the former, the leaves change colour on drying, becoming reddish above and glaucous green below. They are carried on short shoots, with the flowers, which are usually 6-part, the calyx with 6 needle-like lobes, somewhat hairy, with a very hairy receptacle below them. This part swells to an ovoid, somewhat fleshy, edible berry. In *G. ternifolia* (now including both *G. lutea* and *G. triacantha*; Verdcourt 1979–80), the leaves dry brownish-green, and the calyx has 6–7 leafy lobes on a glabrous swollen receptacle. This swells further, to form an ellipsoid, persistent woody fruit. *G. sokotensis*, a shrub species characteristic of rocky hills in the sudan zone, has relatively very small 5-part flowers in pairs, and fruits only 1 cm ⌀, ellipsoid and thin-shelled.

Macrosphyra (Gr. long hammer, the style and stigma). This is a small tropical African genus with only 2 species in West Africa. The more widespread of these is *M. longistyla* (also in Gambia), a climbing shrub of secondary forest and forest inselbergs, with wrinkled, hairy, light-green leaves and hairy, scented, yellow and white contorted flowers in terminal clusters. The fruits are green 1-celled berries with 2 parietal placentae, about 5 cm ∅. The flowers are scented, with a corolla tube at least 2 cm long, and the anthers and stigmas are exserted. This species extends right across Africa to Sudan and Uganda.

Mitragyna (Gr. cap and female, the mitre-like shape of the stigma), is a small Old World genus with only 3 species, all confined to wet places, in West Africa. The large appressed stipules, axillary panicles of pedunculate heads and capsule are distinctive. The 5-part white flowers are attached to the globular top of the peduncle, together with their minute bracts and bracteoles; the corolla lobes are, unusually for this tribe (Naucleae), valvate. The ovary is 2-celled, with axile placentation, the many seeds winged at 1 end. The capsule splits into both outer and inner valves, the former remaining connected to the persistent calyx on top. At this stage the pedicels have lengthened and the capsules are well separated. *M. inermis* (also in Gambia) grows (sympodially) in swampy savanna. It has thin small leaves (up to 13 cm long × 8 cm wide) and calyptrate stipules, which fall early. The flowers are glabrous, with a truncate persistent calyx and exserted pendulous anthers. Another member of the same tribe seen near the same habitat is *Adina microcephala* (now *Breonadia salicina*; Hepper & Wood, 1982), a rather larger tree up to 15 m high, not a shrub. It has narrower whorled leaves in tufts at the ends of the branches. Pedunculate solitary yellow heads, each with a pair of bracteoles, followed by heads of small capsules. *M. ciliata* and *M. stipulosa* (both now *Hallea* spp.; Leroy, 1975) and known as abura, occur in swamp and riverain forest, the former nearer the coast, the latter inland and in forest outliers. Both grow monopodially and have much larger leaves than *M. inermis*, at least 15 cm × 8 cm, with larger and longer persistent only oppressed stipules. *M. ciliata* (now *H. Ledermannii*; Verdcourt, 1985) has a ciliate calyx (hairy along the edges of the lobes), while *M. stipulosa* has a truncate glabrous calyx. This feature can be seen on the capsule. The anthers are included, but the stigma exserted.

Morelia (J. M. Morel, eighteenth-century French botanist). There is a single species, *M. senegalensis*, which grows near streams, mainly in savanna. It is an evergreen shrub or tree up to 12 m, with sticky yellow latex. Numerous short cymes of scented white 5-part flowers appear in the dry season, and again late in the rains, each cyme situated just above the interpetiolar ridge. The corolla tube is short, the lobes contorted and the anthers exserted between them. The berry is small and 4-lobed, as a rule, with 4 cells and 4 placentae, with 2–4 seeds on each. The genus is West African, extending to Sudan and Congo.

Nauclea (Gr. small ship, of the 2-valved fruits of former spp.) (opepe). The 4 West African species of this small, Old World tropical genus, also found in other parts of Africa, have terminal spherical head-like cymes of small whitish flowers, like those of *Adina* and *Mitragyna*, in the same tribe. But in *Nauclea*, the flowers are joined by their calyces, and the fruit is a syncarp. *N.* (now *Sarcocephalus*; Ridsdale, 1975) *latifolia* (also in Gambia) is a savanna shrub, known as Sierra Leone or Guinea peach, with juicy red fruits. The other 3 species are forest trees, of which N. *vanderguchtii*, in riverain forest (Nigeria–Congo), provides domatia for ants in its branches. N. (now *Sarcocephalus*; Ridsdale, 1975) *pobeguinii* is more widespread, and can occur in the same places as *Mitragyna ciliata* and M. *stipulosa*. It has, however, small triangular stipules and large, soft, yellow edible fruits. N. *diderrichii* is at least an upper-storey forest tree, with leaf-like stipules up to 5 cm long and small pitted reddish-yellow syncarps.

Oxyanthus (Gr. pointed flower). The flower buds of this genus are particularly long and slender, only 1–2 mm wide across the tube when flattened. All the West African species are forest shrubs or small trees, with persistent triangular sheathing stipules and leaves which may be asymmetrical. The flowers are in axillary cymes, on 1 side of the stem only. They are scented and white, with well-exserted anthers and style head. The fruit is a berry with 2 parietal placentae, the seeds in a fleshy mass. *O. unilocularis* is the only fringing forest species. This is a shrub (or small tree) with very large asymmetrical leaves with revolute ptyxis on thick branchlets, which may be hollow. It has large cymes of flowers with the corolla tube 13–15.5 cm long × 1–1.5 mm wide, the lobes c. 2 cm long. The fruit is ± spherical, up to 2.5 cm ∅. Further south, 3 other species with asymmetrical leaves are met. These are *O. formosus*, *O. racemosus* and *O. speciosus*. Of these, *O. racemosus* is a rather hairy plant, like *O. unilocularis* but with smaller leaves, only 9–20 cm long × 2–8 cm wide, flowers with a shorter thicker tube (11–14 cm long × 2–3 mm wide) in few-flowered cymes, and yellow spherical fruits 1.5 cm ∅. *O. speciosus*, a chewstick plant, is glabrous, with dense, short cymes with persistent bracts and bracteoles. The flowers are short-tubed, tube 4–4.5 cm long, the fruits stalked, ellipsoidal, shining black and up to 3 cm long × 2 cm wide. *O. formosus* is another glabrous species, also with asymmetrical leaves but with long-tubed flowers, the tube 2 mm wide, in cymes with persistent bracts and bracteoles. The fruits are like those of *O. speciosus* but reddish-yellow in colour.

Psychotria (Gr. life support, from the coffee-like extract prepared from some species). This is a large genus in the warmer parts of the world, but the c. 70 West African species are largely confined to this area, and are often of restricted distribution within it. Most species are small, erect shrubs, sometimes herbs, but rarely climbers or trees. Terminal cymes, without involucres, of white, heterostylous, valvate flowers are produced, each with a 2-celled ovary with an erect ovule in each cell. The fruit is a generally red drupe, with 2 stones. *P. vogeliana* occurs in southern woodland savanna or fringing forest further north (Guinée–Mali to Gabon–Uganda). It has leathery leaves lacking both domatia and nodules, but hairy on the midrib beneath. Young growth is covered with rusty hairs, the stipules 2-lobed and falling early. The inflorescence is a short panicle, made up of head-like cymes, with conspicuous and persistent bracts and bracteoles. The corolla lobes have thickened tips. The fruit is white, the

2 pyrenes grooved on their outer, curved faces. *P. psychotrioides* is widespread by streams in savanna (Sénégal–Sudan). It is not reported to have either domatia or nodules. The flowers are clustered, sometimes without a peduncle, followed by clusters of quite large (1 cm long) ribbed red drupes. The bracts and bracteoles are no more than ridges. The 2 most widespread forest species are *P. sciadephora* and *P. salva*, the latter beside streams. It is distinguished from the other species mentioned by its black

Fig. 24.2. *Oxyanthus speciosus*. A. Flowering shoot × $\frac{3}{4}$. B. Stipules × 2. C. Young inflorescence × 2. D. Tip of young corolla × 2. E. Flower with top half of corolla tube × 2. F. Fruit × 1. (From Hallé, 1970, Pl. 45 *pro parte*.)

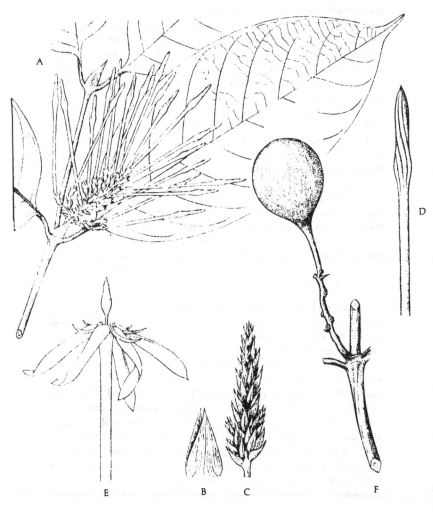

linear leaf nodules alongside the midrib beneath. The drupes are tiny, ribbed and red in short cymes, following the tiny flowers. There are bract and bracteolar ridges only. *P. sciadephora* (now *Chalaziella sciadophora*; Verdcourt, 1977) has stalked ± umbellate cymes of pink flowers, with bract and bracteolar ridges, followed by shining red drupes. The foliage blackens on drying.

Rothmannia (G. Rothmann, traveller in Africa). Forest shrubs and trees in tropical Africa, the 8 West African species extending to other areas to the east and south. The leaves are characteristically carried in 3s, the pair expected plus the first leaf of an axillary shoot. The flowers are scented, large, erect and long tubed, widening apically to a funnel, with the 5 lobes (7–8 only in *R. octomera*) overlapping to the left except in 3 species. The flowers are usually solitary and terminal, i.e. never 2 flowering shoots at 1 node, with a prominent 2-lobed style head at the mouth of the flower. The anthers are similar to those of *Gardenia*. The fruit is a hard berry, at least partially 2-celled, with the numerous seeds in a mass of pulpa on axile placentae. The genus is easily confused with *Gardenia*, but the stipules are quite different, inconspicuous and falling early. The 4 species which are widespread in West Africa tend also to be widespread elsewhere in tropical Africa. *R. megalostigma* occurs nearest the coast, in swamp forest. It has large obovate leaves (up to 35 cm long × 22 cm wide), a 5-toothed ribbed calyx and flowers up to *C.* 30 cm long, brownly hairy outside, white with pink spots inside. The fruit is ellipsoid, up to 12 cm long. *R. whitfieldii* is another forest species, but extends northwards into forest outliers. It has rather smaller, narrower leaves (up to 30 cm long × 11 cm wide) and pendulous flowers, the tube half as long as that of the above species. The fruits are ± spherical, *c.* 7 cm ∅, 10-ridged and yield a black dye. *R. longiflora*, in secondary forest, is very similar to *R. megalostigma*, but smaller leaved (up to 13 cm long × 5 cm wide), the leaves elliptical, and the fruits are ± globose, up to 5 cm ∅. *R. urcelliformis* has leaves similar to those of *R. longiflora*, but the flowers are shorttubed (up to 6.5 cm long) and the calyx splits down 1 side. The corolla lobes overlap to the right. The fruits are ± ellipsoid, up to 7 cm long, with longitudinal lines.

Bibliography

Baker, H. G. (1958). Studies in the reproductive biology of West African Rubiaceae. *Journal of the West African Science Association*, **4**, pp. 9–24.

Hallé, N. (1963). Délimitation des genres *Sabicea* et *Ecpoma* en regard d'un genre nouveau: *Pseudosabicea*. *Adansonia*, sér. 2, **3**, pp. 168–77.

Hallé, F. (1967). Étude biologique et morphologique de la tribu des Gardeniées (Rubiacées). *Mémoire de l'Office des Recherches scientifiques et techniques d'Outre-mer*, **22**.

Hallé, N. (1970). *Flore du Gabon: Rubiácées*, part 2. Paris: Muséum national d'histoire naturelle.

Hepper, F. N. & Wood, J. R. (1982). A new combination in *Breonadia* (Rubiaceae), based on Forskaal's Arabian collection. *Kew Bulletin*, **36**, p. 860.

Leroy, J-F. (1975). Taxogénétique: étude sur la sous-tribu des Mitragyninae (Rubiaceae–Naucleeae). *Adansonia*, sér. 2, **15**, pp. 65–88.

Ridsdale, C. E. (in collaboration with R. C. Bakhuizen van den Brink Jr) (1975). A synopsis of the African and Madagascan Rubiaceae-Naucleeae. *Blumea*, **22**, pp. 541–53.

Robbrecht, E. (1978). *Sericanthe*, a new African genus of Rubiaceae (Coffeeae). *Bulletin du Jardin botanique national de Belgique*, **48**, pp. 3–78.

Robbrecht, E. (1987). The African genus *Tricalysia* A. Rich. (Rubiaceae), 4. *Bulletin du Jardin de botanique national de Belgique*, **57**, pp. 39–208.

Tirvengadum, D. D. (1979). The re-establishment of *Catunaregam* Wolf. *Taxon*, **27**, pp. 513–17.

Verdcourt, B. (1958). Remarks on the classification of the Rubiaceae. *Bulletin du Jardin de botanique de l'État*, Bruxelles, **28**, pp. 209–89.

Verdcourt, B. (1977). A synopsis of the genus *Chalaziella* (Rubiaceae–Psychotrieae). *Kew Bulletin*, **31**, pp. 785–818.

Verdcourt, B. (1979–80). Notes on African *Gardenia* (Rubiaceae). *Kew Bulletin*, **34**, pp. 345–60.

Verdcourt, B. (1985). A new combination in *Hallea* (Rubiaceae–Cinchoneae). *Kew Bulletin*, **40**, p. 508.

25

Compositae (Asteraceae) – *Tridax* family

Although very large on a world basis (20 000 species) the family is only modestly represented in West Africa, by under 300 species. These are mostly annual or perennial herbs or undershrubs, with alternate and/or opposite, simple exstipulate leaves, a few genera having milky juice (*Berkheya* and the Cichorieae tribe).

Members of the family may be recognised by the inflorescence, a racemose head 'composed' of very small flowers (florets), each gamopetalous and, if ♀ or ♂, with an inferior ovary. The axis ('receptacle') of the head is surrounded by an involucre of bracts, a feature also seen in *Protea* (p. 58), which fortunately possesses other distinguishing characters. The heads (in spikes) are ♂ (above) and ♀ (below) only in *Ambrosia maritima*, a whitely hairy undershrub in drier savannas. Compound heads (heads of heads) are seen only in *Echinops* and *Elephantopus* (white, blue or mauve heads), *Sphaeranthus* (red-purple heads) and *Blepharispermum*. *B. spinulosum* is a scrambling shrub (Côte d'Ivoire–Ghana) with greenish-white compound heads, in forest regrowth. It is one of the few West African woody species and it has unique spines, one pointing down the stem from the base of each petiole.

In 'woody' species, no great quantity of wood is developed (see under *Microglossa* and *Vernonia* below). The remaining species which can be considered woody are of restricted distribution, i.e. *Pluchea* spp. in dry savannas from Sénégal to Sudan, Tanzania and Angola; *Erlangea* and *Mikaniopsis* spp. of upland shrubs and shrubby climbers from Guinée–E. Africa; and *Crassocephalum mannii*, a small upland tree, sometimes planted.

There are several species of pyrophytes and numerous weed species, something like one-third of these latter genera having been introduced.

The introduced decorative species include *Cosmos* and *Zinnia* spp. from tropical America, *Tagetes* spp. from Mexico and *Gynura aurantiaca* from Java.

Distinctive vegetative characters, or smaller character combinations, are few in this family, but prickliness is of limited occurrence. Usually, then, the involucral bracts are prickly, often together with the leaves and/or stem wings. Such species include members of the tribe Cynareae, together with *Berkheya*, *Dicoma* and a weed of the sahel, *Geigeria alata*, with whorls of three leaves per node, on three-winged stems. The clearest presentation of floret, stamen and style characters is still that by Bentham (1873).

Flowers (florets) small ⊤ ⊕ or .l. ♂ or ♂ and ♀ or neuter. Calyx absent, or composed of 1 or more whorls of hairs, bristles or scales, then often persistent on the fruit. C(5), basally tubular, but tubular, filiform or ligulate above;

Fig. 25.1. A–C. *Mikania cordata*. A. Flowering shoot × $\frac{1}{2}$. B. Head × 5. C. Fruit × 5. D–H. *Zinnia elegans*. D. Palea × $2\frac{1}{4}$. E. Ligulate floret × $1\frac{1}{2}$. F. Tubular floret × 4. G. Tubular floret in longitudinal section × 4. H. Floral diagram of tubular floret. Neuter and ♀ florets of this type are also found (in other species).

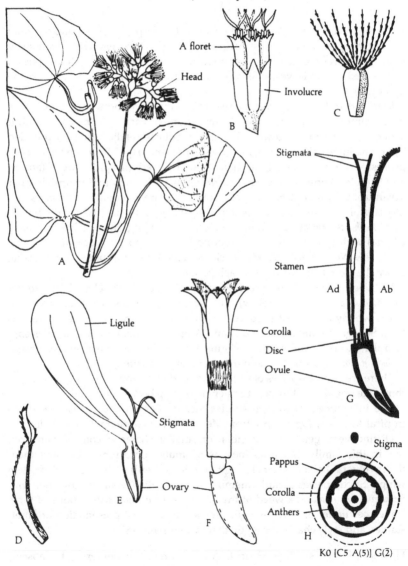

tubular floret ⊕ and ♀ or ♂; filiform floret uncommon, slender, ♀; ligulate floret .l., apparently with 1 petal (actually (5), the ligule), ♀ or neuter (♀ in Cichorieae); when found round the edge of the head these are termed ray florets, the tubular florets in the centre being disc florets. Stamens 5, epipetalous, alternating with the corolla lobes, and joined side to side by the anthers, forming a tube round the style; anther base rounded, prolonged (sagittate) or tailed, the connective often prolonged above. Disc epigynous. G(2) 1-celled, with 1 basal ovule, one carpel adaxial, the other abaxial; style single and terminal with 2 arms, their inner faces stigmatic; the shape of the arms is characteristic for the genus, or tribe of genera, i.e. in the Vernonieae, the style arms are always tapering and hairy all over.

Inflorescence racemose, the oldest florets round the outside. In one tribe, Cichorieae, all the florets are ligulate and the same colour (heads ligulate), while in several other tribes all the florets are tubular and the same colour (heads discoid). A third possibility is that the outer florets are ligulate, the inner ones tubular and another colour (heads radiate); all heads are bisexual except those of *Ambrosia*.

The receptacle may bear bracts (paleae, sing. palea), each with a floret in its axil.

Pollination The massing of very small flowers is one way of constructing a signal large enough to attract a pollinator. As the nectar produced by the disc tends to rise in the corolla tube, it is accessible to both long- and short-tongued insects. The anthers open well before the floret does so, pollen being shed into the anther tube. When the style lengthens later, with the style arms pressed together, pollen is forced out of the top of the tube and the maturing florets pass successively through a ♂ phase. The filaments frequently contract when touched, pollen then being squeezed up the anther tube and presented. The style arms then emerge and separate, and a ♀ phase is established, during which nectar is available. Later the style arms curl back, and self-pollination becomes possible. Although the timing of ♂ and ♀ phases must reduce the possibility of self-pollination within a head, pollination within and between heads on the same plant (geitonogamy, equivalent genetically to self-pollination) must be frequent, unless the heads appear one at a time, or self-incompatibility operates. *Ambrosia* is wind pollinated.

Fruit An achene formed from an inferior ovary, and generally known as a cypsela when crowned by the persistent calyx, which forms a pappus, or as a pseudonut when the pappus is absent.

Dispersal Cypselas with a light-weight pappus of silky hairs are wind dispersed, but often the pappus is made of heavier stiffer bristles that are even barbed or hooked. These are more likely to be animal dispersed, behaving like burr fruits. Sticky glandular pseudonuts (*Adenostemma*) are also animal dispersed. Other pseudonuts may have oil bodies attractive to ants (e.g. *Wedelia*, but yet to be reported in West Africa) or small wings, not likely to be effective in wind dispersal. Mechanical scattering of fruit from the ripe head, especially when this stands on a long stiff peduncle, is probably common.

Economic species The crop weeds are probably the most important economic Compositae in West Africa. The introduced lettuce (*Lactuca sativa*) and perennial Jerusalem artichoke (*Helianthus tuberosus*) provide a salad and a vegetable,

Fig. 25.2. Fruits of weed species. A. *Aspilia africana* × 10. B. *Adenostemma perrottetii* × 10. C. *Struchium sparganophora* × 12. D. *Bidens pilosa* × 5. E. *Emilia coccinea* × 15. F. *Elephantopus mollis* × 9. G–H. *Synedrella nodiflora*. G. Cypsela formed by a disc floret × 5. H. Cypsela formed by a ray floret × 5.

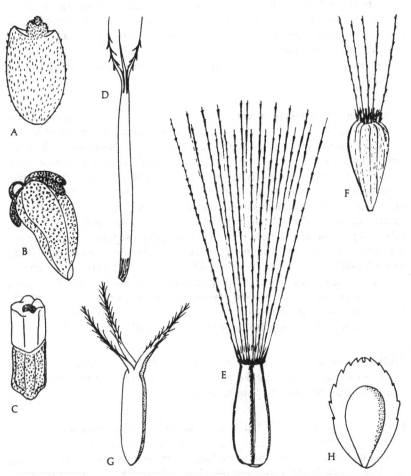

respectively. *Mikania cordata* is a useful cover crop. *Crassocephalum* (now *Senecio*) *biafrae* provides (Yoruba) bologi.

Field recognition *Ethulia* is a small pantropical genus, only *E. conyzoides* occurring in West Africa (Gambia–Indonesia). It is an aromatic herb with numerous, small, pink or mauve discoid heads and ribbed stems bearing alternate oblanceolate serrate leaves with winged petioles. The involucral bracts are in 2–3 series, and there are no paleae. The style arms taper to a point. The pseudonut has an apical rim of hard tissue, rather like a smaller version of this feature in *Struchium*. The species is found in damp grassland and by rivers, both in forest and savanna.

Gutenbergia (? J. Gutenberg, Mainz, inventor of printing) is a small tropical African genus, the most widespread species being *G. nigritana*, an erect grassland herb in derived savanna northwards from Guinée to Nigeria. It has small discoid red/purple/mauve heads with an involucre of 6 series of bracts, but no paleae. The leaves are alternate, wrinkled, ± sessile and narrowed to the base. The pseudonuts are ribbed, and rounded at the top. *G. rueppellii* (N. Nigeria–Sudan–Kenya) is a weed with very small heads (only 5 mm ∅) in large inflorescences.

Gynura (referring to the tapering style arms). *G. sarmentosa* (now *G. procumbens*; Davies, 1978) is a climber of forest edge and regrowth with alternate rather fleshy elliptical leaves, which are often coloured purple beneath when young. The stems are also purple. It has small greenish-yellow discoid heads on short lateral branches, each head with an involucre of *c.* 15 free bracts in 1 whorl, with some smaller scales round the base. There are no paleae. The style arms are long and tapering. The pappus is white. The genus is found in the Old World tropics, with our species occurring from Guinée/Sierra Leone eastwards.

Microglossa (Gr. small tongues, the outer florets). The ligulate florets round the head may indeed be small, and the head appears discoid at first glance. The 4 West African species are mostly scrambling shrubs with alternate leaves and small heads of white and/or yellow florets, the outer ligulate ♀, the inner long-tubular ♂, surrounded by an involucre of 2 series of free bracts. The cypselas are ribbed and bear a pappus of barbed bristles. *M. pyrifolia* is a widespread regrowth species, including upland areas, with petiolate ovate entire leaves and small yellow or white slightly scented discoid heads in corymb-like clusters. *M. afzelii*, a fringing forest species with obvious white ligules, is curiously absent from N. Nigeria, but otherwise widespread in West Africa.

Mikania (J. G. Mikan, eighteenth-century professor of botany in Prague). Of the 2 species, *M. cordata* is a common herbaceous climber in secondary growth opposite leaves and small discoid pink or white heads in corymb-like clusters. Each head has a calyx-like involucre of 4–5 bracts, but there are no paleae. The cypselas are black, 4-angled and with a single whorl of pinkish, barbed hairs forming the pappus. The species is widespread

Fig. 25.3. *Gynura sarmentosa*. A. Base of plant and inflorescence × 1. B. Floret × 2. C. Stamens × 6. (From *Curtis's Botanical Magazine*, **118**, pl. 7244.)

in the Old World tropics, but the genus is almost entirely American, and both West African species may have been introduced.

Vernonia (W. Vernon, seventeenth-century traveller in North America). A genus of 1000 species found in all the warmer parts of the world, represented by *c.* 60 species in West Africa. Most of these are perennial herbs with woody rootstocks, some of them pyrophytes, the remaining ones being softly woody trees or shrubs (rarely climbers). Most species have only a local distribution. All have alternate leaves and discoid heads of red, purple, mauve or white florets, without paleae, with free involucral bracts and tapering style arms. The pappus is persistent and composed of an outer series of scales and an inner series of bristles. There are 3 tree species of regrowth with white heads which are commonly met. In forest regrowth *V. conferta*, in derived savanna *V. colorata* and in forest-savanna mosaic northwards, *V. amygdalina*. *V. conferta* is the tallest-growing, up to 10 m high, with a palm-like habit, little branched and with large leaves up to 90 cm long in tufts. There are large terminal panicle-like inflorescences of small (mauve-to)-white heads *c.* 5 mm ∅ in the dry season. The cypselas have a pappus of 2 series of bristles, the outer shorter than the inner; the achene is glabrous. *V. colorata* (also in Gambia) and *V. amygdalina* are shrubs less than 5 m tall, producing scented white heads in the dry season. *V. volorata* has leaves which narrow abruptly at the base, and the cypselas are glabrous, 10-ribbed and glandular between the ribs, the pappus bristles of 1 kind. *V. amygdalina* leaves taper gradually to the base, and the cypselas are finely hairy with uniform pappus.

Bibliography

Bentham, G. (1873). Notes on the classification, history and geographical distribution of Compositae. *Journal of the Linnean Society (Botany)*, **13**, plates 8–10.

Carlquist, S. (1976). Tribal interrelationships and phylogeny of the Asteraceae. *Aliso*, **8**, pp. 465–92.

Davies, F. G. (1978). The genus *Gynura* (Compositae) in Africa. *Kew Bulletin*, **33**, pp. 335–42.

Heywood, V. H., Harborne, J. B. & Turner, B. L. (1977). *The biology and chemistry of the Compositae*, 2 vols. New York: Academic Press.

26

Solanaceae – tomato family

A tropical and temperate family of herbs and soft-wooded shrubs, a few members being climbers or small trees. Prickles and/or stellate hairs are frequently present. The family is centred in central and south America, and is best known for its economic products, which include tobacco, tomato, Irish potato and egg plant, all introduced, and the indigenous *Solanum nigrum* (ogumo, bologi).

Ornamentals introduced from tropical South America include the shrub *Brunfelsia americana*, which has blue .l. flowers. A number of *Solanum* and *Datura* spp. with ⊕ flowers are weeds. The flowers of *Solanum wrightii* are large and easily examined.

Members of the family have sympodial growth, bearing alternate, simple exstipulate leaves, which are sometimes lobed and usually adnate to the axis above their node of origin. Latex is lacking. The flowers are in cymes, and are generally white or purple and delicate, often with very short free corolla lobes (limbs), the whole folded or pleated. The pistil is of two joined carpels with axile placentation and numerous ovules. Bracts often occur in pairs in the inflorescences, one larger than the other, the larger one remaining adnate to the shoot it subtends.

Flowers ⊹ ⊕ ♂ (but see *Solanum*) 5-part (occasionally more, e.g. tomato), except for the pistil. K(5) persistent. C(5), .l. in only 1 small tribe; ribbed and at least conduplicate (also contorted as in Convolvulaceae in *Datura*) though valvate in *Capsicum*. Stamens epipetalous, 5, 4 or 2 (4 in .l. spp., 2 in *Schwenckia*), opening by slits except in some *Solanum* spp., where the anthers have terminal pores or oblique slits. (G2) oblique, the cross-wall lying at 45° to the adaxial–abaxial plane, a feature which distinguishes the family from all other gamopetalous ⊕ families; the style is terminal and the ovary 2-celled, internally partitioned with false septa to make 4 cells in *Datura* and some *Solanum* spp., often irregularly partitioned in cultivated species.

Pollination The species of economic importance have been best investigated (Free, 1975; Quagliotti, 1979), though observations are often contradictory. Protogyny is common and self-pollination and self-incompatibility is also known. *Capsicum* and *Solanum* have shallow flowers, the anthers standing closely (connivent) round the style (or stigma if the style is short enough), lacking nectar. In (Irish) potato (*Solanum tuberosum*), pollen escapes only through the anther pores and is not available in quantity, though bees are regular visitors, pumping the pollen out (Michener, 1962). In tomato, the anthers open by slits and self-pollination predominates, aided by vibration of the flowers, either in wind or by artificial means. *Physalis* also has shallow

Fig. 26.1. A *Physalis peruviana*, flowering and fruiting shoots × 1. Corolla shallow, rotate. B–D. Other corolla types. B. *Nicotiana tabacum* ×¾. Corolla tube well developed, but lobes (limbs) small. C. *Solanum torvum* ×1½. Corolla lobes well developed, but tube very short. D. *Datura stramonium* floral diagram. An imbricate corolla is more common in the family, and the gynaecium may be twisted to the left in other genera. (A and B by courtesy of F. N. Hepper; D from Rendle, 1938, Fig. 233C.)

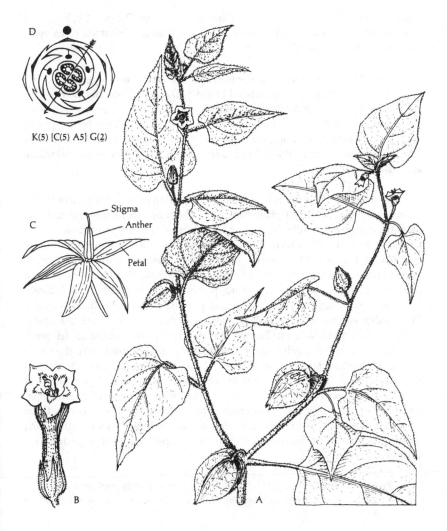

D

K(5) [C(5) A5] G(2)

Stigma

C Anther

Petal

B A

flowers but produces nectar at the base of the ovary. Tobacco differs in having a long corolla tube, but the stigma stands among the anthers and self-pollination is very common, though as a result of the visits of birds and bees, some crossing takes place. In the wild speices of the family, where nectar is usually secreted by glands round the base of the ovary or at the base of the corolla tube, long-tongued insects are likely to be regular visitors; e.g. in *Datura*, given the white flowers, moths might be pollinators. Butterflies and birds might be attracted in other cases. In some *Nicotiana* spp., the stigma secretes a nectar-like fluid which serves to make the bird's beak or moth proboscis sticky, so that pollen grains adhere (Baker, Baker & Opler, 1973).

Fruit Either a capsule or a berry. Capsules occur in *Datura*, 4-valved and prickly, or smooth and indehiscent in *Nicotiana* and *Schwenckia* (2-valved), the other genera possessing berries.

Dispersal The testa of tomato seeds forms an extra juicy layer round the seed, being composed of superficial bladder-like cells. The succulence and colours of the ripe fruits (yellow, red and black, or even green, but with no aroma) is attractive to birds. In *Solanum*, nearly all the West African species fall into this category, though *S. dasyphyllum* has a prickly accrescent calyx (cf. *S. cinereum*, Symon, 1979). In *Datura*, placental flesh forms an elaiosome and dispersal is by ants.

Economic species are important for their production of edible tubers or fruits, or for their production of allelochemicals which are extracted and used as poisons, drugs, etc. While the allelochemicals undoubtedly protect the plant's tissues from herbivore attack, they also make the tissues poisonous to human beings under some circumstances. A sapotoxin, solanine, is made in all green parts of solanaceous plants, and alkaloids, for example nicotine in the tobacco plant, are common. In the Irish potato, patches of green on the tuber indicate the presence of toxins; heavily greened potatoes should be thrown away. The garden egg (*Solanum melongena*) gets its common name from the white-fruited varieties cultivated in Europe in earlier times. The tomato (*S. lycopersicum*) is also commonly cultivated in West Africa. Smoking-tobacco, the cured leaves of *Nicotiana tabacum*, store nicotine and 'nicotine tobacco' (*N. rustica*) is the source of nicotine for pesticides and for nicotinic acid, a vitamin of the B group. Other widely used drugs are atropine (hyoscyamine) and scopola-mine (hyoscine), extracted from *Datura* spp. Ewusie (1971) recognises three species of *Capsicum*, but if *Capsicum chinense* Jacq. (Heiser & Pickersgill, 1969) occurs (as '*sinense*' in Ewusie, 1971), then it is perhaps best regarded as a variety of *C. frutescens*.

Field recognition *Datura* (Hindi name) is known as prickly apple in some languages, from the prickle-bearing fruits developed in some of the 10

Fig. 26.2. A–F. *Datura* spp. A–C. *D. metel*. A. Flower × 1½. B. Capsule × 1½. C. Seed × 6. D–F. *D. stramonium*. D. Flower × 1½. E. Capsule × 1½. F. Seed × 6. G–J. *Brugmansia* spp. G. *B. suaveolens* flower × ¾. H–J. *B. candida*. H. Flower × ¾. I. Capsule × 1. J. Seed × 6. (By courtesy of F. N. Hepper.)

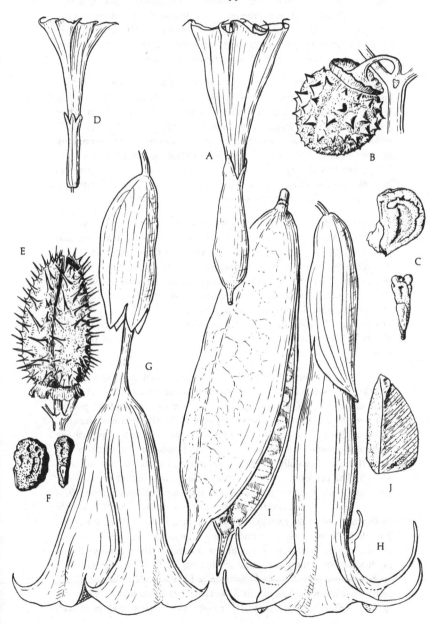

species and by 3 of the 5 species. All these have been introduced, some now occurring as weeds (*D. innoxia* in West Africa, *D. stramonium* also in temperate regions). *D. candida* (now *Brugmansia candida*; Lockwood, 1973), with large pendulous white flowers up to 30 cm long, the corolla ribs each ending in a petaloid tail, is the most spectacular species, occurring only in cultivation. *D. metel*, the most widespread West African species, like *D. innoxia* and *D. stramonium*, has erect flowers and dehiscent capsules covered with blunt prickles or warts. It is an undershrub of regrowth, possibly also an escape from cultivation upon occasion.

Physalis (Gr. bladder). The calyx is inflated and persistent round the berry. Of 4 West African species probably introduced, 2, *P. micrantha* and *P. angulata* (both also in Gambia) have become widespread. They are glabrous annual weeds with solitary yellowish-white flowers. In *P. micrantha* the leaves are almost entire, the corollas only up to 4 mm long and the inflated calyx (± spherical) up to 15 mm long by nearly as wide. *P. angulata* has deeply lobed leaves, and corollas and calyces twice as long though the calyx is ± ovoid. *P. peruviana* (Cape gooseberry) is a hairy perennial cultivated for its fruits in South Africa and Asia.

Schwenckia (M. W. Schwenke, eighteenth-century Dutch botanist). The only West African species, *S. americana*, is a weed of cultivation, also probably introduced from central America, since the other 25 species of the genus occur there. It is widespread, though surprisingly not yet reported from Gambia. It is a slender, usually hairy herb with small leaves, the upper ones sessile. The inflorescence is panicle-like, with narrow, tubular yellowish flowers each about 1 cm long and with only 2 stamens. The fruit is a 2-valved capsule separating from the crosswall, so that this stands like a coin, with seeds on both sides, in the centre of the open fruit.

Solanum (L. sedative) is by far the largest genus both in the family (with *c.* 1700 species out of *c.* 2000) and also in West Africa, although here, there are only *c.* 20 species. Egg plant, *S. melongena*, and Irish potato, *S. tuberosum*, are the well-known edible species. The flowers are distinctive in having 5 stamens with connivent anthers forming a cone and usually opening by pores. The genus is composed of undershrubs, climbers and small trees, often woolly, at least on the younger parts, and often prickly as well. The prickles are found not only on stems, but on the backs of leaf midribs and main veins, and on calyces (*S. dasyphyllum*). Symon (1979) has identified 3 sexual conditions, the commonest being the ♀, i.e. with all flowers similar. In 1 subgenus (*Lepstostemon*; D'Arcy, 1972), however, 2 kinds of flowers may be found, one ♀ the other functionally ♂ with a shorter style, on the same plant (andromonoecism) or on different plants (androdioecism), the latter condition having led to the recognition of 2 species where only 1 exists (Anderson & Levine, 1982). Hossain (1973) recognised andromonoecism in *S. torvum*, *S. wrightii* and *S. melongena* in West Africa. In *S. torvum* and *S. wrightii* the terminal flowers of each inflorescence branch are functionally ♂, the basal ones ⚥, while in *S. melongena*, only the first produced flower of the branch is ⚥. The condition should be looked for in other West African members of the subgenus, e.g. nos. 3, 7, 9, 11, 12, 15 (now *S. anguivi* Lam.; Hepper, 1978), 17, 18 and 20, in the *Flora of West tropical Africa*. Species no. 5 (*S. verbascifolium*), now known as *S. erianthum* (Roe, 1968), belongs to a different subgenus.

S. torvum is one of the commonest West African. Species, appearing as a shrubby weed of cultivation, up to 3 m high. The whole plant is covered with white hairs that appear mealy or felted, but with a hand lens they can be seen to be stellate, occurring even inside the corolla lobes, at the tips. Prickles may also be present. The flowers are generally white, about 2.5 cm ∅ on glandular pedicels in large corymbose cymes. The berrries are 1 cm ∅ ripening through yellow to brown. Approx. 4 other species, all weeds, are also widespread in West Africa.

Withania (H. T. M. Witham, nineteenth-century British palaeobotanist) is a small genus of 10 species, from South America to India, only *W. somnifera* being found in West Africa. It is a stellately woolly undershrub of the drier parts of Africa and India, with shortly petiolate ± elliptic leaves up to 10 cm long. The flowers are small and green, in axillary clusters, such that a dense, leafy, spike-like inflorescence is formed. In the fruit the pedicels elongate and the calyces inflate round red berries.

Bibliography

Anderson, G. J. & Levine, D. A. (1982). Three taxa constitute the sexes of a single dioecious species of *Solanum*. *Taxon*, **31**, pp. 667–72.

Baker, H. G., Baker, I. & Opler, P. A. (1973). Stigmatic exudates and pollination. In *Pollination and dispersal*, ed. N. B. M. Brantjes & H. F. Linskens, pp. 47–60. Nijmegen: University of Nijmegen.

D'Arcy, W. G. (1972). Solanaceae studies 2. *Annals of the Missouri Botanical Gardens*, **59**, pp. 262–78.

D'Arcy, W. G. (ed.) (1986). Solanaceae – biology and systematics. New York: Columbia University Press.

Ewusie, J. Y. (1971). Taxonomic revision of the genus *Capsicum* in West Africa. *Mitteilungen der botanischen Staatssammlung München*, **10**, pp. 253–5.

Free, J. B. (1975). Pollination of *Capsicum frutescens* L., *Capsicum annuum* L. and *Solanum melongena* L. in Jamaica. *Tropical Agriculture* (Trinidad), **52**, pp. 353–7.

Hawkes, J. G. (1978). Systematic notes on the Solanaceae. *Botanical Journal of the Linnean Society*, **76**, pp. 290–1.

Heiser, C. B. & Pickersgill, B. (1969). Names for the cultivated *Capsicum* species (Solanaceae). *Taxon*, **18**, pp. 277–83.

Hepper, F. N. (1978). Typification and names changes of some Old World *Solanum* species. *Botanical Journal of the Linnean Society*, **76**, pp. 290–1.

Hossain, M. (1973). Observations of stylar heteromorphism in *Solanum torvum* Sw. (Solanaceae). *Botanical Journal of the Linnean Society*, **66**, pp. 291–301 with 4 plates.

Lockwood, T. E. (1973). Generic recognition of *Brugmansia*. *Botanical Museum Leaflets, Harvard University*, **23**, pp. 273–84.

Michener, C. D. (1962). An interesting method of pollen collecting by bees from flowers with tubular anthers. *Revista de biologia tropical*, **10**, pp. 167–75.

Roe, K. E. (1968). *Solanum verbascifolium* L., misidentification and misapplication. *Taxon*, **17**, pp. 176–9.

Symon, E. D. (1979). Fruit diversity and dispersal in *Solanum* in Australia. *Journal of the Adelaide Botanic Garden*, **1**, pp. 321–31.

Quagliotti, L. (1979). Floral biology of *Capsicum* and *Solanum melongena*. In *The biology and taxonomy of the Solanaceae*, ed. J. G. Hawkes, R. N. Lester & A. D. Skelding, Linnean symposium series, no. 7, pp. 399–419.

27

Convolvulaceae – sweet potato family

A cosmopolitan family of perennials, mainly twiners and trailers, and often with tuberous roots or roots which readily produce adventitious buds. There are few annual, or erect, or woody species in West Africa. Irvine (1961) named only the liane genera *Bonamia* and *Calycobolus* as being woody, while *Ipomoea verbascoides* is an undershrub which might be thought so. Two introduced species are small trees, but these are seldom seen.

Other introduced species include about one-third of the remaining West African *Ipomoea* spp., e.g. *I. alba*, moonflower, which now occurs naturally in regrowth, but is often seen in gardens; another white-flowered climber, with winged pedicels, peduncles and older stems, *Operculina macrocarpa*, is recorded near the coasts of Ghana and Togo. Most introduced species are native to the Americas, including *Merremia dissecta*, the flowers of which are white with a purple throat opening in the evenings.

Members of the family may be recognised by their alternate, simple exstipulate leaves, which are sometimes digitately lobed, though rarely compound, and latex may be present. The flowers are distinctive, with a funnel-shaped or tubular corolla, often with at most small free lobes but with strongly marked ∧- shaped ribs, pleated and sometimes twisted as well in the bud, usually in some form of cyme. The fruit is generally a small two-celled capsule with four seeds, often surrounded by a persistent calyx.

There are rather few species of distinctive habit in West Africa. The dodders (*Cuscuta* spp.) are exceptional in being leafless, twining parasites of swamp plants, and are often classified in a separate family, the Cuscutaceae. Once past the seedling stage, the dodder becomes rootless and its stems turn yellow. In spite of this appearance, its flowers and fruits are somewhat similar to those of the Convolvulaceae, of which it can be considered a subfamily.

There are two West African genera of erect annuals, *Cressa*, with white flowers, in the Sénégal–Guinée area (not yet reported from Gambia) and *Evolvulus*, with blue flowers, widely distributed in dry savannas (also in Gambia). Both have very small leaves.

Stellate hairs are sufficiently rare in the family as to be distinctive of the genus *Astripomoea*, two species of erect perennials in drier savannas with red-to-purple flowers. Though not often reported in West Africa, the genus occurs right through tropical Africa in similar habitats.

Few genera of Convolvulaceae have distinctive vegetative characters upon which they may be determined. Flowers (and/or fruits) are usually essential for this purpose.

Flowers ⚥ ⊕ ♂ and 5-part, sometimes with an involucre of bracts. K5, often persistent, of similar or dissimilar (*Anisiea, Hewittia, Calycobolus*) size. C(5), the ribs external in the bud (which is not covered by the sepals). Stamens epipetalous, of different lengths, the anthers versatile. Although not a hand-lens character, it may be useful to know that only 4 genera have spiny pollen (*Astripomoea, Ipomoea, Lepistemon* and *Stictocardia*), but see also *Merremia* for other unusual types (Ferguson, Verdcourt & Pool, 1977). Nectar is secreted round the base of the ovary. Pistil characters are important in generic determination. G(2), one cell adaxial, the other abaxial (4-celled in *Stictocardia*), but

Fig. 27.1. A. *Neuropeltis acuminata* fruiting branch × ¾. B–E. *Ipomoea* spp. B. Flower bud × 1½. C. Cross-section of flower bud. D. Cross-section of ovary. The four ovules are basal. E. Capsule × ¾. F. *I. hederifolia* flower × 2; a red-flowered species with awned sepals. G. *Cuscuta australis* flowering and fruiting shoot × 2.

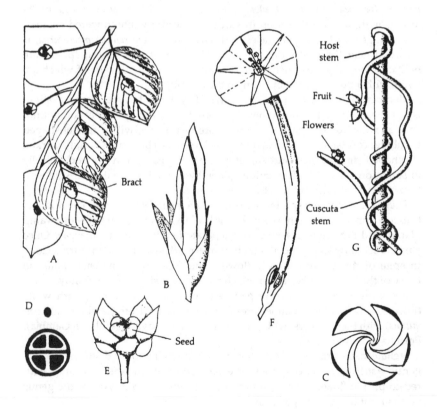

styles and stigmata various, terminal except in *Dichondra*. *Evolvulus* has 2 forked styles (4 stigmata), *Cressa* 2 styles, the remaining genera a single style, unequally forked in *Bonamia* and *Calycobolus* with a stigma to each branch. *Convolvulus* and *Jacquemontia* have simple styles with 2 long stigmata each, while *Ipomoea* and *Merremia* have rounded knob-like stigmata which may be somewhat lobed.

Pollination has been studied mainly in the sweet potato, which is usually self-sterile, cross-pollination being effected mainly by bees in this species. Flowers are protogynous, the stigmas being receptive in the bud the day before opening. The anthers dehisce about midnight, and the flower opens before dawn. White, yellow, blue and red shades are frequent among the flowers of various species, suggesting that both bees and butterflies may take part. Night-flowering white species are pollinated by moths. It would be interesting to know whether the white flowers bear, to us, invisible odour guides, equivalent to the visible nectar guides of other species. The flowers of the temperate *Convolvulus arvensis* do so.

Fruits are generally 2- or 4-valved capsules with 4 basal seeds, rarely dehiscing circularly (a *Cuscuta* sp., *Operculina*), or indehiscent (*Lepistemon*). In *Stictocardia* the persistent calyx becomes fleshy and the fruit berry-like, while in several species with sepals of dissimilar size in the flowers at least 2 of these sepals enlarge round the fruit. In *Neuropeltis* the bract enlarges.

Dispersal in *Neuropeltis* and *Calycobolus* is by wind, the fruits being samara-like, as is the case in so many species of forest lianes. *Stictocardia* 'berries' are undoubtedly animal dispersed, but no other mechanisms have been proposed for other species. The seeds are frequently hairy, but the significance of this feature, if any, is not known.

Economic species Sweet potato, *Ipomoea batatas*, introduced from south America via Europe, is one of West Africa's fastest maturing crops, grown on land liable to flooding, particularly in Guinée, but in general in the sudan and northern guinea zones. The root tubers readily produce adventitious buds, but some of these, or stem cuttings, are used for propagation. Some other *Ipomoea* spp. provide tubers (*I. aquatica*), fodder or spinach crops, and a few other species in other genera provide drugs.

> **Field recognition** *Bonamia* (F. Bonamy, eighteenth-century French botanist). Two species of this tropical genus of forest lianes occur in West Africa, both confined to the area. *B. thunbergiana* is the only widespread species (including Gambia). It has ovate leaves up to 6 cm long × 3.5 cm broad, wrinkled and brownish-hairy beneath. White flowers in dense, 1-sided axillary cymes are produced in the dry season, each pistil with a single but unequally forked style. The seeds are black and glabrous, while those of the other species, *B. vignei*, are red with a black aril, one

of the few cases of the occurrence of an aril reported in the family. *B. vignei* (Ghana) has larger leaves and corymb-like cymes.

Calycobolus (Gr. calyx and throwing). The 6 species (see Lejoly & Lisowski, 1985) of this American and African genus are confined to West Africa. Only 2 species of these are well known. One is *C. africanus* (Sénégal–Congo), with papery long-acuminate leaves and small axillary clusters of white widely tubular flowers, the corolla of each with a small brim. There are no corolla ribs. The style is unequally forked, and the pistil forms an indehiscent fruit. The 2 outer sepals of the flower are larger than the others, and enlarge even further in the fruiting stage, the 1 remaining, however, being only one-third of the size of the other. *C. heudelotii* has shortly acuminate shining leaves, pitted in the axils of the main nerves on the back. The corolla brim is 5-lobed, and, in the fruit, the smaller sepal attains half the width of the larger one.

Ipomoea (Gr. worm-like). This is the largest genus in the family, with *c.* 500 species, also the largest genus in West Africa. Of the species with pink, bluish-red or purple flowers, *I. pyrophila* is a pyrophyte, the only one in the family, while *I. pes-caprae* (now *I. brasiliensis*; St John, 1970) with ± 2-lobed leaves, is common on seashores. As far east as Togo, *I. stolonifera*, with narrow leaves, which may, however, also be notched at the apex, is found in the same habitat. *I. argentaurata* is an erect plant densely covered with silky hairs, its white flowers turning purple with a darker centre. *I. aquatica*, the water potato, also belongs in the red to purple-flowered group. Several species have white flowers, but only *I. obscura* etc., savanna herbs, have yellow flowers (± dark centre). *I. quamoclit* and *I. hederifolia* have scarlet flowers. Some species have petiolar or sepal nectaries, generally both (Keeler & Kaul, 1984). *Ipomoea* spp. are very similar to those of *Merremia*, the only difference being the spiny pollen of the former and the smooth pollen of the latter, plus the flower colour of *Merremia* spp. (white or yellow), and its never more than 4-valved/irregularly dehiscent capsule.

Lepistemon (Gr. scale and stamen). This is a small Old World genus with 2 species in West Africa, 1 of them confined to West Africa. *L. owariense*, which extends to East Africa and Angola, is a forest twiner with ovate cordate leaves and stems and petioles clothed with stiff reflexed hairs. In the dry season there are clusters of white flowers in which the corolla tube is constricted just below the level of the ribs, the stamens being inserted at the base of the tube, each with a scale which bends over the ovary. The pollen grains are spiny. The capsule has yellow bristles and dehisces irregularly (if at all) to reveal 4 seeds. *L. parviflorum* is a more slender twiner (Sierra Leone–Cameroun only) lacking reflexed hairs, with corollas only one-third (1 cm) as long.

Merremia (Blaise-Merrem, early nineteenth-century professor at Marburg) is a genus of trailing and twining plants in the warmer parts of the world, with white or yellow *Ipomoea*-like flowers and smooth pollen

Fig. 27.2. *Lepistemon owariense.* A. Flowering shoot × $\frac{1}{2}$. B. Flower bud × 3. C. Part of corolla, opened × $1\frac{1}{2}$. D. Stamen and scale × 4. E. Fruit × $1\frac{1}{2}$. (From Verdcourt, 1963, Fig. 16 *pro parte.*) (British Crown copyright. Reproduced with permission of the Controller, Her (Britannic) Majesty's Stationery Office & the Trustees, Royal Botanic Gardens, Kew. © 1963.)

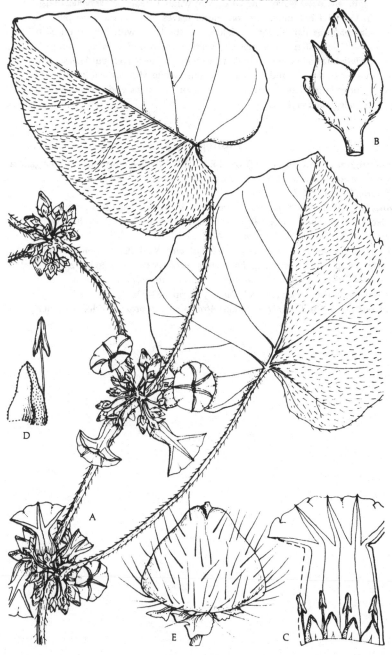

grains. The species can almost be separated on their leaves. Of the annual species, *M. pinnata* (also in Gambia) has deeply pinnately divided leaves (reminiscent of those of quamoclit) and is found in drier savannas, while *M. kentrocaulos* and *M. pterygocaulos* (near water, also in Gambia) have digitately lobed leaves, the latter with winged stems. *M. aegyptiaca* (also in Gambia) has digitately compound leaves and *M. tridentata* (also in Gambia) has narrow leaves, 2-lobed at the base, with 3 teeth on each lobe. The pollen is distinctively multiapperturate. *M. hederacea* and *M. umbellata* (also in Gambia) both have ovate cordate leaves, but *M. hederacea* has small flowers and wrinkled capsules while *M. umbellata* has flowers 3 times larger (corolla 3 cm long), smooth capsules and unusually, 5–6 colpate pollen grains.

Bibliography

Ferguson, I. K., Verdcourt, B. & Poole, M. M. (1977). Pollen morphology in the genera *Merremia* and *Operculina* (Convolvulaceae) and its taxonomic significance. *Kew Bulletin*, **31**, pp. 763–73.

Irvine, F. R. (1961). *Woody plants of Ghana*. Oxford: Oxford University Press.

Keeler, K. H. & Kaul, R. B. (1984). Distribution of defense nectaries in *Ipomoea*. *American Journal of Botany*, **71**, pp. 1364–72.

Lejoly, J. & Lisowski, S. (1985). Le genre *Calycobolus* Willd. (Convolvulaceae) en Afrique tropicale. *Bulletin du Jardin botanique nationale Belgique*, **55**, pp. 27–60.

St John, H. (1970). Classification and distribution of the *Ipomoea pes-caprae* group (Convolvulaceae). *Botanischer Jahrebücher*, **89**, pp. 563–83.

Verdcourt, B. (1963). *Flora of tropical East Africa: Convolvulaceae*. London: Crown Agents.

28

Bignoniaceae – jacaranda family

A tropical family of both Old and New Worlds, though only one or two genera (introductions excluded) occur in both floras. In West Africa, the family is a small one (six genera, nine species) and yields no major economic products. However, its species are commonly seen and are of ecological prominence.

Numerous species have been introduced into West Africa. From the Americas, jacaranda (*Jacaranda mimosifolia*) and yellow tecoma (*Tecoma stans*) are often seen, as well as a red tecoma (*Tecomaria capensis*) from South Africa. This last climbs by twining, and a number of other genera in the family either twine or have claw-like hooks or tendrils (replacing the terminal leaflet), or clasping roots. *Bignonia capreolata*, introduced from America, has tendrils. Only one West African species, *Dinklageodoxa scandens*, reported only from Liberian coastal savanna, is a climber (by twining). Anomalous secondary thickening occurs in these species.

In West Africa, members of the family may be recognised by their large, opposite, compound imparipinnate leaves and conspicuous panicle-like cymes of large, brightly coloured zygomorphic flowers, with darker nectar guides, followed by long, slim two-valved capsules with winged seeds.

Flowers ± .l. σ 5-part. K(5), 5-toothed or spathaceous, in which case the calyx develops in 1 piece and splits at bud-opening, either down the adaxial side (*Markamia*, *Spathodea*) or down the abaxial side (*Newbouldia*). C(5), 2-lipped, but in *Spathodea* with well-developed lobes, splitting adaxially. Stamens 4, epipetalous, with apically attached anthers and of 2 different heights. The staminode is adaxial. There is a large intrastaminal disc. G(2), 2-celled with numerous ovules on axile placentae also in *Kigelia* (Sharma, 1976); the septum is adaxial–abaxial in *Markhamia* (pseudoseptum).

Pollination In the case of *Kigelia africana*, bat pollination has been demonstrated, while *Stereospermum kunthianum* is reported to be particularly attractive to insects. *Spathodea* and *Markhamia* both secrete quantities of liquid into the calyx, that in *Spathodea* with the musty smell generally attractive to bats, that in *Markhamia* 'sweet'. *Spathodea* flowers are robust enough to withstand bat visits, but open by day and are a colour generally attractive to birds, and the same is true of *Markhamia*. *Tecoma stans* is probably pollinated by birds where it occurs naturally.

Fruits Three genera have pendulous, long, slim, 2-valved capsules, compressed at right angles to the septum in *Markhamia* and *Stereospermum*, while *Spathodea* has shorter, thicker erect ones. The seeds of these fruits are winged. *Kigelia* has indehiscent fruits.

Dispersal is by wind for those species with winged seeds, by mammals, baboons in particular, possibly also elephants, in *Kigelia*.

Economic species Many have local uses, the most prominent probably being *Newbouldia laevis*, extensively planted as a fence and boundary marker.

> **Field recognition** *Kigelia* (Mozambique name). *K. africana*, sausage tree (from the shape of its fruits) is the only species. It is a forest tree, persisting by rivers right into the sudan zone (also in Gambia) and even further

Fig. 28.1. A–D. *Newbouldia laevis*. A. Flower in abaxial view × $\frac{3}{4}$. B. The same × $1\frac{1}{2}$, part of calyx and corolla cut away. C. Floral diagram. This is typical for the family, except for the pistil of *Kigelia*. D. Seed × 1. E–F. *Markhamia tomentosa*. E. Calyx × $\frac{1}{2}$. F. Seed × 1. G. *Spathodea campanulata* seed × 1. H–J. *Stereospermum kunthianum*. H. Calyx × 1. I. Seed in face view × 1. J. Seed in side view × 1.

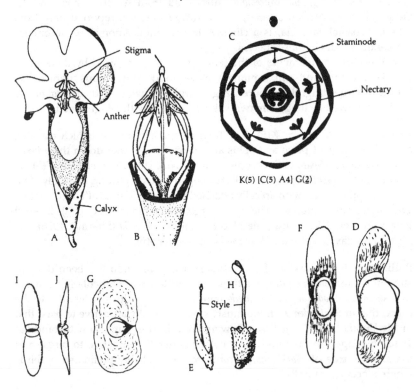

K(5) [C(5) A4] G(2)

north, and is widespread in the rest of tropical Africa. It is a tree up to 20 m high, the leaves with swollen petiole bases covered with light spots (lenticels). In the dry season, on older branches, pendulous panicle-like cymes, up to 90 cm long, of large, dark red-to-yellowish-green flowers, with yellow stamens and style, appear. The pedicels are so twisted that the 2 adaxial petals are still uppermost, and the flower faces horizontally. The fruit starts as a small green berry, but enlarges to a smooth green cylinder about 15 cm ∅ and 45 cm long.

Markhamia (C. R. Markham, nineteenth-century British traveller) is a small, Old World genus, of which we have 2 species. *M. tomentosa* (Sénégal–Angola) is a tree of forest outliers, up to 15 m high, with small pointed pseudostipules (the first pair of leaves of the axillary bud) at each node. These and other young growth are covered with yellow woolly indumentum. The flowers (in the rains) are yellow, marked with red nectar guides, about 5 cm long and in long terminal racemes. The pendulous woolly capsules are narrow and flat, up to 90 cm long, containing 4–6 rows of seeds, each set in a rectangular wing. The pseudoseptum becomes papery. *M. lutea* is very similar though of restricted distribution (Ghana–Cameroun). The pseudostipules are leafy and almost circular, and the leaflets petiolate rather than sessile.

Newbouldia (W. W. Newbould, nineteenth-century British botanist). *N. laevis* is the only species, occurring from Sénégal to Angola (also in Gambia). It is a slender forest tree, up to 12 m high, much planted and often seen near habitation. An ant guard is present, maintained by the secretions of glands, partly at the base of each leaflet blade, partly on the calyces and capsules. The flowers, produced in the dry season, are pink purple and white, in congested terminal erect panicles, and about 5 cm long. The fruits are pendulous and cylindrical, about 25 cm long, with 3 nerves running the length of each valve, and 2 rows of overlapping winged seeds down each side of the woody placentae. The leaves are articulated in the same way as those of *Stereospermum* spp.

Spathodea (Gr. spathe-like, the calyx). The only species is *S. campanulata*, which extends from Sénégal to Angola. It is a deciduous tree up to 20 m high, found in all drier kinds of forest and near rivers, also in fringing forest. The leaves, of which one of the pair is generally smaller than the other, are also distinguished by their slightly asymmetrical, cuneate leaflet bases, each with at least 1 large gland. The flowers are fleshy and bright red, in terminal erect racemes, each flower facing upwards. The fruits are erect, brown and 4-ridged, up to 20 cm long, splitting down the middle of each valve to reveal 2 rows of seeds on each side of the septum. The seed is surrounded by a wing resembling cellophane or plastic foil. The only other species of *Spathodea* is also confined to tropical Africa.

Stereospermum (Gr. hard seed). This is an Old World genus of 24 species, represented in West Africa by *S. kunthianum* and *S. acuminatissimum*. Both are trees with leaves articulated across the rachis between each pair of

Fig. 28.2. *Markhamia* spp. A–D. *M. tomentosa*. A. Flowering shoot × $\frac{1}{2}$. B. Underside of leaf × *c.* 10. C. Anther × 4. Fruit × $\frac{1}{2}$. E–F. *M. lutea*. E. Node with pseudostipules × $\frac{1}{2}$. F. Underside of leaf × *c.* 10. (From Gentry, 1984; A–D, Pl. 11; E–F, Pl. 10.)

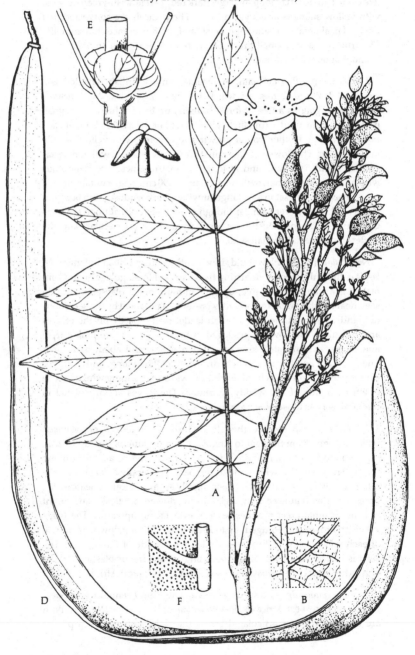

leaflets and again at the base of each leaflet, which has a markedly asymmetrical cuneate base in addition. *S. kunthianum* is a widespread woodland savanna tree (Sénégal and Gambia–Congo, Ethiopia and Moçambique), up to 12 m high, producing drooping panicles of pink and white flowers (with nectar guides) in the dry season, while the tree is leafless. The leaflets are obtuse, notched or acute but never acuminate, like the next species, nor toothed like those of *Newbouldia*. The capsules are pendulous, narrowly cylindrical and about 60 cm long, with a single nerve on each valve, and only 1 row of seeds on each side of the placentae. This is woody, and each seed fits into a notch, by means of a boss. *S. acuminatissimum* is a tall forest tree up to 30 m high, with long-acuminate leaves and erect pyramidal inflorescences of tubular flowers in the rains. It is restricted to the area Guinée–Cameroun.

Bibliography

Gentry, A. H. (1984). *Flore du Cameroun: Bignoniacées.* Paris: Muséum national d'histoire naturelle.

Harris, B. J. & Baker, H. G. (1958). Pollination in *Kigelia africana* Benth. *Journal of the West African Science Association*, **4**, pp. 25–30.

Sharma, B. B. (1976). The bilocular ovary of *Kigelia. Geophytology*, **6**, pp. 63–4 and pl. 1.

29

Acanthaceae – *Thunbergia* family

A large family, mainly of herbs and undershrubs in the tropics and subtropics, and particularly prominent in West Africa in forest undergrowth. Members of the family have opposite exstipulate leaves and lack latex. The leaves are generally shortly petiolate or sessile, and long-cuneate. The inflorescences are often conspicuous; either dense cone-like spikes with prominent overlapping, sometimes forked (or papery – *Rungia*), bracts from between which the flowers appear (Sell, 1969b), or lax and panicle like. The flowers are zygomorphic, with two or four stamens and a two-celled ovary developing into a characteristic capsule.

Introduced species occur as a rule only in gardens. The most prominent are the Indian *Thunbergia grandiflora* and *T. laurifolia*, with blue flowers (the former sometimes with white flowers), and *T. alata*, with winged petioles and yellow-to-reddish flowers with a dark 'eye'. Two *Barleria* spp. have also been introduced.

Several genera contain only, or at least some, prickly species, the prickles often needle like. *Crossandra*, with prickle-edged bracts and pale flowers, and *Acanthus*, with prickly leaves as well, occur in forest, while *Diclipera verticillata* (also in Gambia) is a red-flowered savanna herb, the only one in the genus with a spine at the tip of each bracteole.

Lepidagathis collina is a yellow-flowered savanna herb which can grow as a pyrophyte. When it does so, its globular inflorescences resemble burrs, up to 4 cm ∅, the needle-like tips of the bracts (there are no bracteoles) sticking out all round. The narrow leaves are similarly tipped. Two species of *Blepharis*, white or blue-flowered savanna herbs, have prickly-edged leaves and bracts, the commoner species, *B. maderaspatensis*, with white flowers, also having whorled leaves. The yellow-flowered forest herb *Barleria oenotheroides*, has bracts similar to those of *Crossandra* and *Acanthus*. There are, however, no bracteoles. Two northern Nigerian species, seldom reported, are unique in having paired spines at the nodes. These are perhaps the first pairs of leaves of the axillary shoots.

Flowers ⊹ .l. ♂, corolla 5-part. K(4) or (5) or represented by a ring of tissue in species with large bracts, e.g. *Thunbergia*, where the ring is glandular and attractive to ants. C(5) imbricate or contorted, 2-lipped (1-lipped in *Acanthus*, *Blepharis* and *Crossandra*), ⊕ in *Lankesteria*. Stamens epipetalous, 4 fertile and

at 2 heights in the corolla tube, or 2, abaxial except in *Brillantaisia*; there are
also 2 staminodes in both this genus and *Lankesteria*; anthers introrse, opening
by slits except in *Mendoncia* (apical pores) and *Rungia paucinervia* (now *Ascotheca*) (basal pores); anthers 1-celled in, for example, *Acanthus* and *Physacanthus*;
when 2-celled the cells may be at the same or different heights, or even vertically on top of one another (*Dicliptera*); the anthers are ± a tail; stamen characters can be seen with a hand lens. An intrastaminal disc is usually present.
G(2) 2-celled, mostly with 2 ovules per loculus in species with imbricate corollas, though also in *Thunbergia* (contorted), otherwise several to many in species
with contorted corollas. *Mendoncia* G1 by abortion of the adaxial carpel.

Fig. 29.1. A. Floral diagram. B–F. *Asystasia gangetica*. B. Flower × 2½. C.
Half flower × 5. D. Stamen × 40. E. Fruit × 4½. F. Diagram of cross-section
of ovary. G. *Justicia insularis* fruit × 2½. H–I. *Thunbergia erecta*. H. Fruit
× ¾. I. Diagram of cross-section of fruit. J. *Lankesteria elegans* seed × 4.
(A from Eichler, 1875–8; J from Heine, 1966, Pl. xx.6, by permission of
the Muséum national d'histoire naturelle.)

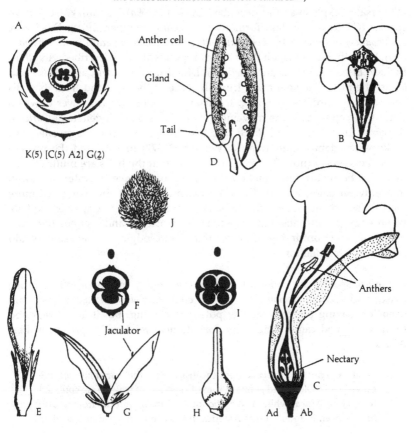

Pollination is by long-tongued insects, whether the corolla is gullet-shaped or tubular.

Fruit Only *Mendoncia* spp., white-flowered ± woody, forest twiners have drupes (1-celled, with 1–2 seeds, standing between paired bracts). Other species have capsules and in most cases these are basally solid ('stalked') and elongated, opening widely and elastically in the adaxial–abaxial plane (loculicidally), the coin-like seeds being thrown out in directions determined by the angle of the curved jaculators, which project below the seeds like a spine, e.g. in *Asystasia*. In *Thunbergia*, the capsules are unusual in being swollen basally, there accommodating 4 seeds, and tapered apically. This genus is sometimes placed in its own family, the Thunbergiaceae. In this and a few other species (e.g. some *Hygrophila* spp. and *Nelsonia canescens*) with more than 4 seeds per capsule, the funicles do not form woody jaculators, but remain very small (papilliform). The seeds may be smooth, scaly (*Crossandra*) or hairy (see below).

Dispersal is explosive, and described by Sell (1969a) for four West African species (*Crossandra nilotica*, formerly *C. massaica*; Napper, 1970); *Dyschoriste* (*Calophanes*) *perrottetii*; *Hygrophila auriculata* (as *H. spinosa*) and *Phaulopsis imbricata*). Five other genera with West African species are also described: *Acanthus*, *Barleria*, *Blepharis*, *Justicia* and *Ruellia*. In *Dicliptera* and *Rungia*, as well as *Phaulopsis*, the placentae split away from the base of the fruit and from its walls and snap upwards on opening, the fruit wall splitting also lower down in order to accommodate the movement. Seed hairs in such genera as *Barleria*, *Blepharis*, *Eremomastix* and *Ruellia* are hygroscopic and become swollen and sticky when damp. Gutterman & Witztum (1977) investigated the process in *Blepharis persica* (now *B. ciliata*), and found that the hairs are multicellular, their tips becoming mucilagenous while the bases take in water and swell unevenly on being wetted. Cells on the inner side of the hair swell more than those on the outer side, causing the hair to erect. Capsules explode when enough rain falls and the seeds are scattered and wetted, the hairs remaining erect on drying. Both scattering and lodging of the seeds would seem to be provided for.

Economic species are of local importance for many purposes, often as fish poisons (*Adhatoda*, *Eremomastix*, *Justicia* etc.). *Blepharis linariifolia* bracts are used for making patterns on clay pots before firing, but it is not reported to provide good camel fodder, as does its near relation, *B. persica*, in East Africa.

> **Field recognition** *Asystasia* (Gr. inconsistency, the variability of the flowers) – an Old World genus of some 40 species, of which 6 are confined to western Africa, while *A. gangetica* occurs all through the tropics, also having been introduced into America. It is a common weed of disturbed

Fig. 29.2. *Rungia grandis*. A. Flowering shoot × 1. B. Bract × 1⅔. C. Calyx × 1½. D. Corolla × ½. E. Corolla opened out × 1½. F. Anther × 4, upper and lower cells distinct. G. Fruit opened × 1⅓. H. Seed × 4. (From Heine, 1966, pl. XLIII *pro parte*, by permission of the Muséum national d'histoire naturelle.)

ground with 1-sided raceme-like inflorescences of small, purple and pale-yellow or white, 2-lipped imbricate flowers with needle-like calyx lobes 5 mm long. There are 4 2-celled tailed anthers at different heights. The seeds resemble half coins, with a ribbed circumference.

Brillantaisia (Brillantais-Morion, who assisted in the compilation of Pali-sot de Beauvois, *Flore d'Oware*, 1805–) – an African and Madagascar genus of glandular herbs up to 3 m high, with winged petioles and strongly 2-lipped contorted blue or purplish flowers in lax panicle-like inflores-cences. Only *B. lamium* and *B. nitens* are widespread in West Africa, and of these, only the former occurs as far east as Uganda. *Brillantaisia* spp. are unusual in having only the 2 adaxial stamens in each flower. The capsule is 2–3.5 cm long and the seeds are hairy.

Hygrophila (Gr. moisture-loving). Swamp herbs, sometimes with 4-angled stems, often creeping and rooting at the nodes; seldom erect and more than 60 cm high. The genus is widespread in the topics, and the best represented in West Africa, with 15 species, but only 2 of those spe-cies, *H. auriculata* and *H. senegalensis* (both also in Gambia) have been so well collected that they may be considered widespread in West Africa. The latter is confined to West Africa, and is well known for its polymor-phism, developing spongy tissue (aerenchyma) at the base of the stem when growing in water, and, when the water level is even higher, deeply pinnatifid leaves under water. The flowers are generally in axillary clusters or spike-like inflorescences, bluish-purple (or pale), contorted and 2-lipped, with 2-celled untailed stamens (2 stamens only in *H. africana*, a seldom-collected hetrophyllous herb of sudan zone swamps). There is a single row of (hairy) seeds in the capsule, as in *Asystasia*.

Hypoestes (Gr. below and ?) is a large Old World genus centred in Mada-gascar, though 4 of West African species are endemic, 2 extend as far as Congo and only 1, *H. verticillaris*, is widespread in West Africa and Africa in general. This is a savanna herb with sessile axillary clusters of widely gaping, 2-lipped purple and white imbricate flowers appearing out of a tube of fused bracteoles. Flowers are either fertile or sterile, 1 fertile accompanying 1 sterile, each with its respective involucre, within a larger involucre of 2 fused outer bracteoles. There are 2 1-celled anthers in the fertile flower.

Justicia (J. Justice, eighteenth-century Scottish gardener) – a large pantro-pical and subtropical genus represented in West Africa by a dozen species, of which 4 are widespread, though only *J. insularis* occurs in Gambia. This species is a spreading herb in savanna, though often also cultivated as a potherb. The *Asystasia*-like flowers occur in clusters at the nodes, and are white, pink or purple, each with only 2 2-celled tailed stamens. The capsules contain only 4 seeds with rough coats. Other species have spike-like or panicle-like inflorescences, and the majority are 1(–3) m high.

Lankesteria (E. Lankester, nineteenth-century British botanist) is a very small genus of tropical Africa and Madagascar, with only 7 species altogether, of which 5 occur in West Africa, all but 1 of these are not found elsewhere. The exception is *L. elegans*, also the most widespread species in West Africa, extending to Congo and Uganda. It has large leaf-like bracts, making the spike-like inflorescences particularly cone like. This is a handsome forest undershrub, with long terminal racemes of tubular, contorted but not 2-lipped, yellow flowers, red inside the throat. There are 2 2-celled untailed stamens, the cells of different lengths. The seeds have hygroscopic hairs.

Bibliography

Eichler, A. W. (1875–8). *Blüthen-Diagramme*. Leipzig: Engelmann.

Gutterman, Y. & Witztum, A. (1977). The movement of integumentary hairs in *Blephairs ciliaris* (L.) Burtt. *Botanical Gazette*, **138**, pp. 29–34.

Heine, H. (1966). *Flore du Gabon: Acanthacées*. Paris: Muséum national d'histoire naturelle.

Morton, J. K. (1978). A revision of the *Justicia insularis–striata* complex (Acanthaceae). *Kew Bulletin*, **32**, pp. 433–8.

Napper, D. (1970). Notes on Acanthaceae 1. *Kew Bulletin*, **24**, pp. 323–42.

Sell, Y. (1969*a*). La dissémination des Acanthacées. *Revue générale de botanique*, **76**, pp. 417–52.

Sell, Y. (1969*b*). Les complex inflorescentiels de quelques Acanthacées. *Annales des sciences naturelles*, sér. 12, **10**, pp. 255–300.

30

Verbenaceae – teak family

A mainly tropical and subtropical family of varied habit, though all with opposite or whorled, exstipulate and sometimes fragrant leaves, often on four-angled stems. Ptyxis is conduplicate as a rule, conduplicate-plicate in *Tectona*. The inflorescences are racemose (less often cymose – *Clerodendrum*, *Premna*, and *Vitex*), but often congested, and made up of small five-part flowers with a two-lipped corolla, four stamens and a two- to eight-celled ovary with a terminal style. The fruit is a drupe, or a schizocarp splitting into four nutlets.

Several ornamental species have been introduced from tropical America, e.g. *Petrea volubilis*, a climber in which the persistent blue calyx lobes act like helicopter rotors in dispersing the fruit. *Duranta repens* (now *D. erecta*: Bromley, 1984), a shrub commonly planted as a hedge, has tubular blue flowers and yellow 'drupes'. *Lantana camara* (wild sage or curse of Barbados) has short spikes of yellow or reddish flowers and does not set fruit in West Africa. In the West Indies it is probably pollinated by butterflies.

From tropical Asia, *Gmelina arborea*, with large flowers suitable for investigation, and *Tectona grandis* have been introduced as plantation trees, while *Holmskioldia sanguinea* (Chinaman's hat) and *Clerodendrum* spp., e.g. *C. fragrans* (now *C. philippinum*; Howard & Powell, 1968), are ornamentals.

Black mangrove (*Avicennia africana*) is now placed in its own family Avicenniaceae, *Congea tomentosa* (with three large persistent bracts and bracteoles per flower) in Symphoremataceae. The former family is monogeneric, the latter trigeneric, both with free central placentation.

Flowers ± .l. 1–2-lipped or ⊕ ♂, stamens 4-part in *Gmelina* and *Lippia*, usually 5-part though 5–7-part in *Tectona*. In 4-part flowers the sepals are diagonal, the petals orthogonal. (K) persistent in several genera. (C) 2-lipped (1-lipped in *Clerodendrum*), not 2-lipped but scarcely ⊕ in *Lantana*, *Stachytarpheta* and *Tectona*. A4 at at least 2 heights (A2 + 2 staminodes only in *Stachytarpheta*, A5–7 in *Tectona*), always epipetalous; the anthers are never apocynaceous or acanthaceous, but the cells may meet only at the top of the connective, cf. Bignoniaceae. A disc is present round the base of the ovary. G(2) placentation axile in the mature ovary; style terminal, ovary simple though lobed, 4-celled by growth of a false septum in the adaxial–abaxial plane (*Clerodendrum*, *Gmelina*, *Premna*, *Tectona*, *Vitex*); G(4) (8-celled) in *Duranta*, G1 (2-celled) in *Lantana*, *Lippia* and *Stachytarpheta*, the adaxial carpel aborting.

Fig. 30.1. A. *Clerodendrum umbellatum* flowering shoot × $\frac{3}{8}$. B–C. *Duranta repens*. B. Floral diagram. C. Diagrammatic cross-section of fruit. D. *Vitex doniana* flower × 2. E. *Lantana camara* diagrammatic cross-section of fruit. F. *Clerodendrum thomsonae* flower × $1\frac{1}{2}$. G–H. *Tectona grandis*. G. Calyx in vertical cross-section × $1\frac{1}{2}$. H. Fruit in cross-section × $1\frac{1}{2}$. (A from Irvine, 1961, Fig. 141; B from Rendle, 1938, Fig. 231B.)

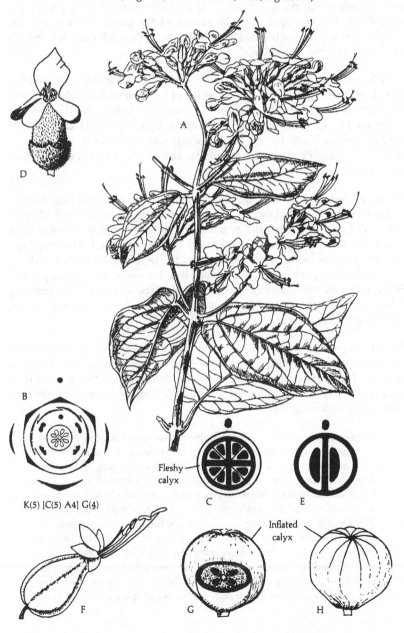

K(5) [C(5) A4] G(4)

Fleshy calyx

Inflated calyx

Pollination Observations on *Tectona grandis* show that flowers have to be cross-pollinated to ensure seed set, and that only two species of bees can perform this function. It is therefore not suprising that fruits are not formed in West Africa, to which the species was introduced. *Lantana camara*, pollinated by butterflies in the West Indies and Central America, has probably also to be cross-pollinated, though it is now sufficiently common to be considered naturalised. Nothing appears to be known of the pollination mechanisms of the other members of the family in West Africa. Protandry and heterostyly have been reported elsewhere.

Fruit The fleshy fruits are either drupes (with 2 1-celled stones in *Lantana*, or a single 4-celled stone in some species of *Clerodendrum*, *Gmelina*, *Tectona* and *Vitex*, or a false drupe in *Duranta*, in which the flesh round the 4 nutlets is provided by the persistent calyx. In schizocarpic fruits, each carpel forms 1 or 2 nutlets, each with 1 or 2 cells. This occurs in most species of *Clerodendrum*, in *Lippia*, *Premna*, *Priva* and *Stachytarpheta*. The calyx persists in a dry and/or coloured form.

Dispersal *Priva* and *Tectona* have hairy nutlets and nut-like fruits respectively, within an inflated calyx. Once this disintegrates, these genera effectively possess burr fruits. In *Lippia*, the two nutlets lie within a tiny globular calyx which is also stiffly hairy, and may also form a burr fruit. *Duranta*, *Vitex doniana* and *Premna* fruits are reported to be dispersed by birds in Africa, a method also reported for *Lippia*, which has, however, very inconspicuous fruits. Baboons also consume fruits of *V. doniana*. It would be interesting to know whether the fruits of *Clerodendrum*, with enlarged contrasting calyces, are also attractive to birds.

Economic species include *Tectona grandis* and *Gmelina arborea*, yielding timber, but the most widespread preserved indigenous species is black plum (*Vitex doniana*). The drupes are edible and also yield ink. Many other species have local uses. *Priva lappulacea*, recently introduced into Ghana from America, threatens to become a weed.

> **Field recognition** *Clerodendrum* (Gr. tree of chance) is a large tropical and subtropical genus, well represented in West Africa by shrubby forest climbers, with about a third of West Africa species being widespread within that area, most of them also in the rest of tropical Africa. The flowers are in cymose inflorescences, 5-part and generally with the petaloid calyx coloured differently from the corolla. The corolla bud is distally globular, but basally tubular, and in this tube the 4 stamens are attached, though they are long and conspicuously exserted, acting as a landing platform for bees. The fruit is generally of 4 black nutlets (or hard drupelets). In forest regrowth are *C. volubile*, appearing first, and *C. buchholzii* both have petiolar spines, but *C. volubile* has dense corymb-like cymes of white flowers in the dry season, the calyx enlarging to a lobed plate

round each fruit. *C. buchholzii* has lax panicle-like cymes on leafless branches of scented white flowers in the rains. The corolla lengthens by about 3 times after pollination, and white calyx enlarges in the fruiting stage. *C. violaceum* and *C. umbellatum* are also confined to forest areas and the southern part of the forest-savanna mosaic. *C. violaceum* has conspicuous, lax, terminal cymes of violet flowers with large corolla lobes, appearing mostly in the late dry season. *C. umbellatum* has lax, terminal and axillary corymb-like cymes of small white flowers in the rains (see also below). The 4 remaining widely spread species also occur in savanna, as does *C. volubile*. *C. splendens* has corymb-like cymes of red flowers in the dry season, while the coarsely hairy *C. capitatum* and *C. polycephalum* flower in the rains, the former having petiolar spines on hollow stems inhabited by ants. Terminal clusters of very long-tubed white flowers are surrounded by an involucre-like group of leaves. The calyces are proment and papery, becoming pale purple in the fruit and enlarged (especially basally). *C. polycephalum* has leaves in whorls of 3, and short but widely branching terminal cymes of small white flowers. *C. thomsonae* flowers at various times of the year in different localities, but is unmistakeable because of its flowers, with large white calyces and red corollas. These are shrubs and lianes with opposite leaves and petiolar spines, producing large terminal and axillary panicles of tubular flowers with K(5) (C(4) A4) G(2). The calyx is generally persistent round the 4-lobed 4-celled fruit, which may break into 4 nutlets at maturity. Although many species are found in forest in forest-savanna mosaic, the following 3 species are found mainly in high forest:

C. violaceum with blue flowers in which one petal is larger than the others. The calyx lobes are rounded, and the fruit does not break up. *C. umbellatum* has white flowers with a pink centre and very long stamens. The calyx lobes are pointed, and the whole calyx becomes red in the fruiting stage, when the fruit breaks up. *C. buchholzii* is the tallest species of the 3, up to 9 m high, producing white, scented flowers, paratly on old wood, in the rains. The calyx is minute, but enlarges and becomes white, the fruits turning red before they break up. Some liane and climbing shrub species of lowland forest flower and fruit in the lower canopy, while others climb much more extensively, and do so in the upper canopy. The fruits of this latter group are often wind dispersed and of distinctive appearance. Only a few of them are cauliflorus, so their flowers are not easily come by. *C. formicarum* has hollow stems which are possibly ant domatia.

Lantana (classical name). Although *L. camara* is so well known as a hedge plant, the genus is both African and American and 2 species of woody savanna herbs are indigenous to West Africa. Of these, *L. rhodesiensis* grows to about 1.5 m high and bears its leaves in 3s. In the rains, short spikes of tiny purplish flowers appear, the bracts quite prominent and becoming more so when the red or purple drupes appear. This is a species of open grassland.

Lippia (Lippi, nineteenth-century French botanist). Another tropical African and American genus, represented in West Africa by savanna undershrubs 3–4 m high, with greyish whorled leaves in 3s and 4s, the foliage strongly scented. The flowers are tiny, 4-part, pale and sessile, in dense silky small spikes (with prominent overlapping bracts) arranged in widely branching panicle. In the fruiting stage the calyx becomes inflated and globular, round the 2 tiny nutlets. *L. multiflora* is the largest-growing species and has smooth upper leaf surfaces, like those of the much smaller (90 cm) *L. chevalieri* (also in Gambia), which is, however, confined to drier savannas. Two other species, also mainly in drier savannas, have rough leaves.

Premna (Gr. tree stump). This is an Old World genus of only 6 species, 5 of them being confined to West Africa but not widespread. *P. angolensis*, a small tree, is found throughout tropical Africa. It has leaves in whorls of 3 or 4, each leaf rounded at the base or at most very shortly cuneate. Small white flowers are produced in large *Tectona*-style cymes in the rains, followed by red or yellowish nutlets.

Stachytarpheta (Gr. dense spike). An American genus, the 3 species of West Africa being herbaceous weeds of farmlands, up to 2 m high, with long rat's tail-like spikes of pale blue 5-part flowers. These are embedded basally in the rather fleshy spike axis, which is covered by spirals of bracts. Each flower has only 2 stamens (and 2 staminodes) and the single carpel forms 2 nutlets. Two of the species are widespread in forest regions, while *S. angustifolia* (also in Gambia) is a savanna species, with toothed narrow leaves.

Vitex (classical name), is a tropical and sub-tropical genus of trees and shrubs, represented in West Africa by a dozen species, most of them confined to West Africa. The 4 most widespread species, however, extend also over much of the rest of tropical Africa. *V. simplicifolia*, a savanna species, has simple leaves in whorls of 3. Its western limit of distribution is Mali–Nigeria. Otherwise *Vitex* spp. have opposite digitately compound leaves and cymes of 4-part flowers followed by black fleshy drupes. The calyx enlarges and becomes plate-like in the fruiting stage, the drupe hanging below it. The 2 most widespread savanna species are *V. doniana* and *V. madiensis*. *V. doniana* is a tree up to 18 m high, with 5-foliolate leathery leaves and dense cymes of pale flowers appearing with the new leaves. *V. madiensis* (also in Gambia) is a smaller tree with 3-foliolate, sometimes crenate, leaves and long-stalked cymes of yellow and purple flowers on new growth. *V. grandifolia* and *V. ferruginea* are under-storey forest trees, the former, which has ant domatia, also in regrowth. *V. grandifolia* has 5 entire leaflets per leaf, the central leaflet being very large (up to 40 cm long × 20 cm wide). The flowers are yellowish, crowded in short-stalked axillary cymes, followed by yellow drupes turning black. *V. ferruginea* is a deciduous tree of similar size (up to 150 cm high) but 5–7 leaflets per leaf, the central leaflet being only up to 15 cm long by half as wide.

The flowers are pink-to-purple, produced in small cymes on leafless branches. *V. thyrsiflora*, a small forest tree or shrub also has ant domatia in its 4-angled stems. The 5-foliolate leaves are distinguished by their covering of yellow glands, and terminal panicle-like cymes.

Bibliography

Barrows, E. M. (1976). Nectar robbing and pollination of *Lantana camara* (Verbenaceae). *Biotropica*, **8**, pp. 132–5.

Bromley, G. L. B. (1984). *Duranta repens* versus *D. erecta* (Verbenaceae). *Kew Bulletin*, **39**, pp. 803–4.

Howard, R. A. & Powell, D. A. (1968). *Clerodendrum philippinum* Schauer replaces '*Clerodendrum fragrans*'. *Taxon*, **17**, pp. 53–5.

Irvine, F. R. (1961). *Woody plants of Ghana*. Oxford: Oxford University Press.

Killick, H. J. (1959). The ecological relationships of certain plants in the forest and savanna of central Nigeria. *Journal of Ecology*, **47**, pp. 115–27.

Rendle, A. B. (1938). *The classification of flowering plants*, vol. 2, 1st edn. Cambridge: Cambridge University Press.

31

Labiatae (Lamiaceae) – Hausa potato family

A cosmopolitan family, mainly of herbs, represented in West Africa by erect perennial herbs and undershrubs rarely over 2 m high, and a very few shrubs, in savanna and upland grassland. Apart from a few weed or culinary species introduced from the New World, our flora has only Old World affinities. Almost all West African genera have at least one species which also occurs in other parts of tropical Africa, sometimes in southern Africa and the Old World tropics in general as well; but within West Africa, most species have only local distributions. Gambia is recorded as having only five species (four genera), only *Leucas martinicensis* being indigenous.

Members of the family may be recognised by their four-sided stems bearing simple, exstipulate opposite or whorled leaves, which are often glandular and generally aromatic when crushed, small, toothed and either petiolate or sessile. Only *Icomum* (one to two species of tiny herbs in Guinée) has alternative leaves. The flowers are small, two-lipped, with four stamens, which usually lie on, or rise from, the bottom lip in West African species. White, pink, blue, violet and purple are common flower colours; yellow, orange and red are quite rare. Inflorescences are short axillary cymes, assembled in various ways into, for example, terminal and axillary interrupted spike-like structures (*Ocimum*), dense cone-like 'spikes' (*Haumaniastrum, Hyptis*), globular nodal clusters (*Leonotis*) or lax, spreading panicles (*Englerastrum, Hoslundia*). There is some doubt as to whether *Haumaniastrum* can be maintained as a genus separate from *Acrocephalus* (Robyns, 1966). Persistent, enlarged or coloured bracts may be associated with the inflorescence, such as the pink bracts on top of spikes of *Hemizygia* spp., grassland herbs. The calyx is persistent, the style usually gynobasic and the fruit schizocarpic, of four nutlets.

In a comparison of the Labiatae and Verbenaceae, El-Gazzar & Watson (1970) maintain that only one subfamily in each family is clearly distinct (the Ajugoideae, e.g. *Tinnea* in the Labiatae, and the Verbenoideae, e.g. *Lantana*, *Lippia* and *Priva*, in the Verbenaceae), and the 25 genera out of the 72 examined are heterogeneous. Codd's (1971) discussion of generic limits for *Plectranthus* and related genera could helpfully be repeated in connection with other groups of genera.

Species introduced from tropical America include several each of *Hyptis* and *Salvia*, while species of *Ocimum* have been introduced from both America and Asia.

There are few species of unusual habit. *Scutellaria paucifolia* is the only pyrophyte, producing blue *Ocimum*-like inflorescences 30 cm high after fires, the leafy shoots following later. A single yellow-flowered species of *Plectranthus* produces brown hairy bulbils in leaf and inflorescence bract axils, and although reported only from upland Liberia, this species may prove to be undercollected. There is also a small group of distinctively bristly species, with spiny-tipped bracts and/or calyx lobes, or bristly stems. *Stachys aculeolata* has markedly bristly stems, *Leucas* (white flowers) and *Pycnostachys* (blue-purple flowers) both have spiny-tipped calyx lobes, while *Leonotis nepetifolia* has both spiny calyces and spine-tipped bracts. It is probably the most often seen species, being common near villages, and sometimes cultivated. The flowers are intensely hairy, whitish or orange-brown, with a long, hooded upper lip. This species has become a tropical weed.

Flowers ± .l. $\male$ 5-part. K(5) various; $\oplus$ with 5 equal teeth, or 2-lipped or with 8–10 or more teeth, or with unequally developed lobes (3 in *Salvia* and *Achyrospermum*), persistent, sometimes inflated or coloured (*Salvia*). C(5), 2-lipped, usually the adaxial 4 corolla lobes forming one lip, the 1 abaxial one the other, the lips equally or unequally developed in different genera. The 2+3 arrangement is less common in West African genera (*Leonotis* etc.). Stamens epipetalous, 4 (2 only, abaxial in *Salvia*, adaxial in *Hoslundia*) seldom accompanied by staminode(s), the filaments usually of two lengths, but rarely joined (*Leocus*, *Plectranthus*, *Solenostemon*); in most West African genera, the stamens and style lie along the lower corolla lip. Disc present, often of 4 or 2 lobes, at least 2 lobes secreting nectar, or of 1 lobe. G(2), 4-lobed and 4-celled, each cell with a single ovule, the style arising basally in the centre, the 4 lobes hardly attached to each other; *Tinnea* is the only West African genus in which the style arises at a higher level, and the 4 lobes are substantially attached to each other.

Pollination In genera in which the stamens and style lie along the lower lip of the corolla (flag flowers), nectar is secreted on the adaxial side of the disc, and insects receive pollen on the legs and underside of the body (or donate pollen from there on the next visit). When stamens and style lie in the upper corolla lip (gullet flowers), the reverse occurs, and pollen is carried on the back and head. Shorter-tubed flowers are usually pollinated by bees, longer-tubed ones by butterflies. In America, some *Salvia* spp. are pollinated by humming birds. The anther connectives are here prolonged and only one anther cell is fertile. The other one is sterile and disc like, and the two together block the entrance to the flower so that a pollinator, pushing against them, causes the two fertile anther cells to swing down, and to touch its back. Gill & Wolf (1977) reported pollination by sunbirds (*Nectarinia* spp.) of *Leonotis nepetifolia* in East Africa.

Most West African genera belong to the subfamily Ocimoideae, where two of the stamens are held under tension in the lower corolla lip, being released explosively when the lower lip is touched by an insect. The genera *Hyptis* (Brantes & de Vos, 1981), *Aeollanthus* (Hedge, 1973) and *Homalocheilos* (now

Fig. 31.1. A–E. *Salvia coccinea*. A. Flowering shoot × 1. B. Floral diagram. The adaxial stamen is never present, but the two indicated by dots are present in many other members of the family, i.e. A4. Disc abaxial, stippled. Abaxial lip formed of the three corolla lobes. C. Flower opened from the abaxial side x *c*. 28. D. Base of flower × 7. E. Calyx × 5. F–H. *Hyptis atrorubens*. F. Flower opened from the adaxial side × 5. G. Flower × 5, the stamens released from the lip as in F. H. Fruiting calyx × 5. c, connective of fertile half-anther; d, disc; l, lower lip. (A, C–H. from Rendle, 1938, Figs. 232 and 234 *pro parte*; B from Eichler, 1875–8.)

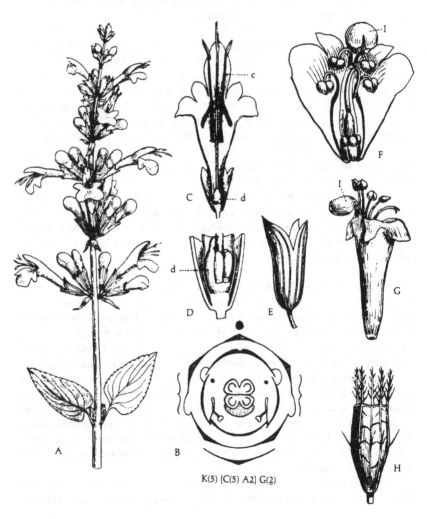

K(5) [C(5) A2] G(2)

Isodon; Codd, 1968) have been reported as having, or being likely to have this mechanism.

Fruit A schizocarp of 4 nutlets, surrounded by the persistent, and often enlarged, calyx. In *Hoslundia opposita,* a common savanna shrub, the ⊕ calyx is persistent and becomes yellow, fleshy and edible, so that the fruit resembles a 4-seeded berry.

Dispersal is generally either aided by animals, or is by ballistic means, or both. The persistent calyx is often either stiffly bristly (*Leonotis, Leucas*) forming a burr, or the fruiting cluster may be disturbed by a passing animal, and the nutlets dislodged. In many of these species the coat of the nutlet contains mucilage and swells up when moistened, becoming sticky and adhesive, cf. Chapter 29, Acanthaceae. In *Ocimum,* rain drops falling on the enlarged, concave adaxial sepal lobe are said to cause the calyx to vibrate and the nutlets to be shaken out. In this genus mucilage is also present. Ballistic dispersal may be followed by ant dispersal in species with elaiosomes. *Salvia* and *Tinnea* have inflated calyces, the latter containing nutlets densely covered with radiating barbed hairs. It has been suggested that hairs such as these may hold the nutlets in place while dispersal takes place. *Hoslundia opposita* false fruits are eaten by baboons.

Economic species The stem tubers of rizga or frafra potato (*Plectranthus esculentus*) and hausa potato (*Solenostemon rotundifolius*) are cultivated in savanna regions. If they are anything like as delicious as the tubers of *Stachys sieboldii* (East Asia, known in Europe as Chinese artichoke or Japanese potato), which store inulin not starch, increased cultivation would be worth while. The family is more widely known for its essential oils, imparting aroma or flavour to whatever they are combined with. Many species have culinary uses, or are used as fumigants or in domestic medicine. The essential oil of *Ocimum gratissimum* contains a high proportion of the antiseptic thymol.

> **Field recognition** *Englerastrum* (G. H. A. Engler, nineteenth-century German botanist) is a small tropical African genus, of which 3 species of small-leaved slender herbs grow in drier savannas in West Africa. All 3 species have blue flowers, in slender axillary racemes with leaf-like bracts, the larger lower corolla lip is keeled. The calyx is 5-toothed, tiny, straight and non-spiny. The 2 widespread, erect, annual species fruit after shedding their leaves, while *E. schweinfurthii,* a creeping perennial with persistent leaves, is more or less confined to the sudan zone.
>
> *Hyptis* (Gr. reflexed, the lower corolla lobe) is a large American–West Indian genus, many species of which are now pantropical weeds or in cultivation. All West African species are erect much-branched strong-smelling herbs with small pale flowers in dense spike- or head-like terminal or axillary cymes. *H. spicigera* has rather furry terminal spikes for example,

H. atrorubens axillary pedunculate heads. The calyx is 10-ribbed, with 5
equal teeth, and the lower lip of the corolla reflexed and quite different
from the upper lobes, marked with mauve in white flowers, not enlarged.
The nutlet coats in some species (e.g. *H. suaveolens*) are mucilaginous.
This species is also unusual in having blue flowers.

Ocimum (classical name) is an Old World genus somewhat similar to
Hyptis in appearance until flowering occurs. The inflorescences are interrupted spikes of white flowers, the calyx 5-lobed but with a larger adaxial
lobe which is either ± ovate when viewed from above or circular (in the
annual species). The stigma is 2-lobed. The calyx enlarges and faces downward in the fruiting stage, and has been shown to play a part in ballistic
dispersal.

Platostoma (Gr. flat mouth). *P. africanum*, a slender, erect annual herb
of damp places, is the only West African species. It is an *Ocimum*-like
plant, but with reflexed white flowers on erect pedicels. The calyx is deeply
2-lipped, the upper lip strongly reflexed, the lower curving upward to
meet it and closing the fruiting calyx round the nutlets. The corolla is
less markedly .l. than in *Ocimum*, and each flower has a tiny, coloured,
reflexed persistent bract.

Plectranthus (Gr. plectrum flower). This is the largest genus in West
Africa (14 species), and includes *P. esculentus* and 4 yellow-flowered species. The inflorescences are spreading panicles or globular nodal clusters
with very small bracts. *P. esculentus* loses its sessile, rather narrow hairy
leaves before producing panicles of relatively large flowers. The calyx
is unequally 5-lobed, the adaxial lobe being larger, the other 4 equally
developed, but the 2 lower teeth not fused for more than half their length.
The corolla is markedly .l., the lower lip hollow and keeled, the 4 stamens,
somewhat joined by their filaments, lying in it, very much as in *Solenostemon*.

Solenostemon (Gr. tube and stamen) is the second largest genus in West
Africa, and consists of bushy erect herbs with white, blue, violet or purple
flowers. *S. rotundifolius* is the hausa potato. *S. monostachyus*, of which
4 subspecies are recognised, is the most widespread species, 2 subspecies
being perennials, another 1 (or 2) being annuals. All have *Ocimum*-like
spikes with very small bracts, violet or purple flowers, with a large adaxial
calyx lobe and reduced abaxial ones (calyx 2-lipped in some other species,
but the 2 lower teeth always fused for more than half their length and
the lateral teeth very short or absent) with a hollow-keeled bottom lip
in which the joined stamens lie.

Bibliography

Brantjes, N. B. M. & de Vos, O. C. (1981). The explosive release of pollen in flowers
of *Hyptis* (Lamiaceae). *New Phytologist*, **87**, pp. 425–30.

Broadie, H. J. (1955). Springboard plant dispersal mechanisms operated by rain. *Canadian Journal of Botany*, **33**, pp. 156–67.

Codd, L. E. (1968). Notes on the genus *Isodon* (Benth.) Kudo. *Taxon*, **17**, p. 239.

Codd, L. E. (1971). Generic limits in *Plectranthus*, *Coleus* and allied genera. *Mitteilungen der botanischen Staatssammlung München*, **10**, pp. 245–52.

Eichler, A. W. (1875–8). *Blüthen-Diagramme*. Leipzig: Engelmann.

El-Gazzar, A. & Watson, L. (1970). A taxonomic study of Labiatae and related genera. *New Phytologist*, **69**, pp. 451–86.

Gill, F. B. & Wolf, L. L. (1977). Non-random foraging by sunbirds in a patchy environment. *Ecology*, **58**, pp. 1284–96.

Hedge, I. C. (1973). The pollination mechanism of *Aeollanthus njassae*. *Notes from the Royal Botanic Garden Edinburgh*, **32**, pp. 45–8.

Rendle, A. B. (1938). *Classification of flowering plants*, vol. 2, 1st edn. Cambridge: Cambridge University Press.

Robyns, W. (1966). On the status of *Acrocephalus* Benth. with some new species from Katanga (Congo Republic). *Botaniska notiser*, **119**, pp. 185–95.

32

Commelinaceae – day flower family

Herbs in the warmer parts of the world, including the New World, represented in West Africa by both annuals and perennials, the latter of quite diverse habit from erect, shrub-like and a few metres high to creeping plants rooting at the nodes or with various kinds of perennating organs (stolons in *Pollia*, rhizomes in *Palisota* spp., bulbs or tuberous roots in *Cyanotis*).

Members of the family may be recognised by their jointed stems bearing entire juicy leaves with convergent (parallel) venation, the blade arising from a closed sheathing base (sometimes with an intervening false petiole), its margins supervolute or involute when young. The flowers are small, three-part, delicate and short lived, in compound cymes, which are often subtended by spathe-like bracts. The fruit is generally a loculicidal capsule, and the seed always has an embryostega or operculum, a patch of testa (opposite the hilum), which is pushed off during germination.

Two species introduced from America are commonly grown in gardens. These are *Rhoeo spathacea*, with a rosette of purple and green sword-shaped leaves and white flowers, and *Zebrina pendula*, a trailing herb with purple-striped leaves and pink and white flowers.

Three genera are unusual in that the closed tubular sheath of the bract subtending the inflorescence is penetrated by the axillary inflorescence, which emerges through a hole. The genera concerned are *Buforrestia*, *Coleotrype* and *Forrestia*. *Commelina* is the only genus with leaf-opposed inflorescences, the remaining genera having the more common kind of terminal inflorescence. The bracteole is lateral in the family, in contrast to the usually abaxial bracteole in other Monocotyledons; it may be deciduous or even absent.

Flowers ± ⊕ (.l. in *Aneilema*, *Commelina*, *Pollia* and *Polyspatha*) ♂, the sepals and petals distinct and free except in *Zebrina*, K(3) C(3), and *Coleotrype* and *Cyanotis*, C(3). Flowers face either vertically upwards or horizontally, but both have fundamentally the same orientation (Faden, personal communication). In some horizontally facing spp. of *Murdannia*, e.g. *M. simplex*, the flower is twisted through 60°, the odd sepal (a lateral one) becoming abaxial. Stamens 3+3 in all ⊕ genera except *Murdannia* (3 outer stamens alternate with 3 staminodes); in .l. genera there are 3(–2) stamens and 0–4 staminodes in separate groups, the stamens abaxial, the staminodes adaxial; joined filaments occur in some species of *Aneilema*, *Coleotrype*, *Floscopa* and *Murdannia* but filaments are generally hairy, with large cells in which cytoplasmic streaming can readily be seen; anthers of various shapes, usually extrorse, opening by slits (basal pores in *Cyanotis*), the connective often wide; staminodal 'anther' 2-lobed (*Aneilema*), 3-lobed (*Murdannia*) or 4-lobed (X-shaped, *Commelina*), or lacking

Fig. 32.1. *Commelina* spp. A. Flowering shoot × ¼. B. Flower in side view, half the spathe removed, × 1. C. Spathe with flower in face view × 1. D. Staminode × 5. E. Lateral stamen × 5. F. Median stamen × 5. G. Seed × 2. H. Floral diagram. I–K. Flower orientation and androecium type. I. *Commelina-Aneilema* type, odd sepal uppermost, flowers face horizontally. J. *Murdannia* spp. (e.g. *M. simplex*)-type, odd sepal lowermost, flowers face horizontally, the flowers having twisted through 60° compared with those of I. K. *Stanfieldiella(–Murdannia* spp.–*Tradescantia*)-type, as J but a fully fertile androecium. (I–K from Brenan, 1966. Figs. 8–10.)

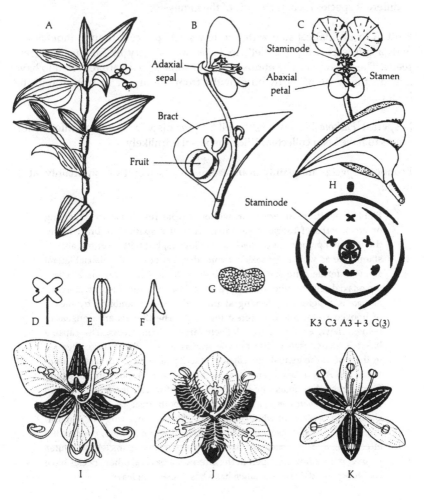

(*Palisota*). G(2) in *Aneilema*, *Floscopa* and *Polyspatha*, but otherwise G(3) 3-celled, 1 to several seeds per axile placenta.

Pollination The flowers are shallow and are visited by a great variety of short-tongued insects, the genera lacking staminodes offering pollen, those with staminodes offering a milky juice which develops in the food anthers of the staminodes. *Palisota* spp. may be strongly scented, and *Aneilema* spp. moderately so. Self-compatibility seems to be the rule, the extreme case being found in *Commelina benghalensis* and *C. forskalaei*, which both produce a proportion of cleistogamous flowers on underground stems. In other species of *Commelina*, and in *Aneilema*, however, some ♂ flowers are produced in each inflorescence, and as these open before the ♀ flowers, some cross-pollination is likely to take place. The brief flowering period at different times of the day in different species is characteristic of the family.

Fruit A loculicidal capsule with 1 to few seeds per cell, though indehiscent with a brittle pericarp ('nutlet') in *Pollia* and a berry in *Palisota*, with few seeds. These have starchy endosperm and some species are reported to have arils (in the sense of fleshy outgrowths from the funicle or from round the hilum).

Dispersal *Palisota* berries might well be bird dispersed, but no information seems to have been collected. Wide dispersal is unlikely.

Economic species are mainly troublesome weeds of cultivation, mainly introduced.

> **Field recognition** *Aneilema* (Gr. without involucre). These are straggling or erect herbs of forest or savanna, without a spathe-like bract round the terminal panicles of *Commelina*-like flowers, but with a central abaxial staminode as well as 3 adaxial staminodes, i.e. only the 2 abaxial lateral stamens fertile. The staminodes are 2-lobed, and the ovary is 2-celled, the adaxial cell aborting, or, if developed, it may be indehiscent. *A. aequinoctiale* is a perennial, rooting at the nodes and scrambling by means of hooked hairs, found in forest regrowth. The flowers are unusual in being yellow, and in being relatively large, 2.5 cm across. The capsule has 1–3 seeds, each with a narrow scar as well as the embryostega (as in the rest of the genus), the capsule with 2 square 'top' corners. *A. beniniense* and *A. paludosum* can be encountered as weeds. Both are straggling fibrous-rooted annuals, the former in lowland forest, the latter in damp savanna. *A. umbrosum* is a similar herb in both areas. *A. beniniense* has close ovoid panicles of white/pink/mauve flowers, opening in the morning; *A. paludosum* dense spike-like panicles of white flowers, open in the afternoon; and *A. umbrosum* loose, little-branched panicles of white/mauve/purple flowers, open in the afternoon. Several other species have root tubers, and *A. pomeridianum* has a basal rosette of leaves.

Commelina (J. and C. Commelin, seventeenth–eighteenth-century Dutch botanists) – day flower. This is a large subtropical and tropical genus, represented in West Africa by about 25 species. These are straggling or prostrate herbs, often rooting at the nodes, with a fibrous root system, and seldom as much as 1 m high. The cymes are leaf-opposed and each is surrounded by a stalked spathaceous bract. The .l. flowers differ from those of *Aneilema* in having 3 abaxial stamens, the central one differing from the 2 laterals, and irregularly lobed or several-lobed staminode anthers. The capsule is 3(–2)-celled, the odd one adaxial as in *Aneilema*. There are several yellow-flowered species, including *C. aspera* and *C. nigritana*, which are widespread in savanna, and *C. africana* and *C. capitata*, both in forest, the latter with red hairs, at least on leaf sheaths. The remaining species have white, blue or violet flowers, e.g. *C. ascendens* is a blue-flowered climber with hooked hairs climbing to *c.* 3 m, occurring especially in riverain forest.

Cyanotis (Gr. blue, the colour of some flowers). These are small herbs mostly with basal rosettes of leaves and fibrous root systems, or with root tubers, stem tubers or bulbs, mostly found in savanna or among rocks. The edges of the leaves are not inrolled with young. The inflorescences each have a spathaceous bract, and the flowers are, in addition, each subtended by a bract. Each flower is white or some shade of blue or purple, ⊕, with 5–6 stamens, the anthers opening basally by pores, and with hairy filaments. *C. lanata* is probably the commonest species. It is a cottony annual or perennial (= *C. arachnoidea*) herb, in the latter case spreading and perennating by means of stolons, which root at the nodes, with purple, pink or white corollas of united petals. This is one of the species without a basal rosette of leaves. *C. longifolia*, a species with tuberous roots and basal rosette, is the only species to be found in Gambia, and also occurs throughout tropical Africa.

Floscopa (L. flower and broom, the inflorescence). Creeping or tufted herbs of swampy places, with terminal cymes not enclosed by their bract. The flowers are ⊕, purple, blue or white with 6 stamens each, and followed by 2-celled capsules with 1 seed in each cell. Three common species of wet habitats are *F. africana*, *F. aquatica* and *F. glomerata*. The leaf of *F. africana* has a false petiole, while that of the other 2 species does not. *F. aquatica* is a creeping herb in swampy ground or free-floating in water. It has leaves up to only 7 cm long, sparsely branched cymes, while *F. glomerata*, which may also appear as a weed, has leaves up to 11 cm long and dense, much-branched cymes. *F. flava*, with a yellow corolla, and *F. axillaris*, with a pink(–purple) one are easily distingiushed.

Murdannia (Murdann Aly, keeper of the herbarium, Saharunpore, Calcutta) – 2 species of pantropical weeds probably introduced to West Africa, and a tropical African species, *M. tenuissima*. Herbs with ⊕ flowers in which the 3 stamens alternate with the 3 3-lobed staminodes, the fruit a 3-celled capsule. Like *Floscopa*, the inflorescence bract is remote and does not enfold it. The 3 species are rather different in habit, and have different flowering times. *M. simplex* is up to *c*. 1 m high, with strap-shaped leaves, up to 40 cm long, at the base of the plant. The flowers are rather large, *c*. 2–2.5 cm Ø, and the pedicels leave bracket-like scars on the peduncle. *M. nudiflora* is an altogether smaller prostrate plant, and *M. tenuissima* is grass like in habit, the flowers lacking staminodes.

Palisota (Palisot de Beauvois, author of *Flore d'Oware*, 1805–). Erect forest herbs, sometimes spreading over large areas, and exceptionally up to 6 m high. Some West African species have basal rosettes, the height of the plant being contributed by the peduncle, while others carry the rosette just under the inflorescence (terminal rosettes). The leaves tend to be large, 30–130 cm long, with a conspicuously silky-hairy margin; a false petiole occurs in most species. The inflorescence is panicle-like, without a spathaceous bract, and the flowers are ⊕ with 3 stamens alternating with 3 staminodes. The fruits are berries, yellow, orange red or purple, often formed in great masses. *P. barteri* and *P. hirsuta* are the most widespread species. *P. barteri* has basal or terminal rosettes and dense inflorescences (usually the former) with white flower and green–red–purple berries. *P. hirsuta* is a taller plant with only terminal rosettes each associated with several loose inflorescences of white-to-purple flowers. The berries are shining and black.

Stanfieldiella (D. P. Stanfield, Senior District Officer and botanist, Nigeria, d. 1971). Four species of creeping forest herbs, rooting at the nodes, with spatheless cymes of white ⊕ flowers with glandular hairs, opening at different times of day in different species. Each flower has its own ± cupular bract, but otherwise resembles *Murdannia*, though with 6 stamens and smaller leaves, up to 10 cm long. There are 2–10 seeds per capsule cell, and in *S. axillaris* the 'capsule' is indehiscent, the seeds being released when it rots. *S. imperforata* is the most widespread species in West Africa. It has glandular hairs 0.5 mm or more long on the outside of the sepals and on the pedicels.

Bibliography

Brenan, J. P. M. (1966). The classification of the Commelinaceae. *Journal of the Linnean Society (Botany)*, **59**, pp. 349–70.

Faden, R. B. & Suda, Y. (1980). Cytotaxonomy of Commelinaceae. *Botanical Journal of the Linnean Society*, **81**, pp. 301–26.

Jackson, G. & Davies, J. S. (1966). Notes on West African vegetation. II. A note on *Cyanotis lanata* Benth. and other possible *Cyanotis* species found on Nigerian inselbergs. *Journal of the West African Science Association*, **10**, pp. 99–102.

Morton, J. K. (1967). The Commelinaceae of West Africa: a biosystematic survey. *Journal of the Linnean Society (Botany)*, **60**, pp. 167–221.

Owens, S. J. (1981). Self-incompatibility in the Commelinaceae. *Annals of Botany*, **47**, pp. 567–81.

33

Zingiberaceae – ginger family

A mainly Old World family, represented in West Africa by only four genera and *c.* 45 species in diverse habitats, but mainly in forest. *Aframomum* and *Kaempferia* are tropical African in distribution, while *Renealmia* and *Costus* are chiefly American.

All are perennial herbs with fleshy rhizomes occasionally with tuberous roots, producing terminal cymose inflorescences of conspicuous ± sweetly scented flowers. The inflorescences are either carried separately on the rhizome from the unbranched leafy shoots or leafy pseudostem (formed, banana-fashion, in *Kaempferia* and two species of *Renealmia*) or they appear as branches at the base of the leafy shoot. In some *Costus* spp., and *Hedychium coronarium*, the inflorescence is terminal on the leafy shoot, while in C. *dinklagei* and four West African species of *Aframomum* (amongst others), the inflorescences are lateral and the shoots bearing them hapaxanthic (Hallé, 1979). The leaves are arranged distichously, those of *Aframomum*, *Alpinia* and *Renealmia* in a plane at right angles to the length of the rhizome, those of *Kaempferia*, *Curcuma*, *Hedychium* and *Zingiber* in line with the rhizome. Phyllotaxy in *Costus* is spiral. Each leaf has an entire blade ± false petiole, and an open leaf sheath. There is often a ligule or a pair of ligules but never a pulvinus. The leaf blade is closely pinnately veined, the two margins symmetrical except perhaps at the very base.

The inflorescences can be congested and cone-like, with prominent spiral bracts. In turmeric the bracts are joined, so that each flower develops in a pocket, while in *Hedychium* and *Renealmia*, the bracteole is tubular.

Costus is sometimes placed in a separate family (Costaceae) on account of, for example, its non-aromatic rhizomes with branched aerial stems bearing leaves (in a spiral) with closed sheaths – fused to the stem – and terminal inflorescences, petaloid stamen etc.

Apart from economic species, the decorative garden ginger lilies, *Alpinia purpurata*, *Hedychium coronarium* and *Zingiber spectabile* have been introduced. All West African genera except *Costus* are liable to be called ginger lilies.

Members of the family may be recognised by their vegetative characters together with their three-part epigynous flowers with one two-locular stamen. The conspicuousness of the flower is provided by the petaloid staminodes, such that the flower may appear to have a one- to four-lobed corolla.

Flowers ÷ .l. ♂ 3-part. K(3) persistent, spathaceous, thin and papery in

Fig. 33.1. A. Floral diagram of a zingiberaceous flower. B. *Kaempferia aethiopica* flower × $\frac{3}{5}$. C–D. *Renealmia cincinnata*. C. Flower in side view, diagrammatic. D. Flower in face view, diagrammatic. E. Fruit × $1\frac{3}{5}$. F–H. *Costus afer*. F. Flowering shoot × $\frac{3}{5}$. G. Two nodes of stem × $\frac{4}{5}$. H. Flower in adaxial view, diagrammatic. I. *Aframomum sceptrum* ligules at junction of leaf blade (above) and false petiole (below) × $\frac{4}{5}$. Close hatching, bract; sparse hatching, bracteoles; stipple, calyx; white, corolla; black, stamen (also staminodes in floral diagram); c, connective; l, lip; ls, lateral staminode; p, petal lobe. (From Koechlin, 1965, Pls. 1, 4, 5, 10 and 20 *pro parte*.)

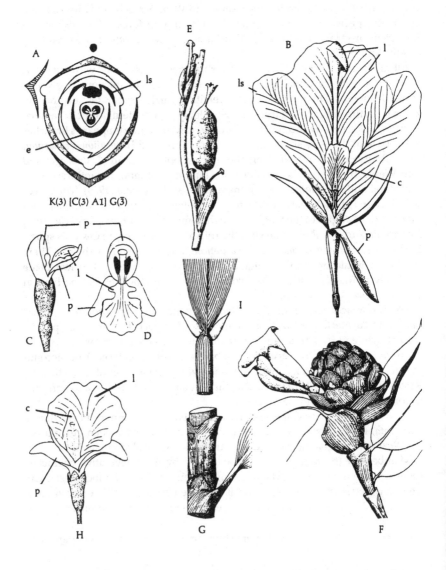

Aframomum. C(3) the adaxial lobe the largest and often the most erect. Stamens basically 3+3, but only the inner adaxial one fertile; this has a large petaloid filament and connective in *Costus*, where the other 5 stamens are represented by a large petaloid abaxial lip (labellum); in the other genera in West Africa, the lip is composed only of the 2 lateral abaxial staminodes of the inner whorl, while the outer whorl is represented only by the 2 lateral adaxial staminodes; these may be small (*Aframomum, Renealmia, Zingiber*) or large and petaloid (*Curcuma, Kaempferia, Hedychium*); the connective is crested in *Aframomum, Kaempferia* and *Zingiber*, basally prolonged into 2 horns in *Curcuma*; the trough-shaped crest embracing the style in *Zingiber* is unique but even where the anther is versatile, as in *Curcuma*, the anther and style are closely associated. There is a pair of epigynous nectaries except in *Costus*, where the nectaries are immersed in the ovary tissue, and *Renealmia*, where the nectary is a ring of tissue round the base of the style. G(3) 3-locular with numerous ovules on axile placentae. The style is terminal and bears a funnel-shaped fringed stigma, which, because the style lies in the groove between the anther loculi, comes to stand above them.

Pollination The flowers are, in general, very short-lived, less than 24 hours, and protandry has been observed in some cases. Bee pollination of *Aframomum* spp. has been observed, while the white flowers of *Hedychium*, which have a very long tube, are more probably pollinated by moths.

Fruit A beaked berry, or a capsule (*Renealmia*, partially dehiscent in *Costus*), with arilloid seeds. The seed reserve is starchy perisperm (also chalazosperm in *Costus*).

Dispersal Some species fruit in, or just under, the soil crust (some *Aframomum* spp., *Kaempferia*) and *Costus* capsules release their seeds into mucilage held in the bract. Both instances suggest dispersal methods different from that of the exposed red berries of *Renealmia* and *Aframomum* spp. The colour would attract birds but the spicy aroma would attract mammals in particular. *Costus afer* germinates from elephant droppings.

Economic species Ginger (*Zingiber officinale*) is much cultivated in West Africa, and turmeric (*Curcuma longa*) is occasionally cultivated along with it, though as a dye plant rather than as a source of spice. *Aframomum melegueta* (alligator (= melegueta) pepper, guinea grains, grains of paradise), is cultivated in forest areas, and the seeds are used as a spice and in medicines, in much the same way as cardamoms in Arabia and further east. The fruits of several other species are gathered for similar purposes.

Field recognition *Aframomum* (L. African *Amomum*, the latter the name of certain eastern spices). Most of West African species are forest herbs. Some species produce separate inflorescences and leafy shoots, others inflorescences at the base of leafy shoots, and the former group has not always been adequately collected. *A. sceptrum* is one of the latter group of species, with large pink or mauve flowers in cone-like inflorescences at the base of leafy shoots up to 2 m high. The leaves are narrow, about 5 times as long as wide. The berries are red, elongated and beaked, containing seeds in mucilaginous arils. Another widespread species is *A. geocarpum*, which has similar flowers but a different habit, and has been confused with *A. sceptrum* and *A. leptolepis* (Lock, 1980). In *A. geocarpum*, flowers are produced at ground level on shoots remote from the leafy ones, and the fruits are globular, in or below the soil crust. Only 3 species occur in savanna. *A. latifolium* (now *A. alboviolaceum*; Lock, 1979) is the widespread savanna species, with leafy shoots up to 2 m high, and separate inflorescences of pink or whitish flowers at ground level. *A. baumannii* (now *A. angustifolium*; Lock, 1980) grows in forest patches and strips in savanna and has red (a rare colour in the genus) and yellow flowers with a narrow lip on either separate shoots or at the base of leafy shoots, and it seems likely that 2 species are involved. *A. elliottii* is also entered in the *Flora of West tropical Africa* as a species of wet ground, from coastal swamp to savanna grassland, but would also appear to confuse 2 species, both with white flowers: *A. elliottii* is a forest species with hairy leaf sheaths, and *A. rostratum* a coastal savanna species with glabrous leaf sheaths. Neither occurs further east than Côte d'Ivoire or Ghana.

Costus (Gr. *Kóstos*, peppery root, Arab. *qust*). There are about a dozen, mainly forest, species. *C. lateriflorus* and *C. talbotii* are epiphytes (S. Nigeria–Gabon). The most widespread species are *C. spectabilis* in rocky savanna and *C. afer* in forest. The former is a rosette herb with up to 4 fleshy ± circular leaves spread on the ground, the inflorescence of reddish-yellow flowers terminal on the short axis. *C. afer* bears cone-like inflorescences of yellow flowers at the tips of leafy shoots 2–3 m high. *C. engleranus* (in moist forest) forms ground cover by means of stolons. *C. dubius* (Liberia–Cameroun) is a species with white flowers in separate inflorescences from the leafy stems, which are up to 2 m high. Nectar is produced in all *Costus* spp. in a pair of pits in the septa of the ovary, and the inflorescences produce mucilage. The arilloid in *Costus* is white, and the seeds of each loculus adhere together.

Kaempferia (E. Kaempfer, seventeenth-century German physician and traveller) (now *Siphonochilus*; Burtt, 1982) (spice lily). The 2 West African species are sometimes known as resurrection lilies, the flowers appearing on the ground in the dry season, before the pseudostem, or tuft of leaves, appears. Both are savanna species, *K. nigerica* in dry savanna, *K. aethiopica* in damp savanna. The former has a short rootstock and tuberous roots, giving rise first to pink flowers, then to leafy shoots. *K. aethiopica* has purple or blue flowers and leafy stems up to 60 cm high arising from a segmented corm-like stem with fibrous roots. The labellum and 2 adjoining lateral staminodes are particularly large in this genus.

Renealmia is the only genus with paniculate inflorescences (of small white, pink or orange flowers), followed by red, later black, capsules. The 6 West African species are of local distribution only, suggesting that they may be undercollected. *R. battenburgiana* (Ghana) and *R. longifolia* (Liberia–Côte d'Ivoire) have the most developed pseudostems in the family, with 'terminal' inflorescences on leafy stems respectively 30 cm and 0.9–1.5 m high. The other species have separate leafy and flowering shoots.

Bibliography

Burtt, B. L. (1972). General introduction to papers on Zingiberaceae. *Notes from the Royal Botanic Garden Edinburgh*, **31**, pp. 155–65.

Burtt, B. L. (1982). *Cienkowskiella* and *Siphonochilus* (Zingiberaceae). *Notes from the Royal Botanic Garden Edinburgh*, **40**, pp. 369–73.

Hallé, N. (1979). Architecture du rhizome chez quelques Zingiberacées d'Afrique et d'Oceanie. *Adansonia*, sér. 2, **19**, pp. 27–44.

Holttum, R. E. (1950). The Zingiberaceae of the Malay Peninsula. *Gardens' Bulletin*, Singapore, **13**, pp. 1–249.

Koechlin, J. (1965). *Flore de Cameroun: Scitaminales (Musacées, Strelitzacées, Zingibéracées, Cannacées, Marantacées).* Paris: Muséum national d'histoire naturelle.

Lock, J. M. (1979). Notes on the genus *Aframomum* (Zingiberaceae) 4. The savanna species. *Bulletin du Jardin botanique nationale belge*, **49**, pp. 179–84.

Lock, J. M. (1980). Notes on *Aframomum* (Zingiberaceae) in West Africa, with a new key to the species. *Kew Bulletin*, **35**, pp. 299–313 with pl. 8.

34

Marantaceae –
(West Indian) arrowroot family

Perennial forest herbs with sympodial fibrous or woody rhizomes and aerial stems bearing distichously arranged leaf sheaths or foliage leaves, both kinds eligulate, each stem ending in an 'inflorescence' (the synflorescence of Andersson, 1976). This is a compound bracteate cyme (synflorescence) with fragile flowers in pairs. This is a mainly New World family, with relatively few species in Africa, Asia and the Pacific area.

The leaves have open sheaths ± a false petiole, but always with a pulvinus (calloused portion) next to the leaf blade. In *Megaphrynium* and *Thaumatococcus* the pulvinus is particularly long, about 10 cm. The leaf blade is entire, ± ovate-elliptic and asymmetrical, one margin being ± parallel to the midrib, the other curved. Venation is close and pinnate, cf. Chapter 33, Zingiberaceae. In the bud, the straight-edged leaf half is rolled round the other half (ptyxis supervolute). Usually, corresponding halves of the leaf blade are either straight or curved (homotropy, *Trachyphrynium*), but occasionally leaves on one side of the stem will show symmetry consistently opposite to that of leaves on the other side (anitropy, *Marantochloa congensis*).

The aerial stems appear forked, often in a zig-zag way, but bear true lateral branches. Each branch bears a two-keeled prophyll (scale) as its first leaf on the adaxial side.

Various developmental patterns exist in the family, according to the relative development of aerial stems and petioles. *Hypselodelphys* and *Trachyphrynium braunianum* are bamboo-like climbers, 2–10 m tall, the aerial stems bearing, basally, leaf sheaths, then relatively small leaves and, terminally, a leafy branched inflorescence. *Haumania* is of similar construction but bears recurved prickles on the stems. Height for the leaf blades is obtained another way in *Thaumatococcus daniellii*, with petioles 2–3 m long, each petiole arising on a short aerial stem, which terminates in an inflorescence very near the ground. In *Afrocalathea* and *Megaphrynium*, leaves and inflorescences arise from separate shoots of the rhizome, as in the Zingiberaceae, a rare condition in the Marantaceae. The remaining genera have intermediate developmental patterns, and, generally speaking, large leaf blades.

Members of the family may be recognised by vegetative characters and the inflorescence of three-part, asymmetrical epigynous flowers with one one-locular anther. Each unit of the inflorescence is, as a rule, a two-flowered cymule in which the left- and right-hand flowers are mirror images of each other, a situation most obvious in the androecium.

West Indian arrowroot (*Maranta arundinacea*), which has erect branched aerial stems, has certainly been introduced, *Thalia welwitschii* possibly so.

Flowers Asymmetrical ♂ 3-part. K3, petaloid in some *Marantochloa* spp., *Megaphrynium* and *Trachyphrynium*. C(3) small. A(3)+(3) epipetalous, but irregular, largely staminodal and petaloid, forming a corona, only the 'adaxial' inner half-stamen represented by a 1-locular anther; the 2 'abaxial' staminodes of the inner whorl form the fleshy and hooded staminodes respectively, the latter sometimes tailed (*Maranta*) or spurred (*Trachyphrynium*); of the outer whorl, only the 2 lateral ('adaxial') staminodes remain, as in the Zingiberaceae (absent altogether in *Thaumatococcus*). G(3̄) 3-locular (1-locular in *Thalia*, 1-locular by abortion in *Maranta*) with a basal ovule in each cell; style terminal, thick and curved, basally fused with the corolla tube, at first held in the hooded staminode under tension.

Pollination Nectar is secreted by septal glands opening on to the top of the ovary, and it accumulates in the bottom of the corolla tube. Pollen is shed

Fig. 34.1. A. *Hypselodelphys* expanded shoot (left) and expanding shoot (right). Shoots bamboo-like, with only bladeless sheaths, leaves with blades confined to lateral branches. B. *Thaumatococcus*. C. *Marantochloa filipes* (basal leaf indicated for *M. purpurea* only). *Ataenidia* is similar, but the inflorescence is congested. D. *Marantochloa cuspidata* is similar to *Thaumatococcus*, but the aerial stem is longer. Also seen in *Megaphrynium* reproductive shoots. E. *Halopegia*, also *Thalia*. *Sarcophrynium* has only one or two basal leaves, but is otherwise similar. Black, vegetative axes and bladed leaves; white, inflorescence axes and buds (also flowers in A) and bladeless sheaths p, prophyll. (Adapted from Tomlinson, 1961, Figs. 2, 4, 6, 9, 11.)

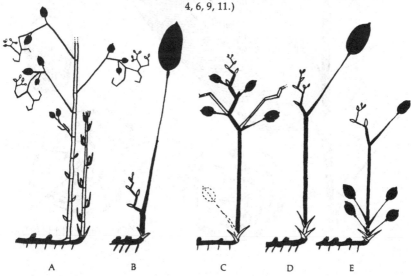

A   B   C   D   E

Fig. 34.2. A–D. *Trachyphrynium braunianum*. A. Flowering and fruiting
shoot × $\frac{3}{4}$. B. Corona opened × 2. C. Fruit × 2. D. Seed × 2. E. Plan
of inflorescence. F. Floral diagram of right-hand flower. Dense hatching,
bract; sparse hatching, bracteoles; stipple, calyx; white, corolla; black,
androecium and position of inflorescence axis; × positions of floral axes;
f, fertile stamen; h, hooded staminode, s, stigma; t, thickened staminode.
(Adapted from Koechlin, 1965, Pls. 21., 21.6, 21.8, 21.9; pl. 2L and M.)

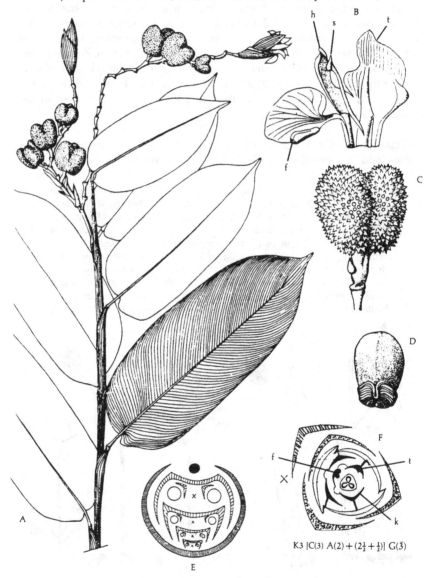

K3 [C(3) A(2) + (2$\frac{1}{2}$ + $\frac{1}{2}$)] G($\bar{3}$)

in the bud on to the style. Long-tongued insects, principally bees, landing on the fleshy staminode, disturb and release the style while probing for nectar, and the stigma springs down on to the insect's back, collecting any pollen that may already be there, while pollen is scattered over the insect from the style. Self-pollination would appear to be virtually impossible.

Fruit In a few genera there is a capsule (*Marantochloa, Megaphrynium, Trachyphrynium*) but a berry is more common in West Africa. Seeds are nearly always arilloid, in *Hypselodelphys* and *Sarcophrynium* with a well-developed pulpa.

Dispersal Birds are the most likely agents. Not only is the arilloid or pulpa a reward, but there is plentiful mealy perisperm and chalazosperm in each seed. Externally the fruits are most often red. The arilloid of seeds in capsular fruits is turgid and may also aid dehiscence.

Economic species include West Indian arrowroot, from the rhizomes of which starch may be washed out. The rhizomes are too fibrous to be used as a vegetable. The arilloid in the three-winged fruits of *Thaumatococcus daniellii* (katemfe) contains the protein thaumatin, a sweetener said to be up to 1600 times sweeter than sucrose, weight for weight. Commercial production should be investigated, since existing sugar substitutes are often unsatisfactory. This species fruits in or below the soil crust.

Field recognition *Halopegia. H. azurea*, the only West African species, occurs in swampy ground as a tufted plant, the ribbed leaf blades very asymmetrical and rather narrow on petioles *c.* 1 m high. The inflorescences are shorter, with pairs of mottled purple flowers each with a yellow staminode. The fruits are 1-seeded and cylindrical, developing within a silky-hairy receptacle. Pneumatophores commonly develop on the roots of the rhizome. This is the only one of the West African genera to extend into Asia.

Marantochloa (*Maranta* and Gr. grass) is a tropical African genus of *c.* 15 species, of which 9 occur in West Africa, constituting the largest genus. There are both branched ± species, 3–4 m high, such as *M. filipes* and *M. purpurea* etc., and unbranched smaller species, such as *M. cuspidata* and *M. holostachya*, in which a large solitary leaf subtends in the inflorescence in the former species, while the leaf arises directly from the rhizome in the latter species. Leaves are strongly asymmetrical and also homotropous, except in *M. congensis*. The inflorescences are erect and panicle or spike like, with white, pink, purple or yellow flowers, followed by smooth red capsules containing 3 dark seeds, each with a whitish arilloid. North of the main forest areas, only *M. leucantha* and *M. purpurea* are found. Both have a branched habit, *M. purpurea* also with a single leaf at the base of the erect shoot. Both species have lax inflorescences,

M. leucantha of white flowers at various times of the year, *M. purpurea* of purple flowers mostly in the dry season.

Megaphrynium (Gr. large and *Phrynium*) is a small genus of equatorial Africa, only 2 of its 3 species extending into West Africa. The more common one is *M. macrostachyum*, which carries large lax inflorescences on stems 2–3 m long, each inflorescence subtended by a ± symmetrical leaf. With only 1 pair of flowers and 1 fleshy bracteole per node, the inflorescence appears sparse, because the bract, though initially extending to the node above, falls when the flower opens. Other leaves with petioles 1–2 m long arise singly from the rhizome, each with a large blade (40 cm × 25 cm), ± asymmetrical and a very long pulvinus (*c*. 10 cm). The red capsules contain black arilloid seeds.

Sarcophrynium (Gr. flesh and *Phrynium*). The 2 West African species of this small tropical African genus are of rather different habit. The more widespread, also north of the lowland forest area, is *S. brachystachys*. It has at most 2 leaves at the base of each erect branch, and another subtending the compact inflorescence of white flowers with persistent bracts. The leaves are up to 30 cm × 15 cm, ± asymmetrical, on petioles up to 1.5 m long with a long (4–6 cm) pulvinus. The purely forest species *S. progonium* is larger in most ways and has lax panicle-like inflorescences. The red berries of both species contain seeds in pulpa. The generic name refers to the fleshy bracteoles accompanying each pair of flowers.

Thalia (after J. Thalius, sixteenth-century German naturalist). *T. welwitschii* is our only representative of this tropical American–African genus of 11 species. It is a tufted swamp plant with leaves up to about 1 m high, the leaf blade ± asymmetrical, up to 40 cm × 20 cm, the margin much thickened. The flowers are purple, each pair subtended by a deciduous bract, and produced mainly in the dry season. The fruits are 1-locular berries. The bases of the petioles and erect branches are spongy, and contain aerenchyma.

Bibliography

Andersson, L. (1976). The synflorescence of the Marantaceae. Organization and descriptive terminology. *Botaniska notiser*, **129**, pp. 39–48.

Holttum, R. E. (1951). The Marantaceae of Malaya. *Gardens' Bulletin*, Singapore, **13**, pp. 254–96.

Koechlin, J. (1965). *Flore du Cameroun: Scitaminales (Musacées, Strelitzlacées, Zingibéracées, Cannacées, Marantacées.)* Paris: Muséum national d'histoire naturelle.

Tomlinson, P. B. (1961). Morphological and anatomical characteristics of the Marantaceae. *Journal of the Linnean Society (Botany)*, **58**, pp. 5–78.

35

Liliaceae – lily family

As defined in the *Flora of West tropical Africa* (1968), this is a family of sympodial herbs with strap-shaped, mainly supervolute leaves ± pseudopetiole, having parallel veins converging at the leaf tip, perennating by rhizomes, stem tubers or bulbs. Flowers are three-part in terminal inflorescences lacking bracteoles and with superior ovaries. There are genera with leafy aerial stems and rosette herbs in which the only aerial axis is a leafless peduncle (scape).

On the basis of a great amount and variety of information (Dahlgren & Clifford, 1982), a new classification is presented (Dahlgren, Clifford & Yeo, 1985)(Table 35.1) in which the Liliaceae as so defined is replaced by six new taxonomic units at the family level. These new families have the advantage of now being as distinct but internally cohesive as for example, the Amaryllidaceae and Iridaceae in Hutchinson & Dalziel's (1927–36) treatment. A synoptic key to the six families is included for convenience (Table 35.2). Each family displays a complex of characters, including easily observed morphological ones.

The largest, most common and widespread genera in the former Liliaceae are the bulbous herbs, now Hyacinthaceae, and two genera with erect 'rootstocks' which otherwise resemble the bulbous genera in many features (now Anthericaceae). These two groups of genera are described in Field recognition, below. The remaining genera are all of distinctive habit, possessing well-developed, persistent and leafy aerial stems.

In the savanna shrubs and climbers of *Asparagus* (Asparagaceae), there is a recurved leaf spine at each node and, in its axil, a cluster of cladodes (leaf-like stems each of one internode's length), together with axillary racemes of small ♂ or ♀ flowers, which are followed by red berries. The perennating part of the plant is a short rhizome bearing fleshy roots.

There are at most 2 species of tendril leaf climbers in the genus *Gloriosa* (Colchicaceae) (*G. simplex* in savanna and *G. superba* in forest). The aerial shoot arises from a deeply buried V-shaped tuber, each arm of the V developing a terminal bud from which a new tuber and aerial shoot arises next season.

Shorter but persistent leafy stems with secondary thickening are seen in *Aloe* spp. (Asphodelaceae), which have sword-shaped fleshy leaves in which the venation is obscured. The leaves are saw-edged and form a rosette, from which axillary panicles of red flowers emerge.

Flowers (in Anthericaceae and Hyacinthaceae) $\pm$ $\oplus$ ⚥ 3-part. Perianth 3+3 petaloid, a stamen opposite each sepal (joined and attached to the perianth tube in *Drimia* and *Dipcadi*); anthers introrse ± basifixed. Nectaries septal. G(3) 3-locular with numerous ovules on axile placentae.

Pollination Nectar is produced and most flowers are also scented. When the sepals are free, a shallow open-access flower, in which nectar is available

Table 35.1 *Comparison of two classifications of the Liliales*

Hutchinson & Dalziel (1931)		Dahlgren *et al.* (1985)	
Order	Family genera	Order	Family
Liliales	Liliaceae Genera[a] *Gloriosa* *Iphigenia* *Wurmbea*	Liliales	Colchicaceae
	Aloe *Kniphofia*	Asparagales	Asphodelaceae
	Asparagus		Asparagaceae
	Eriospermum		Eriospermaceae
	Anthericum		Anthericaceae
	Chlorophytum		
	Albuca		Hyacinthaceae
	Dipcadi		
	Drimia		
	Drimiopsis		
	Scilla		
	Urginea		
	Tecophilaeaceae Genus *Cyanastrum*	Asparagales	Cyanastraceae
	Pontederiaceae Genus *Pontederia*	Pontederiales	Pontederiaceae
	Smilacaceae Genus *Smilax*	Dioscoreales	Smilacaceae
Amarylli- dales	Amaryllidaceae[b]	Asparagales[c]	Amaryllidaceae
	Iridaceae[b]	Liliales	Iridaceae
Dioscoreales	Dioscoreaceae Genera *Dioscorea*	Dioscoreales	Dioscoreaceae
Agavales	Agaveaeae Genera *Dracaena*[d] *Sansevieria*	Asparagales	Dracaenaceae
	Agave *Furcraea*		Agavaceae

[a] In West Africa.
[b] All indigenous genera in West Africa.
[c] Alliaceae is another family in this order.
[d] See Mouton (1976).

to short-tongued insects, is formed, e.g. the flies which visit *Urginea* spp., though *U. indica* is said to be pollinated by moths, *U. ensifolia* by butterflies as well as by bees and flies.

Fruit A loculicidal capsule, the seeds angular or flat, even papery, but black as a result of a layer of a carbon-rich compound, phytomelan, deposited in the testa. Occurrence of this compound is limited to the *Asparagales* as defined by Dahlgren *et al.* (1985).

Dispersal A censer mechanism is most probable, the flat or papery seeds being more effectively dispersed.

Economic species Frequently, local use is made of gathered material.

> **Field recognition** *Albuca* (? L. white, without lustre) (yellow squill). Bulbous plants of savanna with racemes of yellowish bell-shaped flowers with persistent bracts, above a basal tuft of narrow grooved leaves. The capsules are ± ovoid, slightly lobed and narrowed at the tip, with angular black seeds. There are about 50 species in Africa, the most widespread species being *A. nigritana* and *A. sudanica*. Both have slender racemes (of greenish-yellow flowers) at least 30 cm tall, the leaves appearing during and after flowering, which occurs in the dry season. In *A. nigritana* the fruit pedicel elongates to more than 5 mm in length, while in *A. sudanica* it remains at flowering length, under 5 mm. *A. abyssinica* is confined to damp hollows in hill savanna, and is on the western edge of its range. It has pale-yellow flowers in striking racemes, up to 2 m high in the dry season, before the leaves appear. Two more species have been described (*A. scabromarginata* and *A. fibrotunicata*) by Gledhill & Oyewole (1972).
>
> *Anthericum* (Gr. straw, ?the grass-like appearance) (grass lily). This is a large genus of mainly African distribution, with tufts of narrow leaves

Table 35.2 *Synoptic key to the constituent West African families of Hutchinson's Liliaceae and the Alliaceae*

1 Fruit a berry; seeds with phytomelan[a]; cladodes present; rhizomatous with tuberous roots	Asparagaceae
1 Fruit a capsule; cladodes absent	2
2 Phytomelan absent; tuberous plant with hairy seeds	Eriospermaceae
2 Phytomelan present; inflorescence scapose	3
3 Plant bulbous; pedicel not jointed	4
3 Plant with a rootstock or rhizome	5
4 Inflorescence umbel-like (cymose)	Alliaceae
4 Inflorescence racemose	Hyacinthaceae
5 Flowers green or white; with root tubers; pedicels jointed	Anthericaceae
5 Flowers red or yellow; aerial leafy stems (up to 60 cm high) with secondary thickening	Asphodelaceae

Data from Dahlgren *et al.* (1985).
[a] See under Fruit, above.

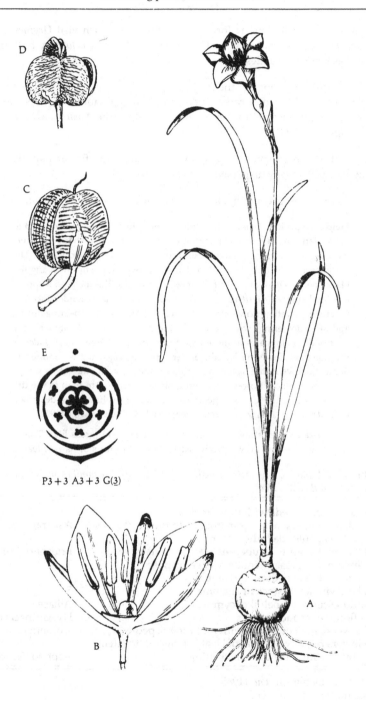

P3 + 3 A3 + 3 G(3̲)

mostly ± mottled red at the base, arising from an erect rootstock with tuberous roots. West African species occur in savanna, and each is of rather local distribution. A. *limosum* (also in Gambia) can, howeveer, be expected right across West Africa in damp areas of rocky savanna. It flowers while in leaf in the rains, producing branched inflorescences *c.* 60 cm high, each with a flattened (winged) main axis. The flowers are in pairs on jointed pedicels, small, white and short-lived, with free spreading tepals. The capsules have ± rounded lobes and contain angular black seeds, which stand horizontally, at right angles to the capsule axis. *A. warneckii* and *A. caulescens* have been transferred to *Chlorophytum* (Marais & Reilly, 1978).

Chlorophytum (Gr. green plant) (green or ground lily) is a genus of over 200 species, of wider distribution than *Anthericum*, and represented in West Africa by more than 20 species of both forest and savanna herbs. The erect rootstock has tuberous roots and produces tufts of leaves, which can be strap-shaped but in some species have broad blades carried on a pseudopetiole. In the latter the convergent venation is particularly clear. The leaf base is never mottled. The inflorescence is mostly a spike-like receme, branched in some species, with small whitish or green flowers on jointed pedicels, the tepals reflexed. The ovary and capsule are both deeply and sharply lobed, the latter with circular black papery seeds orientated vertically, parallel to the capsule axis. There are 3 widespread forest species, *C. alismifolium* (in forest streams) and *C. orchidastrum* with broader, pseudopetiolate leaves, and *C. macrophyllum* (in riverain forest) with lanceolate leaves tapering to the sheathing base (and congested inflorescences, which lengthen when the very large (7–10 cm long) capsules are produced). *C. sparsiflorum* is another forest species sometimes recognisable by the adventitious buds in the inflorescence. At least a third of West African species are found in savanna, among them *C. geophilum*. Here, the almost head-like inflorescence develops on a scape so short that flowering takes place in the soil crust. The fruit pedicels lengthen and the fruits are then buried in the soil.

Dipcadi (S. African name) (awned or tailed squill). These are bulbous savanna herbs with strap-shaped leaves, producing racemes of longlasting greenish flowers in the dry season. The flowers are shortly tubular, with thick joined tepals spreading or reflexed at the tip. The capsules are lobed but do not narrow towards the top, and contain black papery

Fig. 35.1. A. *Zephyranthes citrina* (Amaryllidaceae) flowering plant × $\frac{1}{2}$. B–C. *Anthericum dalzielii* (Liliaceae/Anthericaceae). B. Flower, one inner tepal removed, × 4. C. Fruit × 6. D. *Chlorophytum nzii* (Liliaceae/Anthericaceae), fruit × 4. E. Liliales, floral diagram. (A, from Gooding, Loveless & Proctor, 1965; reproduced with the permission of the Controller of Her Majesty's stationery Office. B, C and E from Hepper, 1968, Figs. 5 and 6 *pro parte*. British Crown Copyright. Reproduced with permission of the Controller, Her (Britannic) Majesty's Stationery Office & the Trustees, Royal Botanic Gardens, Kew. © 1968.)

seeds. This is an African genus of *c.* 55 species, extending to the Mediterranean countries and India. There is perhaps only 1 species in West Africa, though there are then 2 ecotypes, one in rocky savanna, the other in seasonally flooded savanna. At present, *D. longifolium* is recognised by its spike-like racemes up to 60 cm high, produced at the same time as the tufts of narrow twisted leaves. The outer tepal tails are conspicuously reflexed. *D. tacazzeanum* is a smaller plant, producing flowers tinged with yellow or purple in the dry season, but before the leaves appear.

Urginea (?Algerian Arab name) (white squill) is a genus of roughly the same distribution as *Dipcadi* (though *c.* 100 species). Of the 4 West African species, *U. altissima* and *U. indica* are common and widespread in various kinds of savanna. *Urginea* spp. are bulbous plants with rosettes of strap-shaped leaves and racemes up to 2 m high of white or green (or green-striped) flowers, with free tepals which are at least spreading, even reflexed. The pedicels are not jointed. The deeply lobed capsules contain papery black seeds. *U. altissima* produces racemes up to 2 m high in the dry season before the leaves appear. The tepals are only 6 mm long, and the pedicels 1–2 cm long. *U. indica* has racemes up to only 70 cm high, less densely packed with flowers than those of *U. altissima*, but the flowers open at night and they, and the pedicels, are about twice as long. In both species the bracts fall early, while in the 2 remaining species, including *U. ensifolia*, the bracts persist.

Bibliography

Dahlgren, R. M. T. & Clifford, H. T. (1982). *The Monocotyledons: a comparative study.* New York: Academic Press.

Dahlgren, R. M. T., Clifford, H. T. & Yeo, P. F. (1985). *The families of the Monocotyledons.* Berlin: Springer-Verlag.

Gledhill, D. & Oyewole, S. (1972). The taxonomy of *Albuca* in West Africa. *Boletim da Sociedade Broteriana*, ser. 2, **46**, pp. 93–106.

Gooding, E. G. B., Loveless, A. R. & Proctor, G. R. (1965). Flora of Barbados. London: HMSO.

Hepper, F. N. (1968). Notes on tropical African monocotyledons, 2. *Kew Bulletin*, **22**, pp. 449–67.

Hutchinson, J. & Dalziel, J. M. (1931). *Flora of West tropical Africa*, Vol. 2, part 1, 1st edn, pp. 338–52. London: Crown Agents.

Marais, W. & Reilly, J. (1978). *Chlorophytum* and its related genera (Liliaceae). *Kew Bulletin*, **32**, pp. 653–63.

Mouton, J. A. (1976). Identification des *Dracaena* de Côte d'Ivoire à l'état végétatif. *Adansonia*, sér. 2, **15**, pp. 409–13.

36

Araceae – cocoyam family

Perennial herbs in the warmer parts of the world, mainly the tropics, represented in West Africa largely by sympodial rosette herbs with stem tubers or rhizomes giving rise to large, alternate, sometimes solitary and mostly pseudopetiolate leaves, with reticulate third-order venation. The leaf blade is commonly ± hastate or sagittate, seldom peltate (*Caladium*, *Colocasia* and *Remusatia*). Ptyxis is supervolute. Decompound leaves develop by means of irregular growth.

There are some climbing and one facultatively epiphytic genera with leafy, persistent aerial stems, and *Pistia stratiotes* (water lettuce) is the only free-floating aquatic member of the family. Some of the climbing (and epiphytic) genera, as well as a few of the erect cocoyam type, have entire ± elliptic leaves reminiscent of those of the Marantaceae or Zingiberaceae, but there is never a pulvinus or ligule, the leaf base loosely round the stem and passing

Fig. 36.1. A. *Nephthytis afzelii*; 'cocoyam' habit, horizontally growing rhizome × ⅕. B. *Amorphophallus* sp. (diagrammatic); vertically growing stem tubers initiated from below (corms). The two extremes of the range of terrestrial growth habits are represented. (A from Knecht, 1983, Fig. 53; B from Meusel, 1951, Fig. 9.1.)

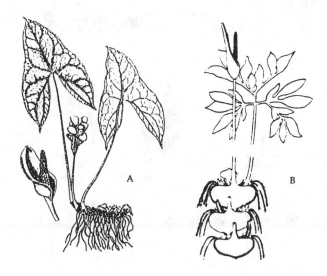

gradually into a channelled pseudopetiole. Where there is close pinnate venation there is also latex (*Anubias, Amauriella*) or reticulate third-order venation
(*Rhaphidophora*). *Culcasia* spp. also lack latex but the venation is not closely
pinnate (see below). *Zantedeschia* has hastate leaves and lacks reticulate venation, but has latex.

The aroids provide an excellent example of modular construction; the shoot
is made up of repeating units (modules), the structure of the unit being so
distinctive as to identify species in at least some cases. There are two kinds
of leaves, assimilating (blade-bearing) ones and cataphylls (scale leaves).
These two kinds are produced in strictly adhered-to patterns and the sympo-
dium (increment of growth) bears in addition a characteristic number of inflorescences (1–12), one of them terminal (Knecht, 1983). A number of the
terrestrial species produce runners or stolons bearing cataphylls (persistent

Fig. 36.2. Growth habit. A. Erect: *Culcasia striolata* (also *C. longevaginata*
and *C. dinklagei*), with stolons. B. Semi-erect: *Culcasia saxatilis*, with stolons.
C. Terrestrial, rarely climbing: *Rhaphidophora* sp. nov., with runners. D–G.
Climbers. D. *Culcasia liberica*. E. *Cercestis stigmaticus*. *C. afzelii* also produces
pendulous branches. F. *Culcasia angolensis*. G. *Rhaphidophora africana*.
Smaller black triangles, cataphylls; larger triangles and other black shapes,
assimilating leaves; the stylised spathe and spadix indicate where flowering occurs. Arrow indicates direction of growth. (From Knecht, 1983, Fig.
24.)

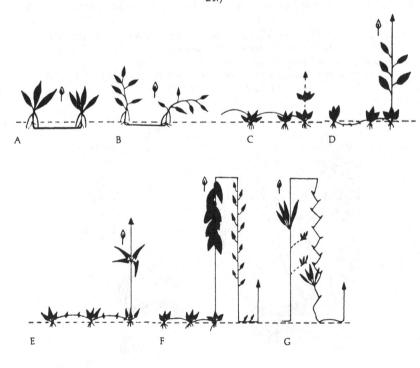

A   B   C   D

E   F   G

in *Cercestis*, deciduous in *Culcasia*) or cataphylls and bladed leaves (*Rhaphidophora africana*).

Members of the family may be recognised by their habit, presence of calcium oxalate crystals, and by their inflorescence. This is a thick axis (spadix) clothed with minute flowers (lacking bracts and bracteoles) and surrounded by a large leaf-like, sometimes strongly coloured spathe. Even *Pistia* has a minute version of this inflorescence hidden in its spirally arranged leaves. The fruit is a berry.

Introduced species, apart from economic ones, include *Caladium bicolor* from tropical south America. It has green and white ± red leaves and flowers readily. In flower characters it resembles cocoyams closely.

Two evergreen forest climbers are easily recognised. One is *Rhaphidophora africana*, which can develop into a liane to 30 m high. This species is not recorded elsewhere in Africa, and the genus is otherwise a large Indo-Malaysian one. The leaves are distichous, entire and asymmetrically oblong-oblaneolate, cuneate at the base. As latex is lacking, they are superficially marantaceous, but have reticulate venation and lack a pulvinus. At the nodes both clasping roots, by which the climber clings to its phorophyte, and longer roots growing into the ground are produced. In the dry season, greenish-yellow spathes appear, each enclosing a spadix of ⚥ flowers with white perianths. If the long roots do not persist, the plant becomes a facultative epiphyte, but seeds can also germinate on the phorophyte. If connection with the soil is unnecessary, these are true epiphytes. In its epiphytic forms, *Rhaphidophora* is found at the tops of trunks and the bases of large branches, in light shade, though without much humus.

Rhektophyllum spp. are smaller climbers than the above. The commoner species is *R. mirabile*, which has latex and unequally pinnately divided decompound leaves, which may also be fenestrated. Both kinds of roots occur, but along the internodes. The spadix has a terminal sterile section and separate belts of ♂ and ♀ flowers, which resemble those of *Cercestis*.

Remusatia vivipara grows both as a facultative epiphyte (in situations similar to those of *Rhaphidophora* but with much humus) and between the rocks and on the ground, but has the cocoyam habit, with peltate leaves which die down at the approach of the dry season. It then produces either spathes or, more commonly, brown stems bearing scale leaves, each with a bulbil in its axil. Each of these is a condensed axillary shoot, the scale leaves of which are hook like, so that the bulbil resembles a burr fruit.

In the most recent account of the family in Africa (Mayo, 1985), *Pistia stratiotes*, *Remusatia vivipara*, *Rhaphidophora africana* and *Rhektophyllum mirabile* are both described and illustrated, and several other genera and species, both indigenous and introduced, of occurrence in West Africa are described.

Flowers ∸ ⊕ and either ♂ and ♀ naked, or ♂ with perianth (*Cyrtosperma* and *Rhaphidophora* etc.), or ♂ and ♀ with perianth in *Stylochiton* only; P4 as a rule. A1–8 free, anthers basifixed and extrorse, or stamens joined into

a columnar synandrium in which the anthers are embedded in its sides and
dehisce apically by pores or slits (*Amauriella*) or laterally (*Anubias, Caladium*
etc.); ♂ flowers stand above the ♀ flowers on the spadix, and there may
be a sterile prolongation of the spadix above the fertile ♂ flowers (*Anubias,
Rhektophyllum* etc.) or a belt of rudimentary ♂ flowers interposed between

Fig. 36.3. A–E. *Caladium bicolor*. A. Spadix, the nearer part of the spathe
cut away (overlapping, lower left) × $\frac{1}{2}$. B. ♂ flower (synandrium) × 7.
C. Synandrium in cross-section × 7. D. ♀ flower (pistil) × 14. E. Pistil
in cross-section × 14. F–H. *Cyrtosperma senegalense*. F. Spathe and spadix
× $\frac{1}{6}$. G. Flower in vertical section × 10, pistil surface view. H. Pistil in
cross-section × 20.

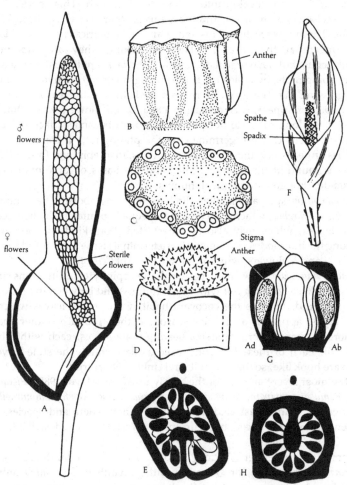

the belts of fertile ♂ and ♀ flowers (*Cercestis*). G1–(2–4), 1–4-celled with basal or parietal placentation and various ovule numbers.

Pollination Aroids are pollinated by beetles and flies, including carrion- and dung-feeders in both groups. These are deceived by the odour of rotting flesh and dung and might well leave the spathe too soon, but for the specialisation of the spathe and spadix as a trap mechanism, which temporarily imprisons the insects. Aroids are protogynous, and the insects are confined in the lower chamber (round the ♀ flowers) by the blocking of their exit at the level of the top of the ♀ flowers; later, after there has been time for any pollen with which they entered to have been deposited, the barrier is removed, and they escape past the pollen which has by then been shed on the surface of the ♂ part of the spadix. *Colocasia* is a good example of this arrangement. Flies enter the spathe at the base (the upper end is closed) and are at first confined to the lower chamber by a constriction of the spathe above. The odour later fades and the entrance closes up, followed by shedding of pollen, and relaxation of the constriction. Next day the spathe opens at its upper end, and the flies, by now covered with pollen having spent the night moving around in the upper chamber, leave. The more lurid colours of such trap inflorescences also suggest rotting flesh, and such species are easily identified by smell. The exact nature of the barrier to insect escape needs investigating in West African species. Not all species, however, are pollinated in the above way. *Cyrtosperma* is reported to smell of blackcurrant jam, and the flies visiting the inflorescence may well feed on special food bodies or cells, since they stay on to breed, their larvae feeding on the fermenting tissues round the berries. Neither is there any indication of a trap in *Amorphophallus variabilis* (tropical Asia), where small beetles feed on special cells lining the base of the spathe, staying on for five days, until pollen falls on them from above. Reports of nectaries, stigmatic secretions and food tissues need checking. In *Typhonium trilobatum* (tropical Asia, introduced into Côte d'Ivoire), food bodies are formed by hook-like projections (rudimentary and modified ♂ flowers?) above the ♀ flowers.

Fruit A berry, usually 1- to few-seeded, and brightly coloured. The outer layer of the testa generally becomes fleshy (sarcotesta) and contributes to the flesh of the fruit. *Pistia* has a drupe of one carpel.

Dispersal It is generally assumed that birds are the only agents of dispersal.

Economic species Both leaves and tubers may be gathered, the former being prepared as a green vegetable, the latter made into fufu. The difficulty in both cases is getting rid of the calcium oxalate crystals and of the poisonous substances in the latex. Prolonged cooking is usually required, and the addition of sodium bicarbonate is sometimes recommended. Cocoyams are one of the principle root crops in forest area, being a useful cover crop in shade between young perennials. *Cocolasia esculenta* (old cocoyam, from tropical Asia) exists in two varieties, one with a larger main corm and a few small cormels, the other with a small central corm and many larger cormels. *Xanthosoma mafaffa* (now *X. sagittifolium*) (new cocoyam, introduced from South America only about 145 years ago), has sagittate leaves and numerous large cormels and makes better fufu than *Colocasia*. Both crops are propagated vegetatively, and rather seldom flower. However, in flower features they resemble *Caladium* closely, though in *Colocasia* the spadix has a sterile tip and in *Xanthosoma* the stigmata protrude and ± touch each other.

Field recognition *Amorphophallus* (Gr. shapeless and phallus) is an Old World genus of herbs with pinnately divided (decompound) leaves. In West Africa there are about a dozen species, some in savanna, some in forest undergrowth. The stem tuber is often large, renewed annually, and produces the inflorescence separately from, and before, the 1 or 2 leaves, in the dry season. The inflorescences are very strong smelling, the spathes purple or reddish in colour, the spadices with a sterile spongy prolongation above the ♂ flowers, the whole on a peduncle up to 90 cm high. The ♂ flower is a group of 3–4 stamens, the anthers opening by apical pores, while the ♀ flower is a 2–4-celled pistil with a basal ovule in each cell. *A. flavovirens* is the most widespread species, distinguished by its yellowish spathes and projecting spadices on a long, spotted peduncle.

Anchomanes (Gr. strangle and rave) is a small tropical African genus, with 2–3 species in West Africa. Single, much-divided leaves on long pseudopetioles are produced from the fleshy rhizome after flowering. The pseudopetiole (and peduncle) carries prickles, and the end section of each part of the decompound leaf blade has 2 tips. The spadices are shorter than the spathes, without a terminal sterile portion, but the flowers are very much as in *Amorphophallus*, though ♂ flowers consist of only 2 stamens, and ♀ ones are 1-celled with 1 basal ovule, and a club-shaped style. The most widespread species is *A. difformis* var. *difformis*, a forest species producing a purple inflorescence on a peduncle 1.5 m long, and a pseudopetiole nearly 3 m long. The savanna var. *welwitschii* (formerly *A. welwitschii*) is smaller, its spathes greenish.

Cercestis (Gr. ?tail, train) is a small West African genus of evergreen climbers and facultative epiphytes in forest, with aerial roots at the nodes, and variously lobed leaves, producing greenish spathes a few centimetres long in the dry season. *C. afzelii*, which shows continuous growth, is the most common and widespread species. The spadix is whitish, with a terminally sterile portion, free but close-packed stamens, a belt of rudimentary ♂ flowers and basally, ♀ flowers, each of a 1-celled ovary with 1 ovule. The species is easily recognised by its 3-lobed or hastate leaves and by the way the spathe breaks off near the base, leaving a cup round the spadix. The pistils, like the stamens, are close packed, and the berries become 5-sided.

Culcasia (Arab. qolquas), is a genus of evergreen forest species, root climbers, facultative epiphytes and some erect herbs (the most common and widespread erect species, *C. striolata*, has stilt roots). The leaves appear somewhat marantaceous, and latex is also lacking, but the second-order veins are seldom closely pinnate, while the third-order ones are reticulate. The sheathing leaf base is prolonged so that the pseudopetiole appears winged (petiolar sheath). The spathes are rather small, at most 7 cm long, with a terminal sterile portion to the spadix, ♂ flowers each of 4 stamens packed back to back, resembling a synandrium, and ♀ flowers each of a 1-celled pistil with a basal ovule, the stigma a disc of 4 segments. Approximately 9 West African species are climbers and most are facultative epiphytes. *C. angolensis* is a liane up to 30 m high, with the largest leaves (up to 50 cm long), bracts (up to 14 cm long), and spathes in the family. The leaf blade and pseudopetiole are at right angles to each other, with a small swelling on the upper side, at the junction. Pendulous stems are also produced. *C. liberica* is a similarly stout climber but without pendulous stems and the leaf blade continues across the junction of the midrib and pseudopetiole. The spathes frequently arise in groups of up to 12, as in *C. angolensis*.

C. scandens is now recognised by Ntépé-Nyamè (1984) as being composed of 2 species, *C. simiarum* with pseudopetioles one-third the length of the leaf blade (illustrated in fig. 360, 1–4 in *Flora of West tropical Africa*, 1968, vol. 3, pt 1), and *C. annetii* with the pseudopetiole half the length of the leaf blade. Both are common wiry-stemmed climbers at the eastern end of our area, with leaves up to 17 cm long, the blade not at right angles to the pseudopetiole but interrupted at its junction with the midrib. The spathes are often in pairs.

Cyrtosperma (Gr. curved seed). *C. senegalense*, the only West African species, is a tall herb with large sagittate leaves on prickly pseudopetioles up to 2 cm high, and even longer prickly peduncles bearing large striped spathes covering shorter spadices. These are purple and bear ♂ flowers (P4 A4 G 1-celled with 2 parietal ovules). The berries are 1- to few-seeded and dark in colour, in contrast to those of most other genera. The perennating organ is a rhizome. The species occurs in freshwater swamps and more open forest areas near rivers as well as in damp savanna areas. It is a common and widely spread species but not yet reported elsewhere in tropical Africa. The main centre for the genus is New Guinea. In West Africa the species is probably evergreen, flowering at various times of the year.

Stylochiton (Gr. pillar and tunic) are small herbs, the leaves scarcely 30 cm high, the spathes produced in the soil crust, the former in the rains, the latter in the dry season. There are 2 widespread savanna species, *S. lancifolius* with sagittate leaves, *S. hypogaeus* with hastate leaves, both species having thick, segmented rather ginger-like rhizomes, but the former with a slim spathe, the latter with a shorter wide spathe (and opening thereto). The ♂ flowers are P4 A2–6, ♀ ones have a perianth enclosing the pistil (the tunic of the name), which is 2–4-celled. The ♀ flowers number only 5, standing in a whorl in the base of the spathe. The other 2 species have 2–3 ♀ flowers each. The genus is confined to tropical Africa and consists of *c*. 18 species.

Bibliography

Bogner, J. (1979). A critical list of the aroid genera. *Aroideana*, 1, p. 63–73.

Bogner, J. (1980). On two *Nephthytis* species from Gabon and Ghana. *Aroideana*, 3, pp. 75–85.

Croat, T. B. (1979). The distribution of Araceae. In *Tropical botany*, ed. K. Larsen & L. B. Holm-Nielsen, pp. 291–308. London: Academic Press.

Crusio, W. (1979). A revision of *Anubias* Schott (Araceae). *Mededelingen Landbouwhogeschool Wageningen*, 79 (14).

Knecht, M. (1983). *Aracées de la Côte d'Ivoire*. Vaduz: J. Cramer.

Mayo, S. J. (1985). *Flora of tropical East Africa: Araceae*. Kew: Royal Botanic Garden.

Meusel, H. (1951). Die Bedeutung der Wuchsform für die Entwicklung des natürlichen systems der Pflanzen. *Feddes repertorium*, 54, pp. 10–172.

Ntépé, C. (1981). Une nouvelle espèce pour le genre *Rhektophyllum* N. E. Brown (Araceae). *Adansonia*, sér. 2, 20, pp. 451–7.

Ntépé-Nyame, C. (1984). Cinq nouvelles espèces pour le genre *Culcasia* P. Beauv. (Araceae). *Bulletin du Muséum national d'histoire naturelle*, Paris, sér. 4, 6, sect. B, *Adansonia*, pp. 313–23.

37

Palmae (Arecaceae) – palm family

Almost all species of this family are confined to the frost-free zone between latitudes 30° N and 30° S. The flora of West Africa is small, with only 12 genera of indigenous palms; of these only *Elaeis* and *Raphia* also have New World species. The palms are, however, important ecologically in West Africa, second only to grasses in economic importance, as well as being prominent by reason of habit.

West African palms are trees or climbers (rattans), all woody and of relatively large dimensions. Species of *Calamus* and *Raphia* provide the largest known climber and leaf, respectively, other genera providing the largest inflorescence and fruit.

Palms lack secondary thickening and depend on the extensive deposition of lignin in existing primary tissue for their rigidity, and on the production of a mass of adventitious roots from the lower nodes to provide increased support and water supply as the stem grows.

Growth is monopodial, infloresences being axillary, and most West African genera show pleonanthic growth, i.e. the stems grow indefinitely and flower repeatedly. There is a small group of genera (*Ancistrophyllum, Oncocalamus* and *Raphia*) in which the stems flower only once (hapaxanthic growth), the equivalent (in effect) of the sympodial growth of most other monocotyledons. These short-lived aerial shoots are replaced by suckers growing from the base of the plant (also seen in *Calamus* and *Phoenix*), but in *Nypa*, the inflorescence is terminal on one of the dichotomies of the rhizome, the leafy shoot being produced separately. *Hyphaene* shows dichotomous branching.

The leaves are alternate, palmately or pinnately (usually paripinnately) lobed, divided or compound (fan and feather palms, respectively) on a woody pseudopetiole arising from a leaf sheath which is, at least initially, closed. The leaf base, pseudopetiole and rachis are fairly commonly armed with long prickles. The leaf blade develops by differential growth and secondary splitting into a large, ribbed plicate structure made up of leaf segments which are V-shaped in cross-section ('induplicate', *Borassus, Hyphaene, Phoenix*) or, more commonly ∧-shaped in cross-section ('reduplicate'). The ligule of the rattans occurs at the junction of the leaf sheath and pseudopetiole and is a well-developed, hood-like structure (ocrea) protecting the terminal bud cf. Chapter 40, Gramineae. In *Ancistrophyllum* and *Oncocalmus* it may be up to 30 cm long and house ants. The hastula is a short-lived ligule-like outgrowth which develops either adaxially or abaxially (or both) at the top of the pseudopetiole.

Fig. 37.1. Palm inflorescences. A–B. *Borassus aethiopum*. A. Basal part of
♂ inflorescence and sheaths × $\frac{1}{6}$. B. ♀ inflorescence branch × $\frac{1}{6}$. C. *Ancistrophyllum secundiflorum*, part of inflorescence branch × 1. D. *Cocos nucifera*
part of inflorescence branch × $\frac{2}{5}$, ♂ flowers above, ♀ flower bud at the
base. E. Diagram of mode of construction of a palm inflorescence. ax^0–ax^4
four orders branching. ax^1 continues, bearing ax^{2-5} branches. The axes
may be very congested as in *Borassus*, or more relaxed as in *Raphia* and
the rattans (though not *R. sudanica*) and *Cocos*. lf, leaf; pr, prophyll; pbr,
sterile peduncular bract; rbr, sterile bract subtending inflorescence branch,
all three kinds encircling the axis (cupular); f, flower; fpr, floral prophyll;
fb, floral bract. (D from Robertson & Gooding, 1963, Fig. 104A; E reprinted
by permission from *Botanical Review*, **48**, pp. 1–69, H. E. Moore Jr &
N. W. Uhl, © 1982, The New York Botanical Garden.)

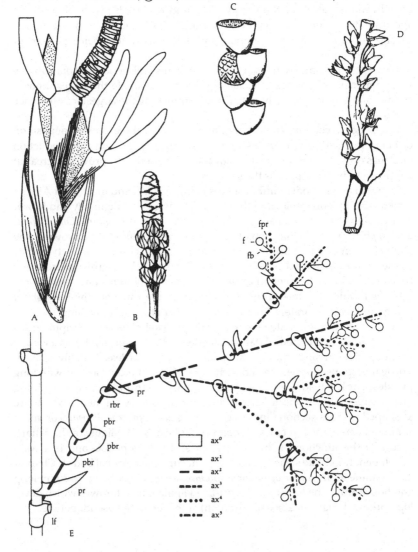

Tree palms are rosette plants with short internodes, steadily producing leaves which live for several years. Initially, the large leaf bases protect the soft tissues of the stem apex, later their remains arm out the trunk for several years. The rattans in contrast, have internodes 1–2 m long sheathed by the prolonged leaf base (variously armed), which ends in the ocrea.

A most useful glossary of the terms used to describe palms has been published recently, together with a summary of generic characters and a field key to nine of the West African genera (Dransfield, 1986).

Fig. 37.2. *Cocos nucifera*. A. Half ♂ flower × 10. B. Floral diagram of ♂ flower. C. ♀ flower from above × 1. D. ♀ flower in oblique section × 1. E. Floral diagram of ♀ flower. A. Adaxial side to right. (From Robertson & Gooding, 1963, Fig. 104 *pro parte*.)

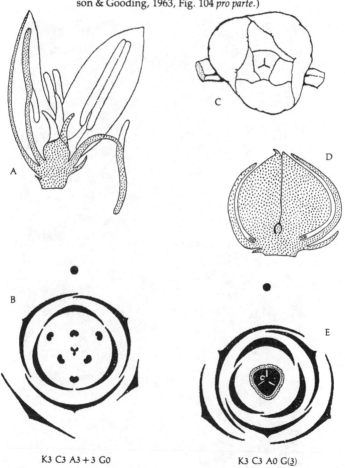

K3 C3 A3 + 3 G0   K3 C3 A0 G(3)

Members of the family may be recognised by their habit, and by their massive panicles or branched spadices emerging from large sheaths of spathe-like bracts.

Species introduced for decorative purposes include the cabbage palm (*Roystonea oleracea*) and the Cuban royal palm (*R. regia*) from Central America. Both make good avenue trees.

Flowers ÷ ⊕, ♂ or ♀, ♀ only in *Ancistrophyllum* and *Eremospatha*, 3-part; tree palms are either dioecious with ♂ and ♀ flowers on separate trees (*Borassus, Hyphaene, Phoenix*) or monoecious, with ♂ and ♀ flowers in different

Fig. 37.3. *Raphia hookeri* in flower. Trunk *c.* 4 m tall. (Courtesy of P. Tuley.)

inflorescences (*Elaeis*), or in the same inflorescence in the remaining species. Flowers may be borne singly (*Phoenix*, *Nypa*) but more often occur in short cymes (cincinnae), ♂ flowers in pits in the spadix in *Elaeis*, in pits formed by bracts in *Borassus* and *Hyphaene*. The receptacle may elongate in such genera. P3+3, similar or dissimilar, sometimes woody or leathery, joined in some species, enlarging and persisting in the fruit (except *Nypa*). A3+3 ((3) in *Eremospatha*, opposite the outer tepals and fused basally in *Nypa*) or more than 6 occur in *Raphia*; anther dehiscence latrorse; staminodes commonly present in ♀ flowers. G 3 in *Phoenix* and *Nypa*, G(3) in all other species, 3-celled or less by abortion, with an ovule in the basal or inner angle of each loculus. Pistillodes may be present in ♂ flowers. Nectar is produced in basal, septal or stylar glands.

Pollination *Cocus*, *Nypa* and *Phoenix* have been well investigated. *Cocos* is protandrous, pollen being shed in the bud, with more ♂ than ♀ flowers. The ♀ flowers have each three hypogynous nectaries, and the effective visitors are thought to be honey bees, though several kinds of insect visitors have been observed. Wind pollination is thought also to occur. *Nypa* is protogynous, and is pollinated by drosophilid flies which also use the inflorescences as breeding sites; it has no nectaries. *Phoenix* has frequently been reported to be wind pollinated, but three small cavities at the base of the style may produce nectar, and several other features, possibly protective of pollen and ovules, suggest that insects may be involved. This is probably also the case in rattans, which also have stylar pockets. In *Elaeis*, where, although nectaries are lacking, the flowers are scented, Syed (1979) reported insect pollination. In *Borassus* and *Hyphaene* there are septal nectaries.

Fruit A berry or drupe. The berries are of 2 kinds, either smooth-skinned (*Phoenix*, a berry-like monocarp) or with pericarpic scales which point towards the base of the fruit (*Raphia* and rattans). There is as a rule only 1 seed. The drupes vary in the nature of the mesocarp and endocarp. In *Cocos* and *Nypa* the mesocarp is thick and fibrous, as it is also in *Borassus*; in *Hyphaene* the mesocarp is dryly fleshy, in *Elaeis* oily. The endocarp is very hard and woody in *Borassus*, *Hyphaene*, *Nypa* and *Cocos*, in which there are 3 'eyes', thin patches, 1 opposite each ovule, and therefore opposite the seed. Seeds always have endosperm and in the rattans and *Raphia* (fruits scale-bearing, lepidocaryoid palms) the seeds may develop a sarcotesta. Germination involves plumule-burying of 1 of 3 kinds: remote tubular, by elongation of the cotyledon sheath (*Borassus*, *Hyphaene*, *Phoenix*); adjacent ligular, the plumule being carried out of the seed only a short distance within the solid cotyledon ligule (*Cocos*, *Elaeis*, *Eremospatha*, *Raphia*); and remote ligular, in which the cotyledon ligule becomes much elongated, a style not yet identified in West Africa.

Dispersal Mammals and birds are undoubtedly the effective dispersal agents. The palm nut vulture (*Gryphjierax angolensis*), rodents and birds are the distributors of oil palm, and the red berries of rattans and *Raphia* are also attractive to birds. Elephants appear to be partial to *Borassus* and *Hyphaene* fruits, since their stones have been found sprouting in elephant dung, and *Borassus* and *Elaeis* stones have been found in baboon droppings. Drills stuff the fruits of *Oncolcalamus* into their cheeks and then crack the stones one by one at leisure. Bats and possibly birds collect dates (*Phoenix dactylifera*), but whether they are interested in the drier fruits of *P. reclinata* is not known. They also eat the fibrous flesh of *Hyphaene* drupes. *Nypa* and *Cocos* are dispersed in water and are both viviparous.

Economic species The list of useful palm products is very long. Leaf and stem materials are used constructionally as well as for tying; sap, rich in sugar, can be fermented to palm wine or vinegar, or distilled into a kind of arrak (also excellent for preserving purposes); oil from the pericarp (*Elaeis*, *Raphia*) or endosperm (*Cocos*, *Elaeis*) is extracted; fibre from the pericarp (*Cocos*) and edible fruits and seeds are all used – the list is long indeed. The terminal bud of the trunk can be eaten raw, but its removal kills the aerial shoot to which it belongs. Sago is prepared from the pith of the trunk.

Field recognition

Tree palms with pinnately compound leaves
  Phoenix (classical name). The only West African species is wild date palm, *P. reclinata*. It is a small spiny palm, up to 8 m, with pneumorhizae, growing in clumps formed by the growth of suckers from the root stock. It is the only West African palm with induplicate leaf segment insertion and leaflet spines at the base of each rachis. The trees are ♂ or ♀, the inflorescences at first being surrounded by a tough spathe, but later expanding into large panicles with wavy, ± distichous branches. Usually only 1 berry-like monocarp develops from each ♀ flower, and its flesh is thin and dry. *P. reclinata* occurs in well-watered ground, from the coast northwards into northern guinea savanna, and throughout most of the rest of tropical Africa and South Africa. The genus is confined to the Old World.

  Raphia (Malagasy name). There are at least 6 species in West Africa, all of which form suckers at the base, and in all of them the aerial shoots are hapaxanthic, possibly truly sympodial, and flower and fruit only once. The formerly huge terminal bud dies after the production of the inflorescence, which carries both ♂ and ♀ flowers. The latter give rise to pointed berries armoured with hard, overlapping scales. Leaf segment insertion is reduplicate, and the petiole is non-spiny. *R. sudanica*, the short-stemmed northern *Raphia*, is a savanna species forming thickets in swampy ground. The fruits are distinctively reddish-brown. This species extends to c. 200 km from the coast and may overlap with *R. hookeri*, the forest wine palm, which forms a belt up to *c.* 250 km deep from the coast northwards, from Guinée to Gabon, growing beside rivers and in swamps. The upper part of the trunk of this species is characteristically covered with black curly fibres and pneumorhizae are formed. In the west, from Gambia to Ghana, the thatch palm, *R. palma-pinus*, with noticeably light-green foliage, suckers very readily to form thickets behind the mangrove zone, and from Dahomey to Congo, the bamboo palm, *R. vinifera*, its trunk covered with straight brown bristle fibres replaces it. *Raphia* occurs in the tropics of both Africa and America, though West African species are seldom found anywhere else.

Tree palms with palmate leaves

Borassus (Gr.) *B. aethiopum*, the only West African species is found in savanna, in swamps and by streams, and occasionally near the coast. It is an unbranched tree with a swelling in the upper part of the trunk in mature specimens; the leaves are palmately lobed. ♂ and ♀ flowers are found on separate trees, in robust branched lateral inflorescences. The fruit is a reddish-yellow drupe, with edible fibrous flesh the same colour and three stones. The genus is an Old World one, but this species is confined to tropical Africa.

Hyphaene (Gr. woven, the fibres of the fruit, used in weaving). *H. the-baica*, the dum or doum palm, is a dioecious palm with a distribution extending across tropical Africa to Arabia and north into Egypt. The genus as a whole is an Old World one, *H. thebaica* occurring in dry savanna in West Africa. Mature specimens usually have forked trunks, but these may in any case be distinguished from *Borassus* by their deeply divided leaves. The inflorescences are similar to, but rather smaller than, those of *Borassus*, and the drupes are greenish-brown, ovoid and about 7 cm long. Each contains a relatively large stone.

Climbing palms (rattans)

Ancistrophyllum (Gr. barbed leaves) (now *Laccosperma*; Dransfield, 1982). *A. secundiflorum*, is by far the commonest of the 3 West African species, especially in wetter forest areas. The hapaxanthic stems are many metres long, and covered with spiny leaf sheaths, bearing blades with paired segments and ending in a cirrus with pairs of hooks so recurved as to resemble 2-pronged forks fixed to the axis. Each leaf sheath ends in a large ocrea. The inflorescences are large and distichously branched, with the lateral branches all hanging down to one side. Each branch is composed of distichously arranged bracts and cupular bracteoles, each of these con-taining 2 white ♂ flowers. These are followed in March–September by red scaly berries, formed in great numbers.

Calamus (L. reed). The only West African species *C. deeratus*, is a fresh-water swamp forest climber (with flagella not cirri) growing from a woody branched rhizome. The leaf sheaths are spiny, particularly round the open-ing, and may also bear a flagellum (equivalent to a sterile inflorescence). The ocrea is small and withers. The inflorescences are adnate to the axis above the node at which they arise, appearing through the leaf sheath of the leaf above that, opposite to the petiole. The inflorescences are long, with short, distichous drooping branches alternately arranged. Each branch is made up of 2 rows of cupular bracteoles containing the flowers, a ♂ only in some inflorescences, a ♂ + ♀ in others. The fruits are small scaly berries. *Calamus* is a mainly tropical Asian genus, with *C. deeratus* isolated in, and confined to, the area of West Africa–Uganda–Angola.

Eremospatha (Gr. single spathe) is represented by 2 species of forest rattans, and the much commoner *E. macrocarpa*, in freshwater swamp forest. This is a smooth-stemmed species bearing leaves with alternate, pleated fan-shaped segments, each leaf ending in a cirrus with paired hooks, but, in this species, no prickles. The inflorescence is lateral, a large yellow panicle on a long peduncle, and bears ☿ flowers. The fruits are scaly berries.

Oncocalamus (Gr. lump and *Calamus*). The 2 West African species are so far reported from Nigeria only, where they occur in freshwater swamp forest or secondary forest, respectively. *Oncocalamus* is hapaxanthic, with long aerial shoots dying after the lateral inflorescences have fruited. The leaf sheath bears caducous flattened spines and the remains of the ocrea may be seen round its top. The leaf blades have rather few segments (5–16 on each side of the rachis) but the rachis ends in a cirrus with reflexed hooks. The flowers are ♂ or ♀, a ♀ with a ♂ on each side (in *O. mannii*, swamp forest), and the fruits are yellow.

Bibliography

Corley, R. H. V., Hardon, J. J. & Wood, B. J. (eds.) (1976). *Development in crop science*, vol. 1, *Oil palm research*. Amsterdam: Elsevier Scientific Publishing Company.

Dransfield, J. (1982). Nomenclatural notes on *Laccosperma* and *Ancistrophyllum* (Palmae: Lepidocaroideae). *Kew Bulletin*, **37**, pp. 445–7.

Dransfield, J. (1986). *Flora of tropical East Africa: Palmae*. London: Crown Agents.

Furtado, C. X. (1972). A new search for *Hyphaene guineensis* Thonn. *The Gardens' Bulletin*, Singapore, **25**, pp. 311–34.

Henderson, A. (1986). A review of pollination studies in the Palmae. *Botanical Review*, **52**, pp. 221–59.

Moore, H. E., Jr & Uhl, N. W. (1982). Major trends of evolution in palms. *Botanical Review*, **48**, pp. 1–69.

Otedoh, M. O. (1975). The distribution of *Raphia regalis* in Africa. *Nigerian Field*, **40**, pp. 172–8.

Robertson, E. T. & Gooding, E. G. B. (1963). *Botany for the Caribbean*. London: William Collins.

Russell, T. A. (1965). The *Raphia* palms of West Africa. *Kew Bulletin*, **19**, pp. 173–96.

Syed, R. A. (1979). Studies on oil palm pollination by insects. *Bulletin of Entomological Research*, **69**, pp. 213–24.

Tuley, P. & Russell, T. A. (1966). The *Raphia* palms reviewed. *Nigerian Field*, **31**, pp. 54–65.

38

Orchidaceae – orchid family

The largest family of monocotyledons with c. 18 000 species, though in West Africa second in size to the grasses. Orchids are perennial herbs found in all but the driest habitats.

The habit is erect (tufted), creeping, climbing or pendulous, and the stems may be sclerified. Erect species seldom become more than 2 m high (including the inflorescence), though creeping and climbing stems may become several metres long. The only West African climber is *Vanilla* with relatively long fleshy internodes and one or two aerial roots at each node. Some species are facultative epiphytes, and the inflorescence is axillary in *Vanilla*.

Nearly half West African genera are (at least facultatively) epiphytic, most of them belonging to the tribe Vandeae *in sensu* Dressler, and growing monopodially. *Ansellia* is an example of a sympodial epiphyte, and forms a bushy plant, with hard stems and short internodes.

Table 38.1 *Classification of West African orchids*

Subfamily	Tribe	Genus number[a]
Orchidoideae	Orchideae	1–6
	Diseae	7–10
Spiranthoideae	Cranichideae	15
	Erythrodeae	16–20
Epidendroideae	Vanilleae	11
	Epipogieae	13
	Gastrodieae	12, 14
	Malaxideae	21, 22
	Epidendreae	25, 28
	Arethuseae	26, 27
		(+ *Spathoglottis*)
Vandoideae	Polystachyeae	24
	Cymbidieae	23, 29–33
	Vandeae	34–58

Data from Dressler (1981).
[a] As in the *Flora of West tropical Africa* (1968).

A second group of monopodial orchids comprises the three genera of leaf-less epiphytes with green assimilating roots (*Microcoelia, Taeniophyllum, Chauliodon*). In these, the stem is extremely short and the tangles of green, rather knotted-looking roots on, or hanging from, the phorophyte are noticeable. *Microcoelia* (including now *Encheiridion*), is the only one of the three genera with several common species, notably *M. caespitosa* (Sierra Leone–Zaïre and East Africa). This is a forest epiphyte with a short creeping stem bearing a great many roots but only a few scale leaves at the tip of the stem. It may be found creeping on, or hanging from, several different species of phorophyte, including cocoa, and flowers most of the year. It seems likely that the roots of these species will prove to have greater chlorophyll concentrations than are possessed by leaves of epiphytic species, but like them, to have CAM photosynthesis (Benzing & Ott, 1981).

Johansson (1977) has produced evidence that epiphytes harm their phorophyte in the case of *Microcoelia exilis*.

The remaining groups of orchids are sympodial, producing annual increments in which leafy shoots and inflorescences are variously associated, but always of limited life span. In the terrestrial 'ground' orchids, the perennating part of the plant is rooted in the soil and seldom visible, though a few species of *Cheirostylis* have creeping rooted stems. Among the epiphytes, the perennating parts, pseudobulbs, are exposed on the phorophyte, renewed growth occurring from a lateral bud on the pseudobulb.

There are two kinds of pseudobulb, though both are green, water-storing stem tubers. In the *Bulbophyllum* type, the pseudobulb consists of the last internode of the annual increment of growth, while in the *Polystachya* type the pseudobulb is made up of several or all of the nodes and internodes of the increment. In both cases the new pseudobulb is either connected with the old one by a length of creeping stem or arises directly upon it. Clump formation results when more than one bud on the pseudobulb develops.

Underground stem tubers, sometimes rhizomatous and somewhat in the style of ginger rhizomes, are common among ground orchids, but in the subfamily Orchidoideae tuberoids, each with a bud concealed in its proximal end, arise round the base of the plant, the bud producing next season's shoot. In *Brachycorythis, Habenaria* spp. etc., fleshy 'furry' roots develop, but these are for storage only.

Among the terrestrial forest species are a few 'saprophytes', small, leafless non-green plants with rootless fleshy underground tubers associated with fungi, and from which the inflorescence arises. This is a scape 10–45 cm high with at most a few scale leaves. Included is a curious species of *Eulophia* (*E. galeoloides*) with yellow flowers (see below), *Auxopus* spp., *Epipogium roseum* and *Schwartzkopffia pumilio*. These orchids remain dependent on their association with fungi (sterile mycelia of various basidiomycete genera) for all their nutrients throughout life while other terrestrial species with green leaves probably show a range of degrees of dependence, according to species and perhaps also according to situation. In the epiphytes, the velamen plays an equivalent role in aiding the uptake of water and nutrients. It consists of one to several layers of dead cells outside the root exodermis. It is generally speaking thin in leafless species but otherwise thicker in species of drier habitats than in one of humid habitats. It is probable that the velamen captures more of the nutrient solution made suddenly or briefly available in rain or dew than the root without a velamen would otherwise manage. It also provides a protective layer on these exposed roots. Terrestrial species also often possess velamen.

Inflorescences arise either terminally on leafy shoots (acranthous sympodia, *Brachycorythis, Polystachya* etc.) or separately from the leafy shoot from a bud on the perennating organ (pleuranthous sympodia, *Eulophia, Bulbophyllum* etc.).

Orchid leaves have parallel (convergent) venation, but this may be obscured as in the narrow leaves, U-shaped in cross-section, of *Cyrtorchis aschersonii*, the terete leaves of *Nephrangis filiformis, Angraecum subulatum* and *Tridactyle tridentata* or the circular ones of *Stolzia*. Venation is also obscured in the fleshy-leaved species, many of which may prove to have CAM photosynthesis.

Leaves are alternate, sometimes distichous, with a basal closed sheath from the whole rim of which the blade may arise, or the sheath may be short and the blade arise on one side only. The blade may be articulate, breaking off from the sheath in age. *Vanilla* has a non-sheathing leaf base and a distinct pseudopetiole. Ptyxis is often conduplicate (and the leaves then leathery) (*Eulophidium* etc.), but supervolute ('convolute') (*Vanilla* etc.), plicate (*Eulophia* spp.) and conduplicate-plicate forms also occur. Leaves with these last two forms are strongly ribbed.

Orchids may be recognised by their habit, their racemose inflorescences of relatively conspicuous $\div$ .l. flowers with prominent bracts and capsules opening by three or six slits. The pedicel is usually short, or even absent, in which case the inflorescence is a spike. Flowering takes place mainly in the rains but some dry- and all-season flowering species are included under Field recognition below.

The position of West African genera in Dressler's classification (Dressler, 1981) is included in order to make the information presented there more readily applicable to West Africa. There are no West African representatives of Dressler's first two subfamilies Apostasioideae and Cypripedioideae, and the family is defined on the same basis as is the *Flora of West tropical Africa*.

Spathoglottis plicata (subfamily Epidendroideae) has been introduced and is commonly available in gardens.

Flowers $\div$ .l. $\male$ 3-part. P3+3 or K3 C3, the inner median tepal differentiated as a lip (labellum), which may also be pouched or spurred; spurred sepals are rare (*Disa, Disperis*). In most genera the flower is resupinate (twisted through 180°), so that the lip lies abaxially, a condition which is achieved in various ways; *Polystachya* and *Satyrium* are exceptions. The single (outer median) stamen is fused to the pistil to form a petaloid column (gynostemium),

which stands opposite the lip; basally it may be fused with the 2 lateral tepals
to form a column-foot (*Bulbophyllum* etc.), while apically it carries the single
anther above the projecting tip of the median stigma lobe (rostellum); the
stigmatic areas are generally found in the hollow under this projection. The
anther may be erect or even reflexed, e.g. *Bulbophyllum*, on the gynostemium,
but often, as in *Spathoglottis*, it is incumbent, at right angles to the gynostemium or bent downwards at some even greater angle; in epidendroid orchids
the bending occurs during growth (incumbent in the strict sense), while in
vandoid orchids anthers dehisce basally and only appear to be incumbent.
Spathoglottis also shows another common feature among the epidendroids,
the operculate (hinged or versatile) anther, which can be lifted like a lid.
Within each of the 2 anther cells the pollen coheres in a pollinium, e.g. the

Fig. 38.1. *Spathoglottis* sp. A. Flower in side view × 1. C and D indicate
the directions from which the column (gynostemium) is viewed in C and
D, respectively. B. Flower in face view × 1. C. Column in ventral view
× 6. D. Column in face view × 6. E. Pollinia × 3. F. Fruit × 1½.

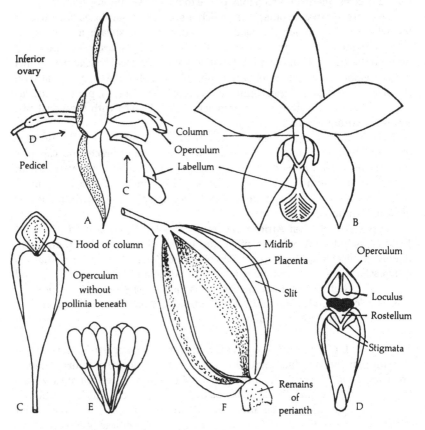

vandoids have 2 pollinia, but in the epidendroids partitioning of the anthers cells results in the formation of 2+2 or 4+4 pollinia; in *Maniella* and *Vanilla* the pollinia are soft (mealy), being composed of pollen grains, but more commonly pollen is present as tetrads; in genera 1–14 and 16–20 (Table 38.1) the tetrads cohere in groups giving a granular (sectile) appearance; in the remaining 2 subfamilies, waxy or hard, bony pollinia are usual; each pollinium has as a rule a stalk, though its origin is various, sometimes within the same genus. G($\overline{3}$), 1-celled (rarely 3-celled) with parietal placentation carrying very large numbers of ovules, and quite immature at pollination. Except in a few genera (*Cymborkis, Hetaeria* etc.) the functional stigma is entirely or mostly developed from the median stigma lobe, the rostellum, which may be elongated into a beak-like projection. Viscidia are sticky areas on the rostellum, to which the pollinia become attached by their stalks at anthesis; they are of diverse morphological origin and not necessarily detachable; pollinia, stalks, and, when detachable, viscidium, form the unit transferred in pollination, the pollinarium.

Pollination is by insects, bird pollination being unknown in West Africa. Most species produce nectar (at the base of the tepals or in a spur or pouch of the lip) though some produce food 'pollen', e.g. *Polystachya*, on the lip,

Fig. 38.2. Diagrammatic relationship of anther and stigma of groups recognised by Dressler (1981). A. Spiranthoideae. B. Neottieae (not West African). C. Orchideae. D. Epidendroideae. E. Vandoideae and some Epidendroideae. F. Vandoideae. Cross-hatching, stigma; stipple, pollinia; black, viscidium. (From Dressler 1981, Fig. 3.19.)

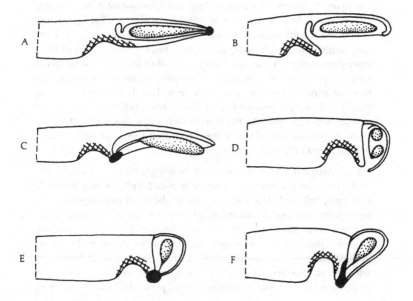

others offering a range of oils or fragrant substances. Deception is frequently used as a means of attracting the appropriate pollinator, which must in any case adopt a precise behaviour pattern. The very large number of pollen grains transferred in each pollinium, and the numbers of ovules developed ensure plentiful seed set from just one successful transfer. Most orchids are self-compatible, but the structure of the flower favours cross-fertilisation. *Spathoglottis* is regularly self-pollinated, hence its fruits are readily available. Lock & Profita (1975) have investigated pollination in *Eulophia cristata*.

Fruit A slow-maturing capsule with 3 or 6 slits, the midribs and placentae remaining attached at the apex. The seeds are minute and contain as a rule no endosperm (double fertilisation does not occur), and only an immature embryo without radicle or cotyledon.

Dispersal The seeds are dust like and fall out as the capsule moves and are carried in air currents. The seeds of some epiphytic species develop hygroscopic hairs, apparently aiding their attachment to phorophyte bark, or there may be hydroscopic hairs between the seeds (Vandeae).

Economic species Vanilla is extracted from the fermented immature fruits (pods) of *Vanilla planifolia*, introduced from central America. This is the only West African potential crop species, though other species have local uses. *V. planifolia* in West Africa has to be pollinated by hand.

> **Field recognition** *Ansellia* (J. Ansell, nineteenth-century explorer in Africa). The 2 species of Ansell's orchids are both large, robust acranthous epiphytes. *A. africana* is the more widespread (Liberia–East Africa, Angola) and is an epiphyte at the base of forest trees, up to 1.5 m high when in flower in the dry season. The erect shoots are pseudobulbs up to 50 cm long, multinodal and bearing scale leaves below, with 2 rows of large lanceolate conduplicate leaves above. The base of the plant is surrounded by a clump of erect (non-attaching) aerial roots. In the dry season a large terminal panicle of yellow, scented flowers, heavily blotched with red, purple or brown, is produced. Each flower has a yellow lip with 2 erect side lobes but no spur. As the flowers wither, the bracts persist as tiny pegs. *A. gigantea* var. *nilotica* is the paler spotted savanna form, known in West Africa only from northern Nigeria.
>
> *Bulbophyllum* (Gr. bulb leaf) is one of the few genera lacking a pollinium stalk. They are common epiphytes with pseudobulbs of one internode bearing apically 1–2 fleshy leaves without sheath or pseudopetiole, the inflorescence arising as a lateral shoot from the base of the pseudobulb. In the flowers, the column is produced downward into a foot and the lip articulates with this, moving freely. This is the largest orchid genus in West Africa, with 65 species recorded, about one-quarter of them being widespread. One species is certainly of restricted occurrence. This is *B. pipio*, the mangrove orchid, with very small pseudobulbs and short thick

spikes of yellowish flowers with purple centres. One of the most wide-spread species, *B. congolanum* (Guinée–Uganda, Congo) is found in north-ern guinea savanna southwards into forest. The pseudobulbs are 3–5 angled, and bear 2 leaves each. The inflorescence axis is flattened and leaf like, ± mottled with red and *c*. 1 cm wide. It bears yellow, greenish or red flowers and is produced at the end of the rains.

Diaphenanthe (Gr. transparent flower). The flowers of these orchids are at least semi-transparent. All are monopodial leafy epiphytes, with clasp-ing aerial roots, one of which emerges above each node through the leaf sheath of, and on the same side as, the blade above. Most are of tufted habit, but the most common species *D. bidens*, an epiphyte often seen on cocoa, has woody or wiry stems with elongated internodes. The shoots bear 2 rows of stiff, ribbed leaves in which all the main nerves are equally visible from above, and the leaf tips are of 2 very unequal jagged lobes. The leaves are pressed to the phorophytes. In the rains the pendulous inflorescence appears with a great many small, reddish-yellow flowers with obtuse sepals and petals, but a decidedly 4-sided lip up to 5 mm long, a little longer than its spur. There is a nodule on the lip, just in front of the opening of the spur. The peduncles and their minute bracts are more or less persistent.

Eulophia (Gr. with a beautiful crest, the callus on the lip). About a third of the 34 West African species are widespread, these also being the species found throughout much of tropical Africa. Only 5 species of *Eulophia* occur in Gambia, from which no other orchids are reported. The West African species are all sympodial ground orchids with pseudobulbs, rhizomes or stem tubers and each growth increment includes 2 or more large often grass-like ribbed leaves, the inflorescence arising separately, often before the leaves appear. The largest species is probably the forest swamp *Eulo-phia, E. horsfalli*, which is up to 2.5 m high when in flower, which is most of the year. It has a thick fleshy rhizome, ending in the swollen base of the flowering shoot, the fleshy roots. Two pseudobulbous *Eulophias* of forest and forest-savanna mosaic are *E. gracilis*, in which the leaves appear with the (green) flowers and are to be seen most of the year, and *E. guineensis* (also in Gambia), which produces its leaves in the rains and flowers towards the end of the rains and in the early dry season. This latter is one of only 2 species with a long spur (up to 25 mm). The pseudobulbs in this genus are of several nodes each; in *E. guineensis* they are conical and carry at the top 2–4 soft, broadly lanceolate leaves up to 45 cm long, tapering basally into a pseudopetiole. The inflorescence is 6–15 large flowers, *c*. 5 cm ∅, and stand up to 90 cm high. Each flower is pink- or purple-green in colour, the sepals and petals similar, the lip a spreading heart or spade shape with a fine point and indistinct side lobes, and without calli. It is generally mauve or pink in colour, darkly veined, and white or greenish towards the column. *E. quartiana* is very similar, but grows in savanna and produces its leaves after its flowers. The other savanna species are *E. cucullata* and *E. cristata* (also in Gambia), the former in seasonally damp savanna, the latter in well-drained savanna. Both these have chains of underground stem tubers which look much like the rhizome of ginger. In both species flowers appear before the leaves. Leaves and flowers also appear separately in *Nervilia*, but here the leaf is solitary and (ovate-)orbicular-reinform, the inflorescences smaller, with smaller spur-less white, yellow or greenish flowers (Hallé & Toilliez, 1971).

Eulophidium (now *Oeceoclades*; Garay & Taylor, 1976). Two species of sympodial pleuranthous orchids in forest, growing either on the ground or on dead wood. The 1–3 leathery conduplicate (not grass-like) pseudopetiolate leaves arise from the top of each pseudobulb. Later, the leaves absciss across the blade, leaving a toothed edge. Separate inflorescences appear at the beginning and again at the end of the growing season, the smaller flowers each with a 4-lobed lip. *E. maculatum* has short pseudobulbs only 4 cm long and bearing a single small variegated leaf with a short but broad and folded pseudopetiole, which looks like a young blade. The lip is white (the inner lobes frilled), veined and spotted with pink and with 2 calli, the spur 4–4.5 mm long and curved forwards. *E. saundersianum* has 1–3 large, dark-green oval leaves with long, cylindrical but channelled pseudostipules up to 20 cm long. The flowers have only a short spur and are yellow, veined with brown or purple, and with two calli.

Graphorkis (Gr. patterned orchid). Of this small mainly African genus, only *G. lurida* has been recorded in West Africa. It is a common widespread epiphyte with yellowish pseudobulbs, occurring in northern guinea savanna and southwards into forest, producing panicles up to 50 cm long of small purple-brown flowers in the dry season, before the leaves appear. The sepals, petals, lip and spur are all about the same length, 6 mm, the sepals purple-brown, the petals brown or yellowish, the lip 3-lobed, the middle lobe yellow with two keels and erect yellow or green side lobes and a purple, forward-bent spur. The column is purple with a white operculum. When the leaves appear, there are 4–6 on top of the pseudobulb, each one stiff, up to 45 cm long × 5 cm wide, ribbed with a hard prominent midrib. Round the base of the pseudobulb is a large tuft of erect white aerial roots, which distinguish the plant at all times of the year.

Habenaria (L. the strap-shaped lip) is a very large (600 species) genus, of pantropical and subtropical distribution, composed of sympodial ground orchids (and only 4 species of epiphytes) mostly of very local distribution in West Africa. Only 3 of the *c.* 50 West African species are widespread, growing in both forest and savanna and extending to East Africa and Angola. *Habenaria*s are erect in habit, with either a small, usually woolly, root tuber or with a fibrous base with fleshy woolly roots. The leaves are cauline; the first and last produced are smallest, with spikes of white or greenish flowers mostly with entire sepals appearing above them in the rains. In all except *H. occidentalis* (with deep yellow flowers) the lip is abaxial, and the anther is attached by its base. The loculi contain

2 pollinia, each divided into granular masses. *H. genuflexa* is a slender tuberous plant of wet patches in grassland or wet pockets among rocks.

Fig. 38.3. *Habenaria zambesina*. A. Base of plant × 1. B. Flower × 2. C. Top of gynostemium, face view × 5. D. Top of gynostemium, side view × 5. E. Inflorescence × 1. F. Floral diagram. a, anther loculus; l, labellum; lr, lateral lobe of rostellum; o, ovary; rp, rostellar projection; s, stigma; sp, spur; st, staminode; v, viscidium. (From Summerhayes, 1968, Fig. 12 *pro parte*.) (British Crown copyright. Reproduced with permission of the Controller, Her (Britannic) Majesty's Stationery Office & the Trustees, Royal Botanic Gardens, Kew. © 1961.)

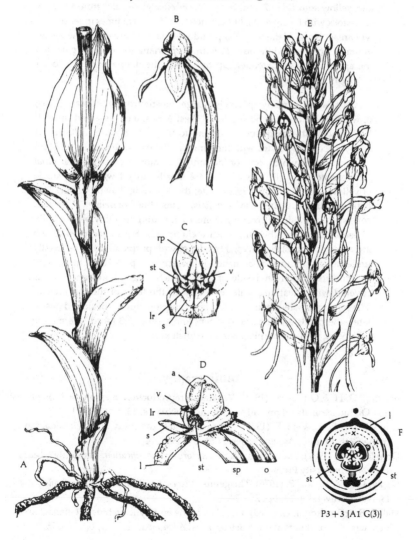

There are a few linear leaves on the stem, and the spike has numerous scented small flowers. Each flower has 3 green sepals, 2 2-lobed petals, the one lobe short and white, the other long and green, and a 3-lobed green lip, with a twisted spur which is swollen at the end. *H. macrandra* is a more robust-looking plant growing to 60 cm high from a fibrous base with fleshy roots. The leaves, except for the last-produced ones, are large, pseudopetiolate with ovate blades, and the flowers, though few in number are *c.* 10 cm Ø in the adaxial–abaxial plane. The sepals are relatively narrow but entire, the outer ones white and up to 3 cm long, the inner whorl of 2 green adaxial tepals and a green spur with 3 green lobes and a green-and-yellow spur up to 7.5 cm long. *H. zambesina* grows also from a fibrous root-stock with fleshy roots, but to a height of 120 cm at times in seasonally wet areas. None of the leaves is pseudopetiolate. There are dense spikes of small, scented, white flowers with short triangular entire tepals, but the spur is long and slender, up to 4 times the diameter of the flower in length, or 6 cm long.

Polystachya (Gr. many spikes) is a large, mostly tropical African genus, of which there are over 50 species in West Africa, 6 of these widespread. Very few West African species extend to the east or south. *Polystachya*s are sympodial epiphytes with long multinodal pseudobulbs, up to 10 cm long, on top of which develop not only 2 or more leaves but a terminal inflorescence as well. The flowers are mostly tinged with some shade of yellow and they are not resupinate, the adaxial lip having a spur. This petal, together with the 2 adaxial sepals, forms a hood or helmet. *P. tesselata* consists of very short, creeping stems each ending in a tiny pseudobulb. These lie close together and form a clump. On top of the pseudobulb are a few oblanceolate leaves, 30 cm long, with purple blotches especially underneath, round a ± branched inflorescence up to 70 cm high, with the flowers carried on 1 side of each branch. The inflorescence axis is covered with sheathing scale leaves up to the level of the top branch. The sepals are pale-green, purple-tipped and sometimes chequered (tesselate), the petals are linear and green, and the lip is 3-lobed, the middle lobe being white or purplish with a reddish keel.

Bibliography

Benzing, D. H. & Ott, D. W. (1981). Vegetative reproduction in epiphytic Bromeliaceae and Orchidaceae: its origin and significance. *Biotropica*, **13**, pp. 131–40.

Cribb, P. J. & Perez-Vera, F. (1975). A contribution to the study of the Orchidaceae of the Côte d'Ivoire. *Adansonia*, sér. 2, **15**, pp. 199–214.

Dressler, R. L. (1981). *The Orchids. Natural history and classification.* Cambridge, Mass.: Harvard University Press.

Garay, L. A. & Taylor, P. (1976). The genus *Oeceoclades*. *Botanical Museum leaflet*, Harvard University, **24**, pp. 249–74.

Hall, J. B. & Bowling, J. C. (1969). Field key to the epiphytic orchids of Ghana, based on characters of shoots and infructescences. *Adansonia*, sér. 2, **9**, pp. 139–75.

Hallé, N. & Toilliez, J. (1971). Le genre *Nervilia* en Côte d'Ivoire. *Adansonia*, sér. 2, **11**, pp. 443–61.

Jaeger, P., Hallé, N. & Adam, J. (1968). Contribution à l'étude des Orchidées des Monts Loma (Sierra Leone). *Adansonia*, sér. 2, **8**, pp. 265–310.

Johansson, D. (1974). Ecology of vascular eiphytes in West African rain forest. *Acta phytogeographica suecica*, **59**.

Johansson, D. (1977). Epiphytic orchids as parasites of their host trees. *American Orchid Society Bulletin*, **46**, pp. 703–7.

Jonsson, L. (1979). The African member of *Taenioiphyllum* (Orchidaceae). *Botaniska notiser*, **132**, pp. 511–19.

Lock, J. M. & Profita, J. (1975). Pollination of *Eulophia cristata* (Sw.) Steud. (Orchidaceae) in southern Ghana. *Acta botanica neerlandica*, **24**, pp. 135–8.

Piers, F. (1968). *Orchids of East Africa*, 2nd edn. Lehre: J. Cramer.

Rasmussen, F. N. (1986). On the various contrivances by which pollinia are attached to viscidia. *Lindleyana*, **1**, pp. 21–32.

Sanford, W. W. (1971). The flowering time of West African orchids. *Botanical Journal of the Linnean Society*, **64**, pp. 163–81.

Sanford, W. W. (1974). The use of epiphytic orchids to characterise vegetation in Nigeria. *Botanical Journal of the Linnean Society*, **68**, pp. 291–301.

Sanford, W. W. & Adanlawo, I. (1973). Velamen and exodermis characters of West African epiphytic orchids in relation to taxonomic groupings and habitat tolerance. *Botanical Journal of the Linnean Society*, **66**, pp. 307–21 and pl. 1.

Segerbäck, L. B. (1983). *Orchids of Nigeria*. Rotterdam: A. A. Balkema.

Summerhayes, V. (1968). *Flora of tropical East Africa: Orchidaceae*, part 1. London: Crown Agents.

39

Cyperaceae – sedge family

Superficially grass-like herbs, in ecological descriptions grouped with grasses as graminoids. The family is a mainly temperate one, occurring also in arctic areas. Most sedges are helophytes or mesophytes growing in seasonally wet places. There are a few hydrophytes in West Africa (*Websteria confervoides* and a few species of *Eleocharis* and *Scirpus*). West African species are mainly perennials, the widespread ones also tending to occur in most of the rest of tropical Africa at least. Less than half West African genera extend to Gambia. Habitat lists of species are to be found in Chapter 2.

Some of the terminology developed for describing grasses is also applied to sedges. The leafy erect aerial shoots (culms) are rosette shoots, the leaves mostly basal and separated from each other by only very short internodes. In perennial species, the culms arise from slender sympodial rhizomes, each culm terminating an increment of growth. The internodes of the rhizome may be long, resulting in a spreading habit, or very short, when a tufted habit similar to that of annual species results. Stem tubers are developed in at least two forms. Terminal tubers on stolons (*Solanum tuberosum* fashion) occur in nut grass (*Cyperus rotundus*), one of West Africa's most pernicious weeds, and the edible tiger nut (*C. esculentus*). Each tuber then produces a culm and several more stolons (but see Lorougnon, 1970). In *Cyperus tonkinensis* and *Mariscus alternifolius*, the base of the culm becomes food-storing, repetition of the process in a series of culms bringing about the formation of a chain of 'bulbous' tubers, a segmented rhizome. Culms are usually three-sided, solid and with three rows of leaves with linear blades arising from closed, usually eligulate sheaths. *Coleochloa abyssinica*, an upland sedge, is unusual in several respects. It has leaves in two rows, the sheaths are open with a ring of ligular hairs at the opening, and there is a joint, like that in grasses, between the blade and the sheath, the blade being deciduous. The flowers are ♂ or ♀.

An outgrowth (antiligule or contraligule) on the side of the sheath opposite to the blade is present in some *Scleria* spp. and *Afrotrilepis pilosa*, the latter a widespread and common sedge of granite outcrops and domes in forest, forest-savanna mosaic and the northern guinea zone. The long, apparently forked, older stems persist to form 'trunks', each one ending in a long-leaved rosette shoot with a tall inflorescence, with separate ♂ and ♀ flowers.

The leaf blade is represented by a small outgrowth (apiculus) in *Eleocharis*, most species of *Scirpus*, some *Cyperus* spp. and *Websteria confervoides*, a plant so soft and fine as to suggest an alga filament at first sight. The development of wide leaf blades, up to 8.5 cm wide, and of a pseudopetiole between leaf and blade, is confined to some species of the forest genus *Mapania*, which thus parallel the forest grass genera closely.

Above the rosette of leaves, the culm develops a scape terminating in an inflorescence, which is usually subtended by one or more leaf-like bracts. The seashore pioneer species *Remirea maritima* is one of the species in which the inflorescence is sessile on the culm. In some species of *Hypolytrum*, the inflorescences are axillary, but the culms are hapaxanthic.

There are quite a number of species in which the rhizome becomes lignified. Apart from *Afrotrilepis* (see above), there is, also on granite outcrops, *Microdracoides squamosus*, with stems which are sclerified and heavily clothed with the persistent bases of the leaves – squared off blade bases as well as sheaths. These 'trunks' can be up to a metre high, but the actual rosettes of leaves are relatively short, and the inflorescences are shorter and stouter than those of *Afrotrilepis*. When young, the plant may well look like a large moss. Although this species is known from Guinée to Cameroun, it is mostly recorded from Guinée and Sierra Leone.

The only regular climber among our sedges is *Scleria boivinii*, as widespread herbaceous perennial species in secondary growth from the wettest forest area northwards into the northern guinea zone. The stems are sharply triangular and there is an antiligule. The stem, leaf blade edges and angles of the leaf sheaths are all armoured with hard reflexed 'hairs', and have a cutting edge.

Sedges may be recognised by their habit, their conical silica bodies in epidermal cells over vascular bundles in the leaves (not *Hypolytrum* or *Mapania*), and by their small, sessile, naked flowers, each in the axil of a floral bract (glume, fertile or flowering glume). The flowers are spirally or distichously arranged on a stalk (rachilla), forming an inflorescence unit. One of these, the spikelet (Koyama, 1961) is of indeterminate growth and has at least one ♂ flower. The other, the cymelet, is of determinate growth and has only unisexual flowers. In *Mapania* and *Hypolytrum*, each cymelet ends in a ♀ flower and the cymelets are agglomerated into partial inflorescences. In *Microdracoides* and *Scleria*, for example, partial inflorescences are very much reduced, with a single ♀ cymelet, or all ♂. *Afrotrilepis* has both kinds of cymelet in the partial inflorescence.

The cymelet-bearing genera are currently placed in the subfamily Mapanioideae (genera 17–23 in *Flora of West tropical Africa*, 1972), and the spikelet-bearing ones in one of the three subfamilies Scirpoideae, Rhynchosporoideae or Caricoideae (genera 1–14, 15–16 and 24, respectively).

Inflorescences are made up of spikelets or partial inflorescences (of cymelets) and are described as 'spikes', 'clusters', 'heads', or 'umbels'. An anthela is similar to the cymose 'cluster' of the Juncaceae.

Flowers ⊹ ⊕ or .l. 2–3-part. Only *Microdracoides* is dioecious, the other West

African species having ♂ or ♂ and ♀ or a mixture of ♂ and ♂ or ♀ flowers on the same plant. Perianth often present in the Scirpoideae and Rhynchosporoideae in the form of 6 hypogynous bristles. The floral bract ± enfolding the ♂ flowers of *Ascolepis* (and floral bract and prophyll in *Lipocarpha*), known as squamella(e), are said to form a utricle, a ± bottle-shaped envelope, while the 6 scales of *Microdracoides* are cymelet bracts, and the utricle of *Carex* is a complete bottle-shaped prophyll round the naked pistil. Stamens 2–3, basifixed; pollen formed in tetrads, but only 1 grain in each survives. G(2) or (3) lens-shaped or 3-sided, respectively, with 2 or 3 styles, but 1-celled with

Fig. 39.1. Diagrams of spikelet structure and floral diagram of Scirpoideae and Rhynchosporoideae. The ♂ spikelet of *Carex* (Caricoideae) resembles that of the Scirpoideae, but the ♀ spikelets are 1-flowered, with a naked pistil in the axil of the enveloping prophyll, the spikelets arranged in spikes. (Spikelet structure from Koyama, 1961, Fig. 1 *pro parte*; floral diagram of *Scirpus* and *Cyperus*, and spikelet diagram of *Lipocarpha* from Kern, 1974, Fig. 1 *pro parte*.)

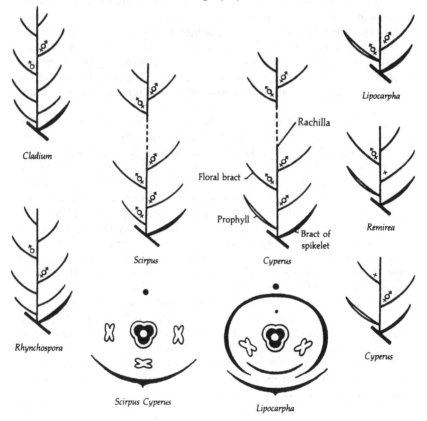

a basal ovule; the stylar base may be swollen and sometimes persists; both 2- and 3-carpelled pistils occur in *Cyperus* and *Fimbristylis*.

Pollination by wind, with dry pollen and, in the species investigated, dry stigmas. Protogyny is general. Entomophily occurs in some (non-West African) species of *Mapania*, possibly also in *Rhynchospora*. Lorougnon (1973) suggests slug pollination for forest species.

Fruit An achene, the production of different above- and below-ground fruits (amphicarpy) being recorded in East Africa (Haines & Lye, 1983). In the Mapanioideae the utricles (of 2 prophylls each) ± adhere to the achene, forming a false fruit.

Dispersal Among helophytes, dispersal by water is most likely and probably all utricles can float. In *Mariscus*, *Remirea* etc. the spikelet is shed whole, and is also likely to float, especially in *Remirea*, where the thick corky rachilla enfolds the achene. Animal-dispersed burr fruits may be formed in *Afrotrilepis* and *Coleochloa*, the achenes of both carrying with them numerous long bristles. *Cyperus papyrus* achenes are found in elephant dung, but seldom germinate from it.

Economic species Sedges are usually of importance for negative reasons, as poor pasture plants and little used as fodder because of the siliceous bodies they usually contain, or as weeds of cultivation. The larger, more cutting species are sometimes known as devil grass. The rhizomes of several species are fragrant and find local use. The tiger nut (*Cyperus esculentus*) and nut grass (*C. rotundus*) are used respectively as food and fodder.

Field recognition *Bulbostylis* (L. globular style (root), now *Abildgaardia pro parte* (Lye, 1974)) and otherwise scarcely separable from *Fimbristylis* (but see Gordon-Gray, 1971)). Annuals or rhizomatous perennials, producing culms with thin wiry leaves (which sometimes resemble long bristles) with hairs round the mouth of the sheath. The base of the style is swollen and persists on top of the achene as a beak, the 3 sides of the achene being decorated with a pattern of wavy lines. The inflorescence is often black, brown or reddish, a 'head' of spikelets or a compound umbel, each branch of which ends in a spikelet. In the heads, the spikelets are sessile, while in the umbels the spikelets are stalked, the stalk sometimes being longer than the spikelet. The inflorescence is subtended by bracts, which are generally hairy at the base. Each spikelet is terete, with several floral bracts (and their ♂ flowers) arranged spirally on the rachilla on which they leave no prominent scar when they fall. The spikelets lack a prophyll. *B. barbata*, an annual of disturbed ground, has heads of sessile spikelets, greenish or yellow-brown in colour. The flowering glumes are boat-shaped and hairy, with a green keel running out at the tip into a recurved point. The following species are also annuals and have

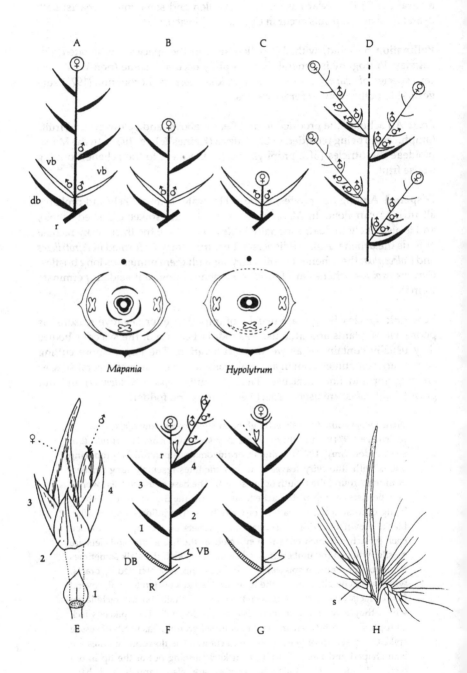

Mapania   Hypolytrum

dark-coloured spreading inflorescences of long-stalked spikelets. They all grow on shallow soils and inselbergs. *B. coleotricha* has few large rounded dark-tipped spikelets (the glumes may have a dark spot near the tip), the upper ones with white-fringed persistent bracteoles round the base. *B. congolensis* (now *Abildgaardia congolensis*; Lye, 1974) has pinkish, papery leaf sheaths round finely and closely hairy stems, and reddish-brown bracts. *B. abortiva* has bracts fringed with at least some long white hairs, glabrous glumes with 3 nerves, and curiously patterned achenes. the surface cells of these are square and raised, giving the appearance of vertical rows of papillae.

Cyperus (Gr. kypeiros) contains 14% of all sedge spp. which are distributed in all the warmer parts of the world. *Kyllinga*, *Pycreus* and *Remirea* are included in *Cyperus* in some classifications. There are 70+ species of both annuals and perennials in a variety of habitats in West Africa, even including 2 upland species of pyrophytes (*C. angolensis*, *C. nduru*). The inflorescence is a compact head of spikelets, or a simple or compound umbel with unequal branches, the spikelets being flattened and made up of 2 rows of floral bracts, each containing a ♂ flower. At maturity, the floral bract and 3-sided achene is shed, leaving the rachilla behind. At least one species of *Cyperus* is likely to be found in any habitat, some of the more obvious ones being mentioned here. *C. maritimus* is always found in sandy situation near the sea, round lagoons and behind mangrove swamps. It is a robust plant with flat, leathery leaf blades of ± equal length. The bracts, long and reflexed, subtend a ± globose congested terminal head of numerous spikelets, each at least 2 mm wide. The style has either 2 or 3 branches. Two of the helophytic species most easily distinguished are *C. papyrus* and *C. articulatus*. The latter has 2 m high scapes which, in the dry state, show the septate nature of the pith as a series of rings. This is not seen in any other sedge. *C. papyrus* (paper sedge) is by far the largest of all sedges, with scapes up to 5 m high. But in any case, the inflorescences are distinctive, with 50 or more subequal branches to the umbel, most of them bearing spikes and spikelets. Two species of sandbanks in rivers are also easily recognised. *C. maculatus* is a rhizomatous perennial with hard round tuberous 'segments', covered

Fig. 39.2. Diagrams of partial inflorescence structure. A–B. *Mapania*. C. *Hypolytrum*. D. Four partial inflorescences of the inflorescence of Mapanieae. Partial inflorescence diagrams for *Mapania* and *Hypolytrum* are shown below. E–F. *Afrotrilepis*. G. Partial inflorescence (♀) of e.g. *Microdracoides* and *Scleria*. H. Fruiting partial inflorescence of *Microdracoides*. db, glume subtending cymelet; DB, glume subtending partial inflorescence; r, partial inflorescence axis; R, inflorescence axis; vb, cymelet prophyll; VB, partial inflorescence prophyll; 1–3, sterile cymelet bracts; 4, fertile cymelet bract; s, squamella (the prophyll in F and G). There are also species of *Mapania* in West Africa with only two stigmata. (A–H from Koyama, 1971, Figs 10–12, 15–17, 19, 21; partial inflorescence diagrams from Kern, 1974, Fig. 1 *pro parte*.)

with dark fibres. The leaves are slim and whitish, and the scape carries an inflorescence of a few subequal branches, each with a compound spike of pale brown or silvery spikelets broader towards the tip. The glumes are at least 2 mm long, obtuse, numerous and closely imbricate on a glabrous rachis. The other most common sandbank species is *C. tonkinensis*, another perennial with a mass of tubers and a narrow inflorescence of silvery, very small spikelets.

Fimbristylis (Gr. fringed style). Tufted annuals or perennials with narrow, sometimes wiry leaves and branched inflorescences, each branch ending in a terete spikelet with spirally arranged floral bracts, each with

Fig. 39.3. A–C. *Pycreus polystachyos*. A. Inflorescence × $\frac{1}{2}$. B. Spikelet × 8. C. Floret dissected × 12. Flowering glume = floral bract. D. *Cyperus* sp. floret dissected × 20. a, spikelet bract; b, prophyll; c, floral bract; d, sterile glume above flower.

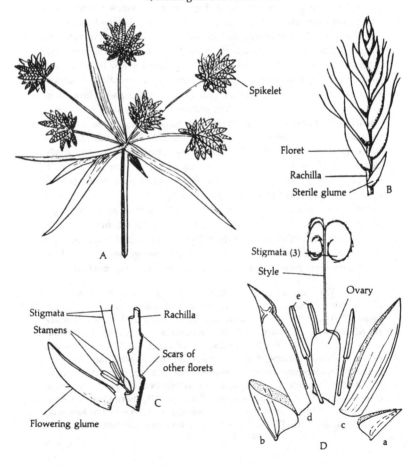

a ♂ flower. The spikelets lack a prophyll. These fall at maturity, leaving a highly scarred rachilla. There may be 2 or 3 style branches (and carpels) but the base of the style, though swollen, is deciduous. The style itself is not fringed in the sense of being hairy but rather in the sense of being covered with cellular projections. In sandy river beds, *F. cioniana* and *F. squarrosa*, the former with G(3̲), the latter with G(2̲), are characteristic. In sandy brackish situations on the coast the perennials, *F. ferruginea* (G(2̲)) and *F. obtusifolia* (G(3̲)) occur, the latter confined to the seashore. *F. ovata*, sometimes included in *Abildgaardia*, is a savanna species, distinctive by reason of its solitary but large (1 cm long) spikelets.

Kyllinga (P. Kylling, seventeenth-century Danish botanist), a genus sometimes included in *Cyperus*, is mainly composed of narrow-leaved mesophytes of various habits in damp places, with heads of flattened spikelets bearing floral bracts in two rows. Above the spikelet bract and prophyll may be a further floral bract lacking a flower then 1–2 fertile ones with ♂ flowers. There are 2 carpels and 2 styles, the achene being flattened in the adaxial–abaxial plane. The spikelet falls when the achene matures, leaving the spikelet bract and prophyll round the prominent rachilla scar. Both species of brackish conditions are perennials. *K. robusta* is a tufted plant of brackish swamps, purple leaf sheaths and roots, and reddish ± hemispherical head of spikelet(s) (usually more than 1) surrounded by bracts broad at the base. Each floral bract has a keel bearing spiny teeth. *K. peruviana* is rhizomatous, and mostly seen in sandy situations on the coast. The scapes are purplish and spongy below, the rhizomes covered with reddish-brown scales. The inflorescence is a single fat spike of crowded spikelets surrounded by a few unequal bracts. Leaf blades may be absent. Of the species of damp, disturbed ground (in forest to driest savanna), *K. squamulata* is representative. It is a small annual, with finely spiny pyramidal spikes subtended by a few broad-based bracts. The midrib of each flowering glume is lobed.

Pycreus (? Gr. thick, firm or anagram of *Cyperus*). A few species are, or can grow as, hydrophytes, but most are savanna plants of various habits. The genus is sometimes included in *Cyperus*. The inflorescence (either head or umbel like) is made up of flattened spikelets with distichous bracts and ♂ flowers with 2–3 stamens and 2 carpels, the achene being flattened in the adaxial–abaxial plane. When they ripen, they and their bracts are shed gradually from the bottom up, exposing the scarred rachilla. *P. polystachyos*, the only coastal species (in rice fields and sandy places) and has either an unequally branched inflorescence with clusters of nearly distichous spikelets, some of which may be over 3 mm wide at maturity (var. *polystachyos*), or the clusters may be sessile (var. *laxiflorus*). Both varieties have anthers under 1 mm long, but in var. *polystachyos* there are generally 2 of them, in var. *laxiflorus* 3. *P. pumilus* is particularly widespread, both in terms of geographical distribution (throughout the Old World tropics, Himalayas and Central America, from the West African coast into the sudan zone) and in disturbed ground. It is a small tufted annual, with yellowish-brown very slim spikelets in a cluster. Each spikelet has floral bracts with a green keel, this extending beyond the bract to a point, the bracts rather spreading and the wide, pale rachilla plain to see.

Bibliography

Goetghebeur, P. (1980). Studies in Cyperaceae, 2. *Adansonia*, sér. 2, **19**, pp. 269–305.

Gordon-Gray, K. D. (1971). *Fimbristylis* and *Bulbostylis*: generic limits as seen by a student of South African species. *Mitteilungen der botanischen Staatssammlung München*, **10**, pp. 549–74.

Haines, R. W. & Lye, K. A. (1983). *The sedges and rushes of East Africa*. Nairobi: East African Natural History Society etc.

Hall, John B. (1973). The Cyperaceae within Nigeria. *Botanical Journal of the Linnean Society*, **66**, pp. 323–46.

Kern, J. H. (1974). *Flora malesiana*, Cyperaceae. Spermaphyta ser. 1, **7** (3), pp. 435–753. Dordrecht: Martinus Nijhoff/Dr W. Junk.

Koyama, T. (1961). Classification of the family Cyperaceae, 1. *Journal of the Faculty of Science, University of Tokyo III, Botany*, **8**, pp. 37–148.

Koyama, T. (1971). Systematic interrelationships among Sclerieae, Lagenocarpeae and Mapanieae (Cyperaceae). *Mitteilungen der botanischen Staatssammlung München*, **10**, pp. 604–617.

Kukkonen, I. (1979). Economic aspects of African Cyperaceae. In *Taxonomic aspects of African economic botany*, ed. G. Kunkel, pp. 72–4. Las Palmas: Perez Galdos.

Kunkel, G. ed. (1979). *Taxonomic aspects of African economic botany*. Proceedings of the IX plenary meeting AETFAT, Las Palmas, 18–23 March, 1978. Las Palmas: Perez Galdos.

Lorougnon, G. (1970). Étude morphologique et biologique de deux variétés de *Cyperus esculentus* L. (Cypéracées). *Cahiers d'Office de recherches scientifiques et techniques d'outre-mer*, sér. Biologique, **10**, pp. 35–63.

Lorougnon, G. (1973). Le vecteur pollinique chez les *Mapania* et les *Hypolytrum*, Cyperacées de sous-bois des forêts tropicales ombrophiles. *Bulletin du Jardin botanique national belgique*, **43**, pp. 33–6.

Lowe, J. (1972). *Flora of Nigeria. Sedges*. Ibadan: Ibadan University Press.

Lye, K. A. (1974). Studies in African Cyperaceae 11. *Botaniske notiser*, **127**, pp. 493–7.

Merxmüller, H. (ed.) (1971). Proceedings of the VII plenary meeting AETFAT, München, 7–12 September, 1970. *Mitteilungen der botanischen Staatssammlung München*, **10**.

Oteng-Yeboah, A. A. (1976). Observations on the genus *Ascolepis*. *Notes from the Royal Botanic Garden Edinburgh*, **35**, pp. 391–7.

40

Gramineae (Poaceae) – grass family

The second largest family of monocotyledons on a worldwide basis, but the first in importance both ecologically and economically, dominating all the 'grasslands' of the world and providing the grain and fodder which are staples of diet for both man and animals. Wheat, rice, maize and guinea corn are reported to cover half the world's land under food crops.

Grasses are annual or perennial rosette herbs, occurring in every kind of habitat. The long-lived bamboos are lignified and of tree-like dimensions, but there are quite a number of species, even annual ones, which reach 5 m or more in height and have cane-like stems. In habit, though plentifully branched, grasses may be compact and upright (tussock-forming), or trailing, creeping or decumbent, or extended by means of axillary shoots, which form stolons above ground or rhizomes underground. Descoings (1972) distinguishes five habits. Annuals lack stolons and rhizomes, but perennials may adopt any of these habits. New axillary shoots may be called tillers, and flowering shoots are known as culms, but all aerial shoots are basally branched and have a limited life span. Growth is sympodial and non-flowering shoots are relatively short lived.

The leaf blade, generally linear to lanceolate in outline, arises from an open sheath which encircles the stem and younger leaves closely. Venation is always parallel (convergent), the wider leaves appearing *Commelina* like. At the adaxial junction of blade and sheath is a ligule of hairs or a scale (absent in some *Echinochloa* spp.) and there may be small outgrowths at the top of the sheath (auricles). A marked pseudopetiole is present in the bamboos and some other species, including *Andropogon gayanus*, where there is also an external ligule abaxial to the blade (Bowden, 1964). There may be a meristem in the leaf sheath base which, by differential growth, can cause flattened or trampled culms to grow upright again.

Aerial shoots grow not only by an apical meristem but also by an intercalary one at the base of each internode, and there is a third kind of meristem, in species without leaf sheath meristems, in the node, having the same function as the leaf sheath meristem. In contrast to sedges, grass stems are cylindrical and the internodes sometimes hollow (except in *Zea* etc.). The leaves are distichous, and each has a bud in its axil.

Forest grasses in general tend to have *Commelina*-like leaves with pseudo-petioles and cross-veins (between the parallel ones). In the subfamily Bambuseae, stomata are confined to the abaxial blade surface and the blade usually breaks off at the base. *Leptaspis cochleata* has, however, stomata on the adaxial surface but the blade is twisted through 180° and the stomata come to lie on the under surface, while *Streptogyna* and *Olyra* have stomata on both sides of the blade.

Savanna grasses are narrow leaved with stomata on the upper side of the blade, which is either permanently folded (conduplicate), pleated (plicate) or rolled or can become so, towards the upperside. The result is, that the stomata open into a chamber or grooves protected from air currents, and stay open for longer than would otherwise be the case. Leaves of *Setaria* spp. may become folded or pleated, those of *Ctenium newtoni* involute and those of *Loudetia arundinacea* supervolute. *Loudetia simplex* is an example of a species with permanently rolled leaves.

Grasses are recognised by their habit and by their small ± naked flowers, each in the axil of a floral bract (lemma), enclosed between this and the adaxial hyaline palea, a prophyll. The flower, lemma and palea constitute a floret. The tissue just under the floret may be enlarged to form a knob, spine or tooth (or two teeth etc.) and is termed a callus.

Florets are arranged distichously on a rachilla to form a spikelet, of which the two lowest bracts, one abaxial, the other adaxial, are (sterile) glumes. Florets are ♂ or unisexual and the distribution of these in the spikelet, and the development of awns, bristles and hooks on lemmas and glumes, provide generic characters.

Fig. 40.1. Diagrams of spikelet structure. A. Indeterminate spikelet, e.g. *Ctenium, Eragrostis*. B. Determinate spikelet, e.g. Andropogoneae, *Loudetia, Panicum*. Black, lower glume and rachilla; l, lodicule; p, palea. (Adapted from Warming, 1933, Fig. 137.)

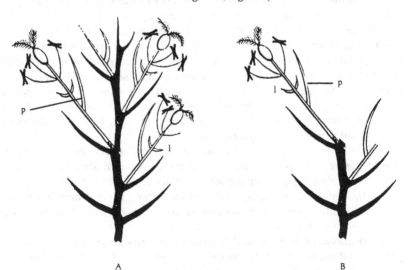

A   B

The inflorescence is composed of spikelets arranged in various ways (panicle, raceme, spike etc.), the spikelet being regarded as the equivalent of a flower as in Cyperaceae. The whole is sometimes aggregated into a leafy false panicle. The spikelet stalk is termed the pedicel, and may carry an involucre of bristles or hairs.

Flowers ÷ ⊕ or .l. ♀ or ♂ or ♀ or barren. The perianth may be represented by 2 or 3 lodicules, or these may be regarded as bracts, and the flower is then naked. A2–3, (3+3 in the Bambuseae, (A) in *Puelia*); anthers dorsifixed, versatile and opening latrorsely, on delicate filaments which become relatively long. G1, 1-celled with 1 lateral and basal ovule, and 2 long, lateral stylodia.

Pollination The family is generally considered to be wind pollinated. Protandry is usual. Among the Bambuseae in South America however, small flies are probably involved (Soderstrom & Calderon, 1971), and such a mechanism would also be possible in the similar conditions of West African forests. Similar observations have been made in Australia and Kenya. Where the lodicules are fleshy, they become swollen at anthesis, forcing the lemma and palea apart and the anthers emerge rapidly on elongating filaments, followed by the styles. Pollination is completed in a few hours.

Fruit Mostly a caryopsis ('grain'), an achene in which the pericarp and testa are fused. The fruit is rarely of another form, e.g. the follicular monocarp of *Sporobolus*. The seed has starchy endosperm, with the embryo placed laterally and outside it in contrast to other monocotyledons. The commonly described form of scutellum, as in *Zea*, for example, is only 1 of 4 kinds (Negbi, 1984).

Dispersal The seed of *Sporobolus*, which is sticky, is squeezed out of the fruit when it swells, but usually the caropsis is shed with at least one other structure, and the fruit is, in fact, a false one. The spikelet itself with two glumes may be shed, or it may break off just above the two glumes, or between each floret, which is thus accompanied by a portion of rachilla, or each floret may be shed, leaving the rachilla behind. Dispersal by water is not well documented but is highly probable for swamp and riverside species. Numerous species are dispersed by wind, having plumed structures of different kinds – hairs on the rachilla, glumes, pedicel, callus or awn. Such species may also float in water, and even adhere to the coats of animals. Internal transport of grain by browsing animals is probably extremely common, by baboons for example, but diaspores with obvious means of external adhesion are fairly common too. Burr fruits have hooks and are sticky, while the springy rachillas of *Streptogyna* act as fur traps, the spikelets being jerked apart as the animal passes. *Leptaspis* has remarkable lemmas (covered in hooked pairs), each enclosing the ♀ flower and with only a small pore at the tip through which the three stigmas emerge. A third forest grass, *Centotheca*, has two elongated tufts of reflexed bristles on the back of the palea. The capacity for penetration of fur or skin is well developed among grasses, both by means of a pungent

Fig. 40.2. A–C. Inflorescence types. A. Panicle. B. Scattered racemes. C. Digitate spikes. Various possible arrangements of spikelets are shown. D–E. Diagrams of floret structure. D. Floret with lemma removed. E. Floret diagram. F. *Eleusine coracana* diagram of spikelet, with a number of similar florets. *Eragrostis* and *Dactyloctenium* have spikelets of this kind. G. *Zea mays* ♂ spikelet 'pair', the sessile floret in section. The 'pairing' of a stalked and a sessile spikelet is constant in the tribes Andropogoneae (four genera of which are described under Field recognition, see the text) and Maydeae. In F and G, the lower glume is black, the upper one striped, the lemma unshaded and the palea stippled.

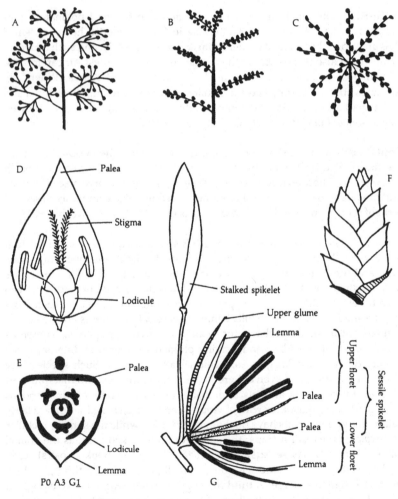

callus and a pungent awn or awns, *Heteropogon contortus*, a shore and savanna species, having both. A weed species also occurring on the shore, *Rottboellia exaltata* (now *R. cochinchinensis*), once the only ant-dispersed species known in West Africa, has an elaiosome formed from the callus of the sessile spikelet (the accompanying spikelet being stalked) which is sunk in the axis of the rachilla. Ants (*Messor* spp.) are now known to disperse grains of various grasses, e.g. *Monocymbium ceresiiforme, Pennisetum hordeoides* and *Andropogon gayanus* (Levieux & Diomande, 1978). The function of the awn appears mainly to be to propel fruit on the ground by means of hygroscopic flexing and coiling, the beard on the callus, when present, preventing backward movement. The activity of the more complex awns can even make the fruit hop. The awn can also turn the fruit so as to favour initial water uptake, and help to anchor it against the thrust of the lengthening radicle.

Fig. 40.3. Animal-dispersed fruits. A. *Tragus berteronianus*, a sahel species, spikelet pair × 10. B. *Cenchrus* sp. inner involucre round spikelet × 4. The outer involucre is of bristles. In West African species, the bristles of the inner involucre are joined only at the base. C. *Pseudechinolaena polystachya* spikelet × 12. D. *Rottboellia exaltata* stalked spikelet and internode of rachilla from both sides, sessile spikelet lower right, × 4. A–C from Clayton, Phillips & Renvoize, 1974, Figs. 108–3, 158–4, 131–3; D from Clayton, 1972, Figs. 461.2, 461.3.) (British Crown copyright. Reproduced with permission of the Controller, Her (Britannic) Majesty's Stationery Office & the Trustees, Royal Botanic Gardens, Kew. © 1961.)

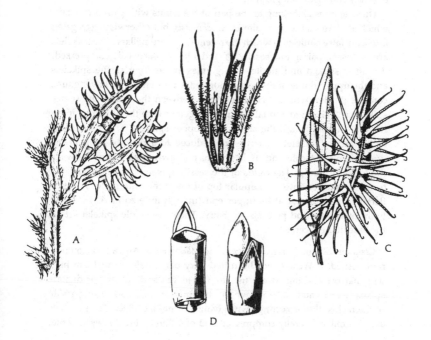

Economic species The important grain crops are bulrush or pearl millet (*Penni-setum americanum*, now *P. glaucum*), guinea corn (*Sorghum bicolor* var. *bicolor*), the 'hungry' rices, *Digitaria* spp. (fonio), all indigenous crops cultivated north of the forest area, and African rice (*Oryza glaberrima*), cultivated wherever the necessary water supply is available, particularly in the southwestern area (Guinée–Côte d'Ivoire). The introduced *O. sativa* from Asia is cultivated in those areas also, while maize ('corn') from America is most often cultivated in the forest-savanna mosaic and northern guinea zones. Another carbohydrate crop is sugar cane (*Saccharum officinarum* from Asia). Many species are suitable as fodder if cut young, and a number of potential pasture and fodder species have been introduced. Many species are used for thatching, woven mats, etc.

> **Field recognition** *Andropogon* (Gr. bearded male). A tropical and subtropical genus with *c.* 35 species of robust annuals and perennials, the latter up to 5 m high, in savanna. At least 1 species may be expected in virtually every habitat. Four very common and widespread species are *A. tectorum* – in shade especially in derived savanna and forest-savanna mosaic, north into the northern guinea savanna; *A. gayanus* (gamba grass) is found in 4 varieties, *A. gayanus* var. *gayanus* in seasonal swamps, var. *tridentatus* (also in Gambia) in the sudan zone, var. *squamulatus* (now var. *polyclados*) and var. *bisquamulatus* (both in Gambia) in derived savanna-sahel; *A. schirensis* on shallow soils and *A. pseudapricus*, an annual up to 1.2 m high, in sandy and disturbed ground.
>
> These species all belong to the part of the genus with paired racemes, which are terminal on the culm in *A. schirensis*, but otherwise aggregated into leafy false panicles in which the raceme pairs are axillary. The spikelets are in 2 (sets, pairs), one sessile, ♂ and adaxially–abaxially compressed, the other stalked and sterile. The glumes are as long as the spikelets and the upper glume of the sessile floret may be awned, the lower glume concave on the back and lacking nerves between the keels. The sessile spikelet is 2-flowered and falls as a unit together with the adjoining internode of the rachis and the stalk of the upper spikelet. The lower floret of the sessile spikelet is sterile and reduced to a delicate lemma. The upper floret is fertile and its lemma has an awn arising from between the lobes at the tip. The callus of the sessile spikelet is blunt, short and bearded, and fits into the cupular top of the rachis internode, which is flattened and thicker at its upper end (though little so in *A. tectorum*). Both internodes and pedicels are hairy. There are sterile spikelet sets at the base of the raceme.
>
> *Ctenium* (Gr. comb) is a small genus in Madagascar, Africa and America, represented in West Africa by tufted wiry perennials up to 1.2 m high with narrow (1–3 mm wide) usually flat leaves and solitary or digitate spikes (which later coil somewhat), the spikelets in 2 rows along 1 side of the rachis, the projecting awns giving a comb-like effect. The spikelets are all similar, laterally compressed and of 4 florets, the 2 lowest sterile,

the middle 1 fertile and the top 1 sterile. The glumes are ± persistent and, at maturity, the spikelet above them breaks up. The upper glume is well developed, pointed and with an oblique dorsal awn. The lemmas are hyaline, with dorsal awns and hairs along the margins, the paleas 2-nerved. *C. newtonii*, with supervolute leaf blades and solitary spikes, is the most common species, found especially on shallow soils from coastal savanna to the sahel. The lower glume has glandular dots. It sometimes grows as a smaller annual. *C. elegans*, also with solitary spikes, has aromatic foliage and yellow glands on the lemmas or paleas of the lowest 3 florets. *C. canescens* (also in Gambia) has digitate spikes and is found on sandy soils throughout the sudan zone. The rhizomes are short and curved and covered with white scales.

Eragrostis (Gr. love grass) is a tropical and subtropical genus of *c.* 300 species, of which about 10% are found in West Africa, mainly in sandy soils and disturbed ground. These are tufted annuals or perennials up to 1.2 m high (though often less) ± glandular leaves and panicles. The spikelets are awnless, and often quite long; new florets are added apically for some time. This is most marked in those species where the spikelet sheds florets from the base upwards at maturity, leaving the persistent ± zig-zag rachilla exposed. A few species shed florets from the top downwards and spikelets in these species have rather fewer florets. The spikelets are stalked, the glumes shorter than the lowest lemma, and the florets are compressed laterally, resulting in the flattened appearance presented by the spikelets. The palea and membranous lemma often behave differently when the spikelet breaks up. The leaves are up to 1 cm wide, with a ligule of hairs, tapered from leaf blade base to apex, and with the ability to become rolled in some cases. *E. aspera* is a weed species in which the spikelet breaks up from the apex downwards, palea and lemma falling together. *E. atrovirens* (also in Gambia) is a species of damp or swampy savanna soils in which the spikelet breaks up from the base upwards, the palea and lemma falling separately but at the same time. The leaves can roll. *E. tremula*, another weed species, is similar but the paleas persist, and there are only 2 stamens in each flower.

Hyparrhenia (Gr. under and male). A mainly African genus with *c.* 30 species of savanna grasses in West Africa. These are almost all perennials, tufted or with short rhizomes, up to 3 m high, and resembling *Andropogon* in many ways, with false panicles of paired racemes made up of the same kind of spikelets and florets. Differences include the generally greater length of the awn, up to 10 cm, and of the sessile spikelet, while the callus of the sessile spikelet, which forms an oblique joint with the rachis internode, these and the pedicels being terete, not thickened; the stalked spikelet lacks a callus. The stalks of the raceme pairs (raceme bases) vary in length, hairiness and shape (flattened or terete) and may become twisted (deflexed) in some species. The lowest spikelets of each raceme are sterile and awnless, and the number of spikelet sets, and therefore of awns, per raceme also varies. The plants in general are rather hairy. *H. cyanescens*

is a tall species of moist savanna in which the upper raceme base is flattened, hardly longer than the lower one, and both are bearded, though
only the lower one has a sterile spikelet set at the base. The awns are
c. 3 cm long, and there are 6–11 of them per raceme pair. *H. rufa* is one
of the most common grasses, especially in damper savanna, but may
grow as a smaller (annual) plant in disturbed ground. It has glabrous
basal leaf sheaths. The spikelets bear reddish-brown hairs but have short
awns as in the above species. The upper raceme base is terete and longer
than the lower one, with a basal sterile spikelet set at the base of the
lower or both racemes. There are 9–14 awns per raceme pair. *H. subplumosa*
occurs on shallow soils, often in woodland. The upper raceme base is
like that of *H. cyanescens*, but both upper and lower bases have 2 sterile
spikelet sets at the base. The awns are long (4.5–7.5 cm) with long hairs,
and there are 3–6 of them per raceme pair, these becoming deflexed.
The annual *H. involucrata*, a 'weed' species, is similar in many respects
and is thought to hybridise with *H. subplumosa* (Olorode & Baquar, 1976).

Loudetia (Loudet, a dentist). These are tufted savanna annuals or perennials, and, with 2 exceptions, up to 1.5 m high. The genus is mainly African
(south of the Sahara) and Malagasy, the 10 West African species representing 25% of the genus. The leaves are under 1 cm wide, tapering at both
ends and rarely rolled, with a ligule of hairs. The brown panicles, which
tend to have branches in whorls, bear small, solitary or paired spikelets
which are all alike, with 2 florets each. The lower floret is sterile, the
upper ♂, with 2 stamens, and the spikelet breaks up beneath each floret.
The glumes are 3-nerved, persistent and unequal; the upper one is the
larger and is longer than the florets. The lower lemma is also 3-nerved
(5-nerved in *L. togoensis*) but both paleas are 2-nerved and hyaline. The
upper lemma is 5–9-nerved with a deciduous flattened awn, which is
(exceptionally) up to 20 cm long, arising between the 2 lobes at the tip.
The callus of the upper floret is distinctive, either truncate or 2(1)-toothed,
and so are the bristles with dark bulbous bases, when present, on the
glumes. A taller species is *L. phragmitoides* on marshy ground. It is up
to 4.5 m high and has larger leaves (2 cm wide by 1 m long), and small,
very shortly-awned spikelets (awns up to 2 cm long), a truncate callus
and hairy upper lemma; *L. arundinacea*, up to 3 m high, is prominent on
inselbergs. It has leaves up to 1.5 cm wide, and rather less than 1 m long,
spikelets about 1 cm long with awns at least 2 cm long. *L. hordeiformis*
(not Sierra Leone) is an annual species of disturbed sandy soils, up to
1.5 m high with narow dense panicles, the 7–12 cm long awn projecting
from the narrow spikelets which are c. 1.5 cm long. The callus is 1-toothed
and pungent. *L. simplex* (including now *L. camerunensis*) has a 2-toothed
callus and spikelets under 1.5 cm long, but is otherwise similar.

Monocymbium (Gr. one, little boat). This is an African genus of only
4 species, of which *M. ceresiiforme* is common and widespread south of
the Sahara in shallow well-drained soils. It resembles *Andropogon* and
Hyparrhenia in a number of features, but has solitary, short, dense racemes,

each with a reddish-brown boat-shaped bract (spatheole). The racemes are made up of spikelets and florets similar to those of *Andropogon* and *Hyparrhenia*. In addition, as in *Hyparrhenia*, the pedicels and rachis internodes are terete and the callus of the sessile spikelet is oblique ± bearded, but the upper glume of this spikelet is awned, i.e. there are always 2 awns from the sessile spikelet. The stalked spikelet has a callus under 1 cm long. *M. ceresiiforme* tends to be relatively tall in West Africa, up to 1.2 m, but the culms are weak and often beaten down.

Panicum (L.) is a very large tropical and warm temperate genus of *c.* 500 species, with nearly 50 species in West Africa. These are annuals or perennials of varied habit (erect and tufted, or with creeping or shortly rhizomatous stems), and flat or rolled (involute or supervolute) leaves which are generally linear (–lanceolate), but ± ovate, often with cross veins, in forest species. The panicles are ± open and bear solitary and similar spikelets each of 2 florets, the upper ♀, the lower ♂. The spikelet is adaxially–abaxially compressed and is shed as a unit. There is no involucre of bristles subtending it. The 2 glumes are membranous and ± ovate, the upper one as long as the spikelet, the lower one generally shorter. The lower lemma and palea are membranous, the upper lemma and palea brittle, varying in texture between papery and bony. About half the West African species are no more than 60 cm high although, among the *c.* 10% widespread species, some much taller ones are included. A *Panicum* species may be expected in virtually every habitat. *P. maximum* (up to 3+ m) is the tallest West African species. It is usually a perennial and has linear-narrowly lanceolate leaves generally 12–40 cm long × *c.* one-tenth as wide, the plant lacking club-shaped hairs. The spikelets are oblong, *c.* 3 mm long and rounded on the back, the lower glume 3-nerved, the upper glume 5-nerved. The lower lemma is also 5-nerved and there is a well-developed lower palea; the upper lemma and palea are wrinkled (rugose). This is a southern savanna species, of forest-savanna mosaic extending somewhat northwards in suitably damp places. *P. fluviicola* (1 of the 4 species recorded for Gambia) is widespread, occurring near water or in flood plains and in damp savanna. It is up to 2 m high, with leaves up to 30 cm long and *c.* 1 cm wide. The spikelets are ovate, under 3 mm long and the glumes have pointed recurved tips, the lower glume also being as long as the spikelet. The upper lemma and palea are pale and glossy.

Schizachyrium (Gr. split and chaff, hull). Apart from *S. pulchellum*, a stoloniferous seashore pioneer, this is a savanna genus in West Africa, with *c.* 20 mostly tufted annuals or perennials. Like *Monocymbium*, the racemes are solitary and may be assembled into loose false panicles, but the spatheoles are not boat-shaped and the racemes are slender, the spikelet sets spaced out. The genus is otherwise similar to *Andropogon* in spikelet, floret and callus characters, but, unlike *Andropogon*, *Hyparrhenia* and *Monocymbium*, the lower glume of the sessile spikelet is concave on the back and has several nerves between the keels. There are a few creeping species such as the small annual *S. brevifolium*, found with other, taller grasses, but the common erect species (1–3 m high) is *S. sanguineum*, which has reddish culms, hence its name. The sessile spikelet in this species is linear, compressed between the rachis internode on one side and the pedicel on the other. On disturbed stony soils, the rather smaller annual *S. exile* appears. The raceme is partly covered by a spatheole, which turns reddish-brown.

Bibliography

Bowden, B. N. (1964). A note on the external ligule and ligule of *Andropogon gayanus squamulatus* (Hochst). Stapf. *Botanical Journal of the Linnean Society*, **59**, pp. 77–80.

Brown, W. V. (1977). The Kranz syndrome and its subtypes in grass systematics. *Memoirs of the Torrey Botanical Club*, **23**, pp. 1–97.

Clayton, W. D. (1972). *Flora of West Tropical Africa*, 2nd edn, vol. 3, part 2. London: Crown Agents.

Clayton, W. D., Phillips, S. M. & Renvoize, S. A. (1974). *Flora of tropical East Africa: Gramineae*, part 2. London: Crown Agents.

Descoings, B. (1972). Méthode de description des formations herbeuses intertropicales par la structure de la végétation. *Candollea*, **26**, pp. 223–57.

Jacques-Felix, H. (1962). *Les Graminées d'Afrique tropicale*. Paris: Institut des recherches agronomiques tropicales et des cultures vivrières.

Levieux, J. & Diomande, T. (1978). The nutrition of granivorous ants. 1. Cycle of activity and diet of *Messor galla* and *Messor regalis* (*Hymenoptera, Formicideae*). *Insectes sociaux*, **25**, pp. 127–40.

Negbi, M. (1984). The structure and function of the scutellum of the Gramineae. *Botanical Journal of the Linnean Society*, **88**, pp. 205–22.

Niklas, K. J. (1987). Aerodynamics of wind pollination. *Scientific American*, July, pp. 72–7.

Olorode, O. & Baquar, S. R. (1976). The *Hyparrhenia involucrata-H. subplumosa* (Gramineae) complex in Nigeria: morphological and cytological characterization. *Botanical Journal of the Linnean Society*, **71**, pp. 212–22.

Peart, M. H. (1984). The effects of morphology, orientation and position of the grass diaspores on seedling survival. *Journal of Ecology*, **72**, pp. 437–53.

Rose Innes, R. (1977). *A manual of Ghana Grasses*. Surbiton: Ministry of Overseas Development.

Soderstrom, T. R. & Calderon, C. E. (1971). Insect pollination in tropical rain forest grasses. *Biotropica*, **3**, pp. 1–16.

Stanfield, D. P. (1970). *Flora of Nigeria. Grasses*. Ibadan: Ibadan University Press.

Warming, E. (1933). *Frøplanterne*, 2nd edn. Copenhagen: Borgen.

General bibliography

Agnew, A. D. Q. (1974). *Upland Kenya wild flowers*. Oxford: Oxford University Press.

Berhaut, J. (1971–6). *Flore illustrée du Sénégal*. Dakar: Ministère de Developpement rural (incomplete).

Brunnel, J. F. *et al.* (1984). Flore analytique du Togo. Phanerogames. *Englera*, 4.

Burger, W. C. (1967). *Families of flowering plants in Ethiopia*. Stillwater, Oklahoma State University Press.

Burkill, H. M. (1985–). *Useful plants of tropical West Africa*. Kew: Royal Botanic Gardens.

Busson, F. (1965). *Plantes alimentaires de l'ouest africain*. Marseilles: Leconte.

Cole, N. H. Ayodele, (1968). Review of the classification of vegetation in Sierra Leone. *Journal of the West African Science Association*, **13**, pp. 81–92.

Drummond, R. & Palgrave, K. (1973). *Common trees of the high veldt*. Salisbury, S. Rhodesia: Longman.

Durand, J-R. & Leveque, C. (1981). *Flore et faune aquatiques de l'Afrique sahelo-soudanienne* 1. Paris: Office de la Recherche scientifique et technique outremer.

Edberg, E. (1982). *A naturalist's guide to the Gambia*. Channel Is., St Anne: J. G. Sanders.

Fanshawe, D. & Hough, C. (1967). Poisonous plants of Zambia. *Forest Research Bulletin, Zambia* 1.

Flora of tropical East Africa (1952–). Ed. R. M. Polhil *et al*. London: Crown Agents; from 1982, Rotterdam: A. A. Balkema.

Flora of West tropical Africa (1954–72). Ed. F. N. Hepper *et al*., 2nd edn. London: Crown Agents.

Flora zambesiaca (1960–). Ed. E. Launert *et al*. London: Crown Agents.

Flore du Cameroun (1963–). Ed. R. Letouzey *et al*. Paris: Muséum national de l'histoire naturelle.

Flore du Congo etc. (1958–). Ed. W. Robyns *et al*. From 1972, *Flore d'Afrique centrale* (Zaïre–Rwanda–Burundi). Meise: Jardin botanique national de Belgique. See also *Flore du Rwanda*.

Flore du Gabon (1961–), Ed. A. Aubréville *et al*. Paris: Muséum national de l'histoire naturelle.

Flore du Rwanda (1978–85). Ed. G. Troupin. Vol. III *Spermatophytes* in *Annales du Musée royal de l'Afrique centrale* (Tervuren, Belgium), Sciences économiques, sér. in 8., nos. 9, 13 and 15.

Geerling, C. (1982). Guide de terrain des ligneux sahéliens et soudano-guinéens. *Mededelingen Landbouwhogeschool, Wageningen*, **82**(3).

Good, R. (1974). *The geography of flowering plants*. London: Longman.

Hall, J. B., Pierce, P. & Lawson, G. W. (1971). *Common plants of the Volta Lake*. Legon: University of Ghana.

Harlan, J. R., de Wet, J. M. J. & Stemler, A. B. L. (1976). *Origins of African plant domestication*. The Hague: Mouton Publishers.

Hedberg, I. (ed.) (1979). *Systematic botany, plant utilization and biosphere conservation*. Stockholm: Almquist & Wiksell.

Hopkins, B. & Stanfield, D. P. (1966). *A field key to the savanna trees of Nigeria*. Ibadan: Ibadan University Press.

Houssain, M. & Hall, J. B. (1969). *A field key to the trees of the Mole Game Reserve, Damongo, Ghana*. Legon: University of Ghana.

Irvine, F. R. (1961). *Woody plants of Ghana*. Oxford: Oxford University Press.

Irvine, F. R. (1969). *West African crops*. Oxford: Oxford University Press.

Ivens, G. W., Moody, K. & Egunjobi, J. K. (1978). *West African weeds*. Ibadan: Ibadan University Press.

Johansson, D. (1974). Ecology of West African epiphytes in rain forest. *Acta phytogeographica suecica*, **59**.

Keay, R. W. J., Onochie, C. F. A. & Stanfield, D. P. (1960–4). *Nigerian trees*. Lagos: Federal Government Printer.

Moriarty, A. (1975). *Wild flowers of Malaŵi*. Capetown: Purnell.

Morton, J. K. (1961). *West African lilies and orchids*. London: Longman.

Myers, N. (1979). *The sinking ark*. Oxford: Pergamon Press.

Palmer, E. & Pitman, N. (1972–3). *Trees of southern Africa*. Capetown: Purnell.

Pillinger, J. S. & Kitchin, A. M. (1982). *Trees of Malawi*. Blantyre: Blantyre Print and Publishing.

Purseglove, J. W. (1968–72). *Tropical crops*. London: Longman.

Purseglove, J. W. *et al.* (1981). *Spices*. London: Longman.

Schnell, R. (1977). *La flore et la végétation de l'Afrique tropicale*. Paris: Gauthier-Villars.

Soule, M. E. (ed.) (1986). *Conservation biology: the science of scarcity and diversity*. Sunderland, Massachusetts: Sinauer Associates.

Soule, M. E. & Wilcox, B. A. (eds.) (1980). *Conservation biology: an evolutionary perspective*. Sunderland, Massachusetts: Sinauer Associates.

Troupin, G. (1982). Flore des plantes ligneuses du Rwanda. *Annales du Musée royal d'Afrique centrale*, Sciences economiques, sér. in 8., no. 12.

Verdcourt, B. & Trump, E. (1969). *Common poisonous plants of East Africa*. London: Collins.

Vickery, M. L. & Vickery, B. (1979). *Plant products of tropical Africa*. London: Macmillan Press.

Voorhoeve, A. G. (1965). *Liberian high forest trees*. Wageningen: Centrum voor Landbouwpublikaties en Landbouwdokumentatie.

White, F. (1962). *Forest flora of Northern Rhodesia*. Oxford: Oxford University Press.

Index of family, generic and common names

Anthericaceae 279–80
Anthericum 27, 280–1, 282
Anthocleista 6, 37, 60
Anthospermum 214
Antiaris 6, 57, 165
Antidesma 125
Antrocaryon 185, 187
Anubias 286, 288
Apocynaceae 6, 11, 34, 37–8, 40–1, 43, 47, 52, 59, 61, 196
Aptandra 37, 50
Araceae 3, 39, 41, 56, 285
Arachis 151—4
Araliaceae 49
Arecaceae 293
Aristolochiaceae 38
Artocarpus 9, 164–5, 166
Asclepiadaceae 14, 26, 41, 43, 56, 59, 205
Asclepias 205, 206, 207
Ascolepis 314
Ascotheca 247
Asparagaceae 280–1
Asparagus 279–80
Asphodelaceae 280–1
Aspilia 224
Asteraceae 221
Astripomoea 233, 236
atemoya, (hybrid), see Annona
Attractogyna 210
Aubregrinia 191
Aubrevillea 53, 144
Auxopus 302
Avicennia 29
Avicenniaceae 29
awned squill, see Dipcadi
awusa nut, see Tetracarpidium
Azadirachta 6, 171, 173
Azanza 118, 121, 122

Baissea 199, 200
Balanitaceae 64
Balanites 12, 64, 65, 68
Balanophoraceae 1
balloon vine, see Cardiospermum
balsa, see Ochroma
Bambara groundnut, see Voandzeia
bamboo, see Bambuseae
Bambuseae 39, 321
baobab, see Adansonia
Baphia 152
Baphiopsis 136
Barleria 246, 248
Barteria 14, 15, 16
batchelor's button, see Gomphrena
Bauhinia 69, 133, 135
Beaumontia 196
Benincasa 87

Bequaertiodendron 190
Berkheya 219
Berlinia 34, 136, 137–8
Bidens 224
Bignonia 241
Bignoniaceae 6, 11, 14, 34, 37, 49, 53, 57, 59, 61, 64, 68, 178, 241
Bixa 13
black mangrove, see Avicennia
black plum, see Vitex
black walnut, see Mansonia
Blepharis 246, 248
Blepharispermum 221
Blighia 68, 181
blood flower, see Asclepias
Boletus 7
bologi, see Crassocephalum and Solanum
Bombacaceae 6, 11, 13, 35, 40, 51, 53, 64, 68, 114
Bombax 61, 114–15, 116
Bonamia 235, 237
Boraginaceae 11
Borassus 54, 293, 294, 296–9
Borreria 211
Boscia 64, 69
Bosquiea 6, 165, 167, 168
Boswellia 64, 66
bottle gourd, see Lagenaria
Brachiaria 24, 51
Brachycorythis 303
breadfruit/nut, see Artocarpus
Breonadia 216
Breynia 124
Bridelia 41, 57, 125, 127, 129
Brieya 74
Brillantaisia 247, 250
Brunfelsia 228
Brugmansia 231, 232
Bryophyllum 48, 54
Buchholzia 12
Buforrestia 264
Bulbophyllum 302–4, 306
Bulbostylis 45, 315
bulrush millet, see Pennisetum
Burkea 58, 66, 133, 134
Burmannia 47, 54
Burmanniaceae 7, 54, 55
Burseraceae 12, 66
Bussea 133
butter tree, see Pentadesma
Butyrospermum 48–9, 57, 64, 68, 190, 192
Byrsocarpus 43, 61
Byttneria 107, 109

cabbage palm, see Roystonea
Caesalpinia 135, 137